Protein–Protein Interactions
in Plant Biology

Annual Plant Reviews

A series for researchers and postgraduates in the plant sciences. Each volume in this annual series will focus on a theme of topical importance and emphasis will be placed on rapid publication.

Editorial Board:

Titles in the series:

1. Arabidopsis
Edited by M. Anderson and J. Roberts

2. Biochemistry of Plant Secondary Metabolism
Edited by M. Wink

3. Functions of Plant Secondary Metabolites and their Exploitation in Biotechnology
Edited by M. Wink

4. Molecular Plant Pathology
Edited by M. Dickinson and J. Beynon

5. Vacuolar Compartments
Edited by D.G. Robinson and J.C. Rogers

6. Plant Reproduction
Edited by S.D. O'Neill and J.A. Roberts

7. Protein–Protein Interactions in Plant Biology
Edited by M.T. McManus, W.A. Laing and A.C. Allan

Protein–Protein Interactions in Plant Biology

Edited by

MICHAEL T. McMANUS
Institute of Molecular BioSciences
Massey University
Palmerston North
New Zealand

WILLIAM A. LAING
and
ANDREW C. ALLAN
HortResearch
Mt Albert Research Centre
Auckland
New Zealand

Sheffield
Academic Press

CRC Press

First published 2002
Copyright © 2002 Sheffield Academic Press

Published by
Sheffield Academic Press Ltd
Mansion House, 19 Kingfield Road
Sheffield S11 9AS, UK

ISBN 1-84127-229-9
ISSN 1460-1494

Published in the U.S.A. and Canada (only) by
CRC Press LLC
2000 Corporate Blvd., N.W.
Boca Raton, FL 33431, U.S.A.
Orders from the U.S.A. and Canada (only) to CRC Press LLC

U.S.A. and Canada only:
ISBN 0-8493-9790-1
ISSN 1097-7570

Printed on acid-free paper in Great Britain by
Antony Rowe Ltd, Chippenham, Wiltshire

British Library Cataloguing-in-Publication Data:
A catalogue record for this book is available from the British Library

Library of Congress Cataloging-in-Publication Data:
A catalog record for this book is available from the Library of Congress

Preface

The purpose of this volume is to review protein-protein interactions in plant biology. This is a rapidly emerging research theme in plants which has not been addressed previously in a single volume. In animal cell biochemistry, the importance of protein-protein interactions is well established—particularly in signal transduction, but also in many other areas of cellular metabolism. The knowledge gained from these animal studies suggests the importance of protein-protein interactions in the plant cell. Consequently, several homologous systems in plants have been identified [e.g. metabolons, proteosomes, chaperonins], which are considered in this volume along with the unique biochemistry associated with photosynthesis in higher plants.

The volume is divided into three sections. The first considers examples of single protein-single protein (binary) interactions and concerns the regulation of enzymes by protein effectors. Chapter 1 describes the plant thioredoxins and interactions with the thioredoxin-dependent enzymes. Chapter 2 considers the Rubisco activase-Rubisco system, one of the key protein-protein interactions that regulates photosynthesis. Chapter 3 deals with the emergence of 14-3-3 proteins as important regulators of both metabolic and non-metabolic enzymes in higher plants, while chapter 4 surveys the increasingly diverse range of proteinase inhibitors in higher plants—well characterised examples of single protein-protein interactions.

The second section deals primarily with multi-protein complexes, including situations in which the protein-protein interaction results in cooperation of the complex in terms of function. Chapter 5 deals with multienzyme complexes in higher plants, with particular reference to complexes in the Benson-Calvin cycle and fatty acid metabolism. Chapter 6 is concerned with the proteolytic complexes in higher plants, including coverage of the proteosome, the large ATP-dependent multisubunit protease that mostly degrades polyubiquitylated proteins, the ATP-dependent Clp proteases and the Lon proteases. The final chapter in this section, chapter 7, describes the higher plant chaperonins, the term reserved for the well characterised 60 kDa heat shock protein (hsp60) family, a sequence-related family of the chaperone proteins.

The third section concerns protein-protein interactions as they occur as part of signal perception, transduction and elicitation of cellular response. The extensive literature associated with the elucidation of signal transduction and the importance of protein-protein interactions in animal cells has provided plant biologists with a framework with which to investigate whether similar signalling pathways occur in plant cells. The advent of sophisticated technologies such as two-hybrid systems has significantly advanced this area of plant biology research. The first chapter in this section, chapter 8, considers the receptor-based kinases, of which the serine/threonine kinases predominate. In plants, little is known of the putative ligands of these proteins and so they are more accurately termed receptor-like kinases (RLKs). A second class of receptor kinase, the histidine kinases, is also discussed. Chapter 9 deals with the cytoplasmic-based kinases, the function of which is to transduce ligand-binding events into a cellular response. The emphasis is on MAP kinases (again serine/threonine kinases), but coverage includes cyclin-dependent kinases. Chapter 10 considers protein-protein interactions regulating transcription of plant genes, including the plant DNA-binding transcription factors and the combinatory interactions of these proteins with other regulatory proteins. Finally, chapter 11 discusses the calcium-binding protein calmodulin and calmodulin target proteins. In this chapter, emphasis is placed on the Ca^{2+}-dependent interactions involving calmodulin.

Where possible, the focus of each chapter is on examples from higher plants. However, authors have included examples from the animal and microbial kingdoms. In many instances, the study of protein-protein interactions in higher plants is in its infancy. We expect the volume to stimulate interest in this emerging field of research. It is directed at all who are researching or teaching plant protein biology.

Michael T. McManus
William A. Laing
Andrew C. Allan

Contributors

Dr Andrew C. Allan

HortResearch, 120 Mt Albert Road, Private Bag 92169, Auckland, New Zealand

Dr Abdussalam Azem

Department of Biochemistry, Tel-Aviv University, Tel-Aviv 69978, Israel

Dr Teerapong Buaboocha

Department of Biochemistry, Chulalongkorn University, Bangkok, Thailand

Dr Robert J. Ferl

Program in Plant Molecular and Cellular Biology, 2109 Field Hall, University of Florida, Gainesville, FL 32611, USA

Dr Brigitte Gontero

Institut Jacques Monod, CNRS-Universités Paris VI and VII, 2 place Jussieu, Tour 43, 75 005 Paris, France.

Ms Emmanuelle Graciet

Institut Jacques Monod, CNRS-Universités Paris VI and VII, 2 place Jussieu, Tour 43, 75 005 Paris, France.

Dr Heribert Hirt

Institute of Microbiology and Genetics, Vienna Biocenter, Dr. Bohrgasse 9, A-1030 Vienna, Austria

Mr Keith Hudson

Genesis Research and Development Corporation, 1 Fox Street, Parnell, Auckland, New Zealand

Dr Christina Ingvardsen

Royal Veterinary and Agricultural University, Department of Plant Biology, Thorvaldsensvej 40, 1871 Frederiksberg C, Denmark

Dr Jean-Pierre Jacquot

UMR 1136 Interactions Arbres Microorganismes, INRA-Université Henri Poincaré, BP 239, 54506 Vandoeuvre Cedex, France

Ms Claudia Jonak

Institute of Microbiology and Genetics, Vienna Biocenter, Dr Bohrgasse 9, A-1030 Vienna, Austria

Dr William A. Laing   HortResearch, 120 Mt Albert Road, Private Bag 92169, Auckland, New Zealand

Dr Sandrine Lebreton   Institut Jacques Monod, CNRS-Universités Paris VI and VII, 2 place Jussieu, Tour 43, 75 005 Paris, France

Ms Galit Levy-Rimler   Department of Biochemistry, Tel-Aviv University, Tel-Aviv 69978, Israel

Dr Michael T. McManus   Institute of Molecular BioSciences, Massey University, Private Bag 11222, Palmerston North, New Zealand

Dr Yves Meyer   Laboratoire de Physiologie et Biologie Moléculaire des Plantes, UMR 5545 Université de Perpignan, 52 Avenue de Villeneuve, 66860 Perpignan Cedex, France

Dr Myroslawa Miginiac-Maslow Institut de Biotechnologie des Plantes, Bât. 630 Université Paris-Sud, 91405 Orsay Cedex, France

Ms Adina Niv   Department of Biochemistry, Tel-Aviv University, Tel-Aviv 69978, Israel

Professor Archie R. Portis Jr USDA-ARS and Department of Crop Sciences, University of Illinois, Urbana, Illinois 61801, USA

Professor Peter Schürmann   Laboratoire de Biochimie Végétale, Université de Neuchâtel, CH 2007 Neuchâtel, Switzerland

Dr Paul C. Sehnke   Program in Plant Molecular and Cellular Biology, 2109 Field Hall, University of Florida, Gainesville, FL 32611, USA

Dr Rajach Sharkia   Department of Biochemistry, Tel-Aviv University, Tel-Aviv 69978, Israel

Dr Bjarke Veierskov   Royal Veterinary and Agricultural University, Department of Plant Biology, Laboratory of Plant Physiology, Thorvaldsensvej 40, 1871 Frederiksberg C, Denmark

Dr Paul Viitanen   The Molecular Biology Division, Central Research & Development Department, E.I. DuPont de Nemours and Company, Experimental Station, Wilmington, Delaware 19880-0402, USA

Dr Celeste Weiss   Department of Biochemistry, Tel-Aviv University, Tel-Aviv 69978, Israel

Professor Dao-Xiu Zhou   Institut de Biotechnologies des Plantes, CNRS/UMR 8168, Université de Paris XI, F-91405 Orsay, France

Professor Raymond E. Zielinski Department of Plant Biology and the Physiological and Molecular Plant Biology Program, University of Illinois, Urbana, IL 61801, USA

Contents

6 Self-compartmentalizing proteolytic complexes
BJARKE VEIERSKOV and CHRISTINA INGVARDSEN

7 The higher plant chaperonins
RAJACH SHARKIA, PAUL VIITANEN, GALIT LEVY-RIMLER, CELESTE WEISS, ADINA NIV and ABDUSSALAM AZEM

11. Calmodulin   285
TEERAPONG BUABOOCHA and RAYMOND E. ZIELINSKI

Index   315

1 Protein–protein interactions in plant thioredoxin dependent systems

Yves Meyer, Myroslawa Miginiac-Maslow,
Peter Schürmann and Jean-Pierre Jacquot

1.1 Introduction

Thioredoxin from *Escherichia coli* was discovered originally as an electron–proton donor to ribonucleotide reductase (RNR), acting through a dithiol–disulfide interchange reaction involving its own dithiol and an internal disulfide bridge of the RNR (Holmgren *et al.*, 1968). The electrons docked on the dithiol of the reductase are used later for the reduction of deoxyribonucleotides and the disulfide bridge is reformed. In plants, the first thioredoxins discovered were those that served as regulators for the chloroplastic enzymes: fructose-1,6-bisphosphatase (FBPase) and NADP malate dehydrogenase (NADP-MDH). It was found that these enzymes are inactive when isolated from plants maintained in the dark, but that they become active in the light after being reduced by thioredoxin. This finding opened the concept of redox regulation of enzymes via dithiol–disulfide interchange reactions. This regulatory mechanism has gained ever more importance in the last decade in plant and animal systems after the discovery of redox regulation of transcription factors. Those redox mechanisms have been described as playing a very important role in various cellular mechanisms such as stress responses, apoptosis and viral recognition. In oxygenic photosynthetic plants, an additional and specific role for these reactions is to regulate the rate of photosynthesis. Several catalysts can perform oxidation–reduction of disulfide bridges of various enzymes (known as target enzymes). They include the oxidizing protein disulfide isomerases (PDIs) and the reducing proteins thioredoxin and glutaredoxin. All these proteins possess a so-called thioredoxin fold (i.e. a central pleated β sheet surrounded by α helices. The sequence of the active site (CysGly/ProProCys, in thioredoxin, CysProTyrCys, in glutaredoxin, and CysGlyHisCys in PDI) of those catalysts is directly related to their oxidizing–reducing properties and to their redox potential.

A very interesting feature of the redox systems is that they function as regulatory cascades that depend often on extremely specific protein–protein interactions. In this chapter, we describe what is known today about the molecular recognition of the various redox protein partners in the plant thioredoxin systems. Knowledge in this area derives from experiments *in vivo* that involve trapping target enzymes using monocysteinic mutants of thioredoxin, use of two hybrid systems in yeast and biochemical approaches that include expression of recombinant enzymes, site directed mutagenesis and elucidation of three-dimensional (3D) structures.

1.2 Description of the chloroplastic and cytosolic systems in eukaryotic plants

It is clear that eukaryotic plants possess several thioredoxin systems, one located in the chloroplasts, one in the cytosol, and one less well documented system in the mitochondria (Laloi *et al.*, 2002). Extensive evidence indicates that the cytosolic system uses NADPH as an electron donor and a flavoenzyme, NADPH thioredoxin reductase (NTR) that reduces a class of thioredoxins known as thioredoxin h (figure 1.1a). However, in the chloroplastic system the reducing power is provided by visible light through the photosystems to reduced ferredoxin, and transmitted through a ferredoxin-dependent thioredoxin reductase (FTR) and specific thioredoxins called thioredoxin m or thioredoxin f (figure 1.1b). A series of biochemically identified target enzymes is also listed in table 1.1.

The recent progress in the generation of expressed sequence tags (ESTs) and genome sequencing shows that the situation is far more complex. Most proteins of the thioredoxin cascade are encoded by a gene family, with each

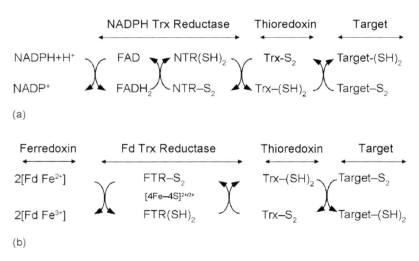

Figure 1.1 The two thioredoxin dependent reduction cascades acting in higher plants: (a) The cytosolic NADPH dependent thioredoxin reductase transfers the reducing power through a flavin to its disulfide, forming a dithiol which reduces cytosolic thioredoxin h. (b) The chloroplastic ferredoxin-dependent thioredoxin reductase transfers the reducing power through its 4Fe–4S center to its disulfide which reduces chloroplastic thioredoxins m, f, and possibly some of the recently discovered chloroplastic thioredoxins.

Table 1.1 Biochemically identified thioredoxin target proteins[a]

Target enzyme	Plant	Localization	Activator
FBPase	Spinach	Chloroplast	Trx f
SBPase	Wheat	Chloroplast	Trx f
PRK	Spinach	Chloroplast	Trx m(f)
ATP synthase	Spinach	Chloroplast	Trx f
(γ-subunit of CF1)			
NADP-MDH	Sorghum	Chloroplast	Trx f(m)
G6PDH	Potato	Chloroplast	Trx m
Rubisco activase	*Arabidopsis*	Chloroplast	Trx f
Thiocalsin	*Hordeum*	Cytosol	Trx h
Storage proteins	Wheat	Cytosol	Trx h
Apurinic endonuclease	*Arabidopsis thaliana*	Nucleus	Trx h

[a] Most of the references concerning the studies on target enzymes can be found in Schürmann and Jacquot (2000) and in this chapter (for FBPase and NADP-MDH), with the exception of Babiychuk *et al.* (1994) for apurinic endonuclease, and Geck and Hartman (2000) for PRK.
Abbreviations: FBPase, fructose-1,6-bisphosphatase; SBPase, sedoheptulose-1,7-bisphosphatase; PRK, phosphoribulokinase; NADP-MDH, NADP malate dehydrogenase; G6PDH, glucose-6-phosphate dehydrogenase.

member differentially expressed (Brugidou *et al.*, 1993). In addition, genes encoding novel thioredoxin types are present in the genome (Mestres-Ortega and Meyer, 1999, Meyer *et al.*, 2001). For example, table 1.2 contains a list of thioredoxins and thioredoxin reductases in the *Arabidopsis* genome. In most cases the genes encoding the new thioredoxins are expressed at a low level and consequently the corresponding proteins have not yet been detected in previous studies.

1.3   Mechanism of thioredoxin catalysis and consequences

Thioredoxin is a very powerful reductant of disulfide bridges in proteins, and its regulatory function is mainly related to this property. There are only a very few cases where it exerts a strictly structural function, independent of its redox properties, e.g. phage assembly in *E. coli* (Feng *et al.*, 1999) and vacuole formation in yeast (Xu *et al.*, 1997). Its reductive activity involves both cysteines of the active site, which play very dissimilar roles (Geck *et al.*, 1996; Lancelin *et al.*, 2000). The most reactive and exposed *N*-terminal cysteine of reduced thioredoxin's active site disrupts the disulfide bridge of the target by nucleophilic attack (figure 1.2a–b), reduces one cysteine of the disulfide bridge, and forms a bridge with the second cysteine of the target protein (figure 1.2c–d). The resulting mixed disulfide is disrupted by the attack of the second cysteine of thioredoxin (figure 1.2e–f), yielding a reduced target enzyme and an oxidized thioredoxin. This mechanism for disulfide bridge reduction is not specific for thioredoxins. Glutaredoxins, the other protein disulfide reducers,

Table 1.2 Compilation of thioredoxin and thioredoxin reductase genes present in the *Arabidopsis thaliana* genome. The proposed names are indicated in the first column, the positions on *Arabidopsis* chromosomes or chromosome sections (available only for chromosome 2 and 4) in the second column, and their presence in particular bacterial cosmids used during the sequencing process in the third column. The putative localization was inferred for the deduced protein sequence using Psort (Nakai and Kanehisa, 1992)

Classical thioredoxin types			
Thioredoxins m			
m1	1	F21B7	chloroplastic
m2	4-9	F9H3	chloroplastic
m3	2-91	F9O13	chloroplastic
m4	3	MJK13	chloroplastic
Thioredoxins f			
f1	3	F13F9	chloroplastic
f2	5	MQK4	chloroplastic
Thioredoxins h			
h1	3	F24M12	cytosolic
h2	5	MYH19	cytosolic
h3	5	MBD2	cytosolic
h4	1	F14P1-F6F9	cytosolic
h5	1	F27F5	cytosolic
h7	1	F23H11	cytosolic
h8	1	T17F3	cytosolic
h9	3	F17O14	cytosolic
New thioredoxin types			
Thioredoxin x			
x	1	F14I3	?
Thioredoxins ch2			
ch2-1	1	F28O16	chloroplastic
ch2-2	1	T10P12	chloroplastic
Thioredoxins o			
o1	2-192	F19I13	mitochondrial
o2	1	F17F8	?
Monocysteinic thioredoxins			
CxxS1	1	T23J18	cytosolic
CxxS2	2-220	T7D17	cytosolic
Thioredoxin reductases			
NTR			
NTRa	2-100	F5J6	mitochondrial/cytosolic
NTRb	4	F15J1	cytosolic
NTRc	2-225	T32G6	cytosolic
FTR			
FTRa	5	F8L15	chloroplastic
		K19M13	
FTRb	2-23	F28I8	chloroplastic

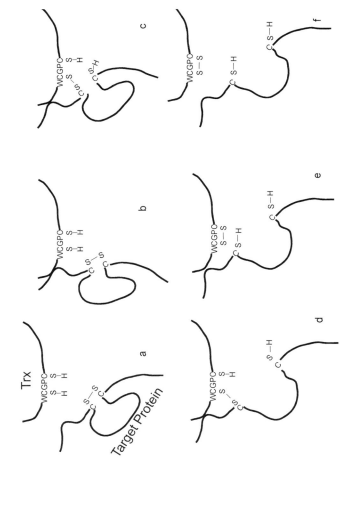

Figure 1.2 Reaction mechanism of thioredoxin mediated reduction of target proteins. Reduced thioredoxin interacts with the oxidized target protein (a,b), then the most *N*-terminal cysteine attacks the disulfide of the target, forming a mixed disulfide between thioredoxin and the target. This disulfide is reduced by the *C*-terminal cysteine of thioredoxin, releasing the oxidized thioredoxin and the reduced target protein.

react in a similar way (Gan *et al.*, 1990). Even NTR and FTR reduce thioredoxins using a dissymmetrical dithiol attack and PDI, which is responsible for disulfide isomerization (i.e. disulfide displacement), attacks the original disulfide bridge by the *N* terminal cysteine of its dithiol active site and forms an intermediate disulfide bridged complex with its target. The PDI reaction differs in the second step during which the complex is attacked by a thiol from the target, releasing the reduced PDI and the target protein with a displaced disulfide bridge (Darby and Creighton, 1995). This strongly suggests that the mono-cysteinic thioredoxin homologues present in *Arabidopsis* (table 1.2) and other plant genomes could act as PDIs or even disulfide reductases, if the environment is sufficiently reducing to break the intermediate complex. For a reason that is still poorly understood, the chemical dithiol DTT, widely used for reducing the disulfide bridge of thioredoxins, is a very poor reductant for some thioredoxin targets, suggesting that besides its effect as a disulfide reductant, thioredoxin also has a structural function. Indeed, the redox potential of DTT ($-330\,mV$ at pH 7.0) is quite adequate for the reduction of the disulfides for those chloroplastic target enzymes where the redox potentials have been determined (Hirasawa *et al.*, 1999, 2000).

1.4 Molecular properties of the thioredoxin dependent enzymes and protein–protein recognition properties

Thioredoxins are reduced by two different reductases (figure 1.1a, b) and they in turn regulate many enzymes, especially in the chloroplast. We have chosen in this chapter to address only the most documented cases of protein–protein interactions, namely the thioredoxin reductases and the target enzymes NADP-MDH and FBPase. Information about the other thioredoxin dependent catalysts is contained in other recent reviews (see for example Meyer *et al.*, 1999; Ruelland and Miginiac-Maslow, 1999; Schürmann and Jacquot, 2000; Schürmann and Buchanan, 2001).

1.4.1 Thioredoxin reductases

1.4.1.1 NADPH dependent thioredoxin reductase
In *Arabidopsis thaliana*, the cytosolic NADPH dependent thioredoxin reductase is a flavoenzyme which catalyzes the transfer of electrons from NADPH through a flavin moiety to a redox active site which contains two cysteine residues with the sequence CysAlaValCys (Jacquot *et al.*, 1994). The plant enzyme is more closely related to the *E. coli* protein than to the mammalian enzyme which possesses a *C* terminal extension and a catalytically active selenocysteine. The *A. thaliana* and *Chlamydomonas* NTRs preferentially reduce thioredoxin h, then chloroplastic thioredoxin m with less efficiency, and then chloroplastic thioredoxin f very poorly, if at all (Huppe *et al.*, 1991; Jacquot *et al.*, 1994).

1.4.1.2 The three-dimensional structure of NTR
The X ray crystallographic structure of thioredoxin h reveals that the enzyme comprises a dimer of identical subunits that are inversely oriented in respect to eachother (Dai *et al.*, 1996). Each subunit contains two separate domains: the FAD domain, which is constituted by the *N* and *C* termini of the polypeptide, and the NADPH binding domain, which corresponds to the central part of the sequence. Each of the subunits contains a bound FAD and a disulfide. The redox active disulfide bridges are located extremely close to the isoalloxazine rings of the FAD molecules, a feature which facilitates electron transfer. Since the NADPH binding domain is far removed from the disulfide/FAD centre, it has been postulated that there must be a rotational movement of the subdomains that allows the NADPH molecule to become postioned closer to the active site. This was demonstrated recently to be the case for the *E. coli* enzyme which was crystallized in the two conformations (Lennon *et al.*, 2000).

1.4.1.3 Interaction with thioredoxin h
After reduction, the disulfide bridge of NTR reduces the disulfide bridge of the substrate, thioredoxin. In order to do that, there must be a precise positioning of the two protein partners. How the recognition is achieved in plant systems is poorly documented, but the fact that higher plant and algal NTR interacts with thioredoxin h and m, but very poorly with thioredoxin f and *E. coli* thioredoxin, indicates some specificity. There is now a structure of a complex between *E. coli* NTR and thioredoxin available (Protein Data Bank access number 1F6M) (Lennon *et al.*, 2000). This complex was obtained by creation of a mixed disulfide between a C35S thioredoxin mutant and a C135S NTR mutant. In the *E. coli* system, the recognition site seems to be largely hydrophobic, involving, respectively, the thioredoxin residues $G_{71}I_{72}G_{74}$ (seq GIRG) situated in a loop between the secondary structures, the $\beta3$ and $\beta4$ sheets, spatially close to the active site, and two phenylalanine residues of NTR (F_{141}, F_{142}). Another region that could participate in the recognition is the sequence *N* terminal of the active site of thioredoxin. It has, however, been shown in plant systems that changing the tryptophan residue adjacent to the active site into an alanine residue (mutation *W35A* in *Chlamydomonas reinhardtii* thioredoxin h) does not hamper recognition of NTR; on the contrary, the K_m for thioredoxin h of the *W35A* mutant is threefold lower than the K_m of the wild-type thioredoxin, suggesting that this residue is not involved in protein–protein recognition but rather slows down this process (Krimm *et al.*, 1998; and unpublished data). Overall, NTR/thioredoxin recognition seems to proceed mainly through hydrophobic contacts, and does not seem to depend much on ionic interactions, as is observed for the FTR/thioredoxin interaction site. It is not clear yet whether this *E. coli* model can be transposed to the plant system as *E. coli* NTR does not use thioredoxin h as a substrate and *vice versa*. This could be due to the absence of the two phenylalanine

residues in NTR (F_{141}, F_{142}) that are replaced by two alanines in the plant sequence.

1.4.2 Ferredoxin–thioredoxin reductase

Ferredoxin–thioredoxin reductase (FTR) is a unique enzyme of oxygenic photosynthetic organisms catalyzing the reduction of thioredoxins with electrons provided by the photosynthetic light reactions. FTR is a relatively small protein of about 25 kDa with a yellowish-brown color. It has been purified and characterized from different organisms: spinach (Schürmann, 1981; Droux *et al.*, 1987), maize and *Nostoc* (Droux *et al.*, 1987), soybean (P. Schürmann, unpublished data) and *Chlamydomonas reinhardtii* (Huppe *et al.*, 1990). All known FTRs are composed of two dissimilar subunits—designated the catalytic and variable subunits. The catalytic subunit of the enzyme, which is highly conserved in different organisms with a constant molecular mass of about 13 kDa, contains a [4Fe–4S] cluster and a redox-active disulfide bridge that functions in the reduction of thioredoxins. The variable subunit ranges from 8 to 13 kDa when primary structures from different species are compared, and appears to have only a structural function.

The FTRs from spinach (Gaymard & Schürmann, 1995; Gaymard *et al.*, 2000) and *Synechocystis* (Schwendtmayer *et al.*, 1998) have been cloned and expressed as perfectly functional enzymes in *E. coli*.

1.4.2.1 Three-dimensional structure of ferredoxin–thioredoxin reductase

FTR is an unusual enzyme with respect to its 3D structure and its mechanism, since it is able to reduce a disulfide bridge with the help of its Fe–S cluster. The recombinant *Synechocystis* FTR has been crystallized (Dai *et al.*, 1998) and its 3D structure solved by X-ray analysis (Dai *et al.*, 2000a, 2000b). The variable subunit is a heart-shaped β-barrel structure on top of which sits the essentially α-helical catalytic subunit. Together, they form an unusually thin molecule with the shape of a concave disc measuring only 1 nm across the centre. The active site is located in this centre with the functionally important structures, a cubane [4Fe–4S] cluster and, in close proximity, a redox active disulfide bridge. The Fe–S cluster is liganded by four cysteine residues, which are not present in the usual consensus motif seen in other [4Fe–4S] proteins. Instead, the cluster is arranged in an entirely new pattern, consisting of two CysProCys and one CysHisCys motifs. Two iron atoms are liganded by the cysteines in the $Cys_{74}ProCys_{76}$ (*Synechocystis* FTR numbering) motif, whereas the other two iron atoms are coordinated by Cys_{55} in the $Cys_{55}ProCys_{57}$ motif and by Cys_{85} of the $Cys_{85}HisCys_{87}$ motif. The remaining two cysteines, Cys_{57} and Cys_{87}, form the active site disulfide bridge, ensuring that it is located close to the Fe–S centre. The Fe–S cluster is located on one side of the flat molecule close to the surface of the ferredoxin docking site, whereas the disulfide bridge is accessible from the opposite side, i.e. the thioredoxin docking site. This arrangement allows simultaneous docking of ferredoxin, the electron donor, and thioredoxin, the electron acceptor protein, on opposite sides of the FTR, and the transfer of an electron across the centre of the molecule. This is a pre-requisite for the proposed reaction mechanism, which envisions the formation of a transient mixed disulfide between FTR and thioredoxin (Holmgren, 1995; Staples *et al.*, 1998; Dai *et al.*, 2000b). The docking sites on both sides of the FTR molecule are well adapted for their function and their surface residues are highly conserved.

1.4.2.2 Interaction with ferredoxins
It had been shown that FTR forms a stable 1:1 complex with ferredoxin and that this complex is stabilized by electrostatic forces (Hirasawa *et al.*, 1988). This rather strong interaction has been successfully exploited for the affinity purifica-tion of FTR by chromatography on ferredoxin–Sepharose (Droux *et al.*, 1987; Schürmann, 1995). The structural analysis has now shown that the potential ferredoxin interaction area, which is on one side of the disk-like FTR, contains three positively charged residues. Together with hydrophobic residues around the Fe–S cluster they form a docking area for negatively charged ferredoxin. Furthermore, this ferredoxin docking site has shape-complementarity with the ferredoxin molecule (Dai *et al.*, 2000a). Conversely, there are two clusters of negative charges located respectively in helix 1 and the *C*-terminal helix 3 of ferredoxin that are essential for the FTR–ferredoxin recognition (de Pascalis *et al.*, 1994). The essential role of one of the *C*-terminal glutamate residues (Glu$_{92}$ in spinach) has been confirmed both by chemical derivatization and site directed mutagenesis (de Pascalis *et al.*, 1994; Jacquot *et al.*, 1997a). The opposite charges of the two proteins and the structural fit together contribute to a rather strong interaction between FTR and ferredoxin, as well as forming the basis for a certain specificity of this interaction. There is clearly a lower affinity observed in the interaction between heterologous reaction partners, i.e. spinach ferredoxin and *Synechocystis* FTR, than between ferredoxin and FTR from the same species (Schwendtmayer *et al.*, 1998; and unpublished observation).

1.4.2.3 Interaction with thioredoxins
The thioredoxin interaction area is less charged, contains more hydrophobic residues as well as three histidines, with one of these very close to the active site. In *Synechocystis* FTR, there is one negatively charged residue, Glu$_{84}$, about 1.2 nm away from the disulfide and it fits with a positively charged residue in thioredoxin. The absence of other charged groups makes the thioredoxin interaction area less selective. This seems to be important *in vivo*, where FTR reduces different thioredoxins. For example, in spinach chloroplasts, two types of thioredoxins, m and f, are reduced indiscriminately. We have found that

Synechocystis FTR reduces spinach thioredoxin f with equal efficiency to that observed for spinach FTR (unpublished observations).

These properties make the FTR a versatile thioredoxin reductase, capable of accepting electrons from diverse ferredoxins, although with different efficiencies, and reducing the disulfides of various thioredoxins.

1.4.2.4 Mechanism of thioredoxin reduction

For the full reduction of the disulfide bridge of thioredoxin, two electrons are needed. However, ferredoxin, the electron donor for this reduction, is a one-electron carrier. Therefore, the essential function of FTR is to catalyze the reduction of the disulfide in two consecutive steps and to stabilize the transiently formed one electron reduced intermediate. Unlike the NADP-dependent thioredoxin reductase, which is a flavoprotein, the FTR is a Fe–S protein which uses its [4Fe–4S] cluster to cleave the active site disulfide. This appears to be possible due to the close proximity of the active site disulfide and the cluster. One of the cysteines of the disulfide bridge, Cys_{57}, is at the molecular surface and, when reduced, acts as the attacking nucleophile. The second active site cysteine is protected from the solvent, and is so close to the Fe–S centre that it can interact with the cluster and probably never exists as a free thiol until the final reduction step.

In a first reduction step, initiated by an electron delivered by ferredoxin on the ferredoxin docking side, the active site disulfide is cleaved forming a thiol, located on the surface exposed Cys_{57} and a thiyl radical (Cys_{87}). The latter is stabilized by ligation to the cluster, thereby oxidizing it from the $[4Fe–4S]^{2+}$ to the $[4Fe–4S]^{3+}$ state. While the X-ray analysis of the oxidized FTR positions a cluster Fe atom closest to the Cys_{87}, suggesting their coordination in this one-electron reduced intermediate state (Dai *et al.*, 2000a), recent structural analyses of the reduced FTR (Dai *et al.*, unpublished) suggests a ligation between Cys_{87} and a sulfide ion of the cluster as originally proposed (Staples *et al.*, 1998). The exposed Cys_{57} now acts as attacking nucleophile and cleaves the disulfide of thioredoxin, forming a mixed disulfide between FTR and thioredoxin linked through a covalent intermolecular disulfide bond. In a second reduction step, again driven by an electron delivered by a second ferredoxin molecule, the bond between the cluster and Cys_{87} is cleaved and the cluster is reduced to its original 2+ oxidation state. The Cys_{87}, which has now become a nucleophilic thiol, attacks the intermolecular disulfide releasing the fully reduced thioredoxin, and returning the FTR active-site to its disulfide state.

This proposed reaction scheme, which is based on spectroscopic analyses of native and active-site modified FTR (Staples *et al.*, 1996, 1998), is entirely compatible with the structural model derived from the crystallographic data (Dai *et al.*, 2000a). The described sequence of interactions is possible because of the particular disk-like structure of the FTR, enabling the simultaneous docking

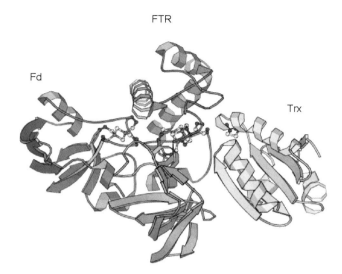

Figure 1.3 Modeling of the interaction between ferredoxin (red, to the left), ferredoxin:thioredoxin reductase (catalytic subunit in blue and variable subunit in green) and thioredoxin (yellow, to the right). The thin, disk-like structure of the FTR allows simultaneous docking of ferredoxin and thioredoxin on opposite sides of the molecule. The Fe–S centers and the disulfide bridges are shown in ball and stick representation. (Reproduced with permission from Dai *et al.*, 2000a. Copyright Cambridge University Press.)

of ferredoxin and thioredoxin on opposite sides of the molecule and the rapid transfer of electrons across its centre (figure 1.3).

1.4.3 Target enzymes

1.4.3.1 NADP malate dehydrogenase

One of the most strictly thioredoxin-dependent target enzymes is the chloroplastic NADP malate dehydrogenase (EC 1.1.1.82) (Miginiac-Maslow *et al.*, 2000). This enzyme is involved in the reduction of oxaloacetate into malate using the NADPH generated by photosynthetic electron transfer. It has no activity at all in the oxidized form (in the dark). Its activation in the light by reduced thioredoxin is slow (ca. 10 min) with a distinct lag phase, and is known to be inhibited by $NADP^+$ (Ashton and Hatch, 1983; Scheibe and Jacquot, 1983). In contrast with the strong interaction between ferredoxin and FTR, there is no stable complex formed between NADP-MDH and thioredoxin. Both proteins are separated easily from each other by size-exclusion chromatography, and none of the partners can be retained on an affinity column made with one or other of the two proteins. Almost any thioredoxin isoform is able to activate NADP-MDH (chloroplastic m and f, but also cytosolic h, the latter, however, with a somewhat lower efficiency). Nevertheless, reduced thioredoxin cannot be replaced with dithiothreitol (DTT) in standard conditions, and the invariant thioredoxin's active site tryptophan residue (**Trp**CysGlyProCys) is required for activation. The *W35A* mutant thioredoxin h is unable to activate wild-type MDH (Krimm *et al.*, 1998), while mutant thioredoxin is reduced very efficiently with NTR. Hence, this residue seems to be important for protein–protein contacts between MDH and thioredoxin.

Unlike other thioredoxin-dependent enzymes, whose activation requires the reduction of one disulfide per monomer, this homodimeric protein requires the reduction of two different disulfides per monomer. These disulfides have been localized by site-directed mutagenesis and chemical derivatization, and their position has been confirmed recently when the 3D structure of the oxidized enzyme has been solved (Carr *et al.*, 1999, for the *Flaveria* enzyme, Protein Data Bank file 1CIV; Johansson *et al.*, 1999, for the sorghum enzyme, Protein Data Bank file 7MDH). They are located respectively at the N-terminal and C-terminal ends of each monomer, in sequence extensions specific for the thioredoxin-regulated NADP-dependent forms and absent from the extra-chloroplastic, constitutively active NAD-dependent forms (EC 1.1.1.37). In sorghum NADP-MDH, the N-terminal bridge links Cys_{24} and Cys_{29} located in a ca. 43 amino acid long N-terminal extension. The C-terminal bridge links Cys_{365} and Cys_{377}. Only the latter cysteine belongs to the approximately 17 amino acid long unstructured C-terminal extension. The Cys_{365} residue is located on the conserved last α-helix of the protein core. These disulfides are strictly conserved among all the higher plant NADP-MDHs sequenced so far. These include sequences from two C3, two C4 and one CAM-type plants, as are the four more internal cysteines at 175, 182, 207 and 328 (sorghum numbering) (see Johansson *et al.*, 1999, for sequence alignments). All eight cysteines are also conserved in mosses (*Selaginella*), but both cysteines of the N-terminal extension are missing in the NADP-MDH of green algae (Ocheretina *et al.*, 2000).

1.4.3.2   *Role of the* C-*terminal disulfide*

Early mutagenesis studies (Issakidis *et al.*, 1992, 1994, 1996; Riessland and Jaenicke, 1997) led to the conclusion that the C-terminal extension was involved in obstructing the active site. Indeed, mutation of the C-terminal cysteines conferred a weak catalytic activity on the oxidized enzyme, with a high K_m for oxaloacetate, suggesting that the mutation opened access to the active site for the substrate. This conclusion is fully supported by the structural data that show that the C-terminal bridge pulls the extension towards the core of the protein, where its C-terminal end occupies the space devoted to substrate binding. Further site-directed mutagenesis studies have identified the penultimate C-terminal glutamate residue (Ruelland *et al.*, 1998) and the active-site-located Arg_{204} (Schepens *et al.*, 2000a) as being the interacting residues preventing the binding of oxaloacetate. Indeed, Arg_{204} is also one of the two arginines (along with Arg_{134}) directly involved in oxaloacetate binding (Schepens *et al.*, 2000a). From the structure, it is clearly apparent that the *C*-terminal end forms interactions with this Arg, and also with the catalytic His_{229} residue, thus stabilizing the extension inside the active site. The puzzling inhibitory effect of $NADP^+$ on activation was finally solved from the structural data. Site-directed mutagenesis showed that mutation of either of the cysteines of the *C*-terminal bridge to alanines (Issakidis *et al.*, 1994) or of the penultimate glutamate residue to a glutamine (Ruelland *et al.*, 1998) suppressed the inhibitory effect, suggesting an interaction of $NADP^+$ with the negatively charged *C*-terminal tip. The site-directed mutation of the cofactor specificity for catalysis towards NADH also changed the cofactor specificity for inhibition towards NAD^+ instead of $NADP^+$: clearly, the inhibitory effect involved the cofactor binding site (Schepens *et al.*, 2000b). The availability of the structure reconciled both observations. A bound $NADP^+$ molecule is visible in the structure of the *Flaveria* enzyme. Its positively charged nicotinamide moiety is facing the penultimate glutamate residue (Carr *et al.*, 1999) and interacting with its side-chain carboxyl group. This interaction strengthens the locking of the active site, and logically should delay the release of the *C*-terminal end upon the reduction of the *C*-terminal disulfide bridge.

No crystallographic data are available yet for the reduced (active) enzyme. However, two-dimensional proton nuclear magnetic resonance (NMR) spectra obtained with a monomeric form truncated at the *N*-terminus show that in the reduced enzyme, the 15 amino acid long *C*-terminal stretch (Ala_{375} to the *C*-terminal valine) acquired a mobility similar to the motion of a free peptide, indicating that it has been released from the active site (Krimm *et al.*, 1999). The way the peptide is rebound upon oxidation is still unknown, but the deactivation of the enzyme is also thioredoxin-dependent. This observation is consistent with the redox potential value of thioredoxin disulfide that is less electronegative than the redox potential of the disulfides of NADP-MDH (-280 mV compared with -330 mV for the *C*-terminal disulfide; Hirasawa *et al.*, 2000). However, it also suggests that thioredoxin might exert a structural effect on this process.

1.4.3.3 Role of the N-*terminal disulfide*
Whereas the role of the *C*-terminal extension as an 'internal inhibitor' is well-documented, warranting the classification of NADP-MDH among the enzymes regulated by 'intrasteric inhibition', such as some protein kinases (Kobe and Kemp, 2000), the role of the *N*-terminal disulfide is much less clear. Nevertheless, it is well established that the events triggered by its reduction are rate-limiting for activation: its elimination by site-directed mutagenesis greatly accelerates the activation rate, which becomes completed within one minute. This observation led to the hypothesis that the reduction of the *N*-terminal disulfide is followed by a slow structural change at the active site towards

better catalytic efficiency. The precise nature of the structural change is still unknown. However, the high K_m for oxaloacetate observed in oxidized mutant proteins devoid of the C-terminal bridge is markedly lowered upon activation. It has also been reported that mutation of the N-terminal cysteines, or the deletion of the N-terminal extension, loosened the interaction between subunits, which could then be dissociated more easily into monomers by increasing the ionic strength of the medium. Moreover, one of the conserved four internal cysteines, namely Cys_{207}, appears to be implicated in the activation process: its mutation also accelerates the activation rate. Combined cysteine mutations showed that a thioredoxin-reducible disulfide can be formed between Cys_{24} and Cys_{207} (Ruelland et al., 1997; Hirasawa et al., 2000). The Cys_{207} residue was also shown to be accessible to thioredoxin. Here, a mixed disulfide forms between this cysteine and a single cysteine mutant thioredoxin when the N-terminal disulfide is missing (Goyer et al., 1999). These observations suggest that a disulfide isomerization may occur during the activation process and that thioredoxin might act as a disulfide isomerase. The 3D-structure shows that the Cys_{207} of one subunit is sufficiently close to Cys_{24} of the other subunit to be able to form an intersubunit disulfide. However, the structure also shows that the N-terminal bridge stabilizes a short α-helix in the extension and that this helix is bound by many hydrophobic interactions to both subunits. The disruption of these interactions and release of the N-terminal extension could also be rate-limiting for activation. The postulated conformational change might then result in an increased flexibility of the active site, necessary for a high catalytic competence.

A sequential view of the activation mechanism can be drawn from the redox potentials of the individual disulfides of NADP-MDH (Hirasawa et al., 2000). The N-terminal disulfide is the less electronegative ($-290\,mV$) and would be almost in equilibrium with reduced thioredoxin. It is also the less structurally demanding since a mutant MDH, where only this disulfide remains, can be activated by the W35A mutant thioredoxin (though less efficiently than with the wild-type thioredoxin). On the contrary, the reduction of the C-terminal disulfide ($-330\,mV$) would require an excess of structurally unaffected reduced thioredoxin. Then, the N-terminal disulfide could be the 'preregulatory disulfide' postulated by Hatch and Agostino (1992). Its reduction would lead to a rapidly activatable form of the enzyme, immediately functional when the reducing power of the chloroplast is increased. It has been reported that the active site unlocking by elimination of the negative charge of the penultimate glutamate is facilitated when there is no N-terminal bridge present (Ruelland et al., 1998). This suggests a synergy between the reduction of the N-terminal bridge, and the release of the C-terminal extension from the active site.

Although many aspects of the activation of NADP-MDH have been clarified, the structural role of thioredoxin is still somewhat obscure and requires the co-crystallization of both partners in order to fully understand the protein–protein interactions involved. The 3D structure suggests that Cys_{29} and Cys_{377} are the most accessible cysteines of the regulatory disulfides, and should be the primary targets of a nucleophilic attack by reduced thioredoxin. The observation that MDH activation is markedly accelerated by increasing the ionic strength of the medium underlines the importance of hydrophobic interactions in the activation process.

A 3-D structure of the active form of the enzyme would certainly contribute to the understanding the exact nature of the modification of the interaction between monomers and the consequences of this on the structure/flexibility of the active site. A critical aspartic acid residue implicated in dimerization has been identified at the monomer interface area but its function is poorly understood (Schepens *et al.*, 2000c). It has been observed that in the oxidized enzyme, there is a slight rotational shift of the catalytic site of one of the monomers, compared with the structure of the active, NAD-dependent forms. This might result in a catalytically unfavorable conformation (Dai *et al.*, 2000a).

The highly sophisticated regulation of NADP-MDH (figure 1.4) in higher plants enables these organisms to adjust enzyme activity very precisely and rapidly in response to metabolic needs and to cope with changing environmental conditions. Indeed, transgenic approaches have shown that the activation level of the enzyme is higher in lines expressing lower levels of the transgene and is lower in overexpressors. This feature allows a change in enzyme activity within minutes, whereas hours would be needed to synthesise new enzyme *de novo* (Trevanion *et al.*, 1997; Faske *et al.*, 1997).

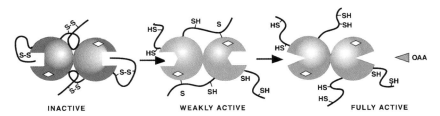

Figure 1.4 Schematic representation of the activation steps of NADP-MDH by reduced thioredoxin. The enzyme is a homodimer of two asymetric subunits. In its inactive form (left) its *C*-terminal ends (on the outside) are pulled back by disulfide bridges linking Cys_{365} and Cys_{377} and obstruct the active sites. The *N*-terminal ends form short α-helices stabilized by disulfides linking Cys_{25} and Cys_{29} and are inserted at the dimer interface area. The active site is in a catalytically unfavourable conformation (square). Upon reduction of the disulfides by reduced thioredoxin (middle), the *C*-termini are expelled out of the active site. The *N*-terminal bridges are isomerized and form intersubunit disulfides linking Cys_{24} of one subunit with Cys_{207} of the other subunit (this step is still somewhat hypothetical). The active site starts changing conformation but is only weakly active. Upon reduction of this intermediary disulfide (right), the active site reaches its final highly active conformation (triangle). The cofactor binding site is denoted by a diamond.

1.4.4   Fructose-1,6-bisphosphatase

Fructose-1,6-bisphosphatase (FBPase) is an enzyme which is present in several cell compartments in plants. The reaction catalyzed by this protein is the cleavage of the substrate fructose-1,6-bisphosphate (FBP) to fructose-6-phosphate (F6P) and inorganic phosphate. One isoform present in the cytosol is involved in the synthesis of sucrose and its subsequent export to non-photosynthetic tissues. The chloroplastic enzyme on the other hand is involved in the Calvin cycle responsible for the fixation of CO_2 into organic compounds. Both enzymes depend on Mg^{2+} for their function. In addition, it has long been known that the chloroplastic eukaryotic protein is also subject to redox regulation (for a review of the historic elucidation, see Buchanan, 1980). A difference between FBPase and the NADP-MDH is that FBPase can also be activated in the oxidized state by incubating it with high Mg^{2+} concentrations, in a highly cooperative process (Chardot and Meunier, 1991). The redox-dependent regulation of eukaryotic chloroplast FBPase is strictly dependent on thioredoxin f. The enzyme is very poorly activated by thioredoxin m, with the exception of chloroplastic pea thioredoxin m that shows some reactivity (Schürmann et al., 1981; Hodges et al., 1994; Geck et al., 1996; Lopez-Jaramillo et al., 1997). On the other hand, the cyanobacterial FBPase can also be activated by glutathione and lacks the regulatory insertion (see below) (Udvardy et al., 1982).

1.4.4.1   Redox active site

The redox active site of FBPase has been elucidated from amino acid sequence comparisons and from site-directed mutagenesis. The primary structure of the chloroplastic protein has been determined both by direct amino acid sequencing for spinach (Marcus et al., 1987) and from cDNA sequences from wheat (Raines et al., 1988; Tang et al., 2000), pea (Carrasco et al., 1994) and spinach (Martin et al., 1996). The plant chloroplastic enzyme is not only homologous to the cytosolic plant enzyme but also to the mammalian gluconeogenic enzyme, despite the fact that the proteins do not cross-react with the antibodies directed against one another (Fonolla et al., 1994; Hur and Vasconcelos, 1998). Sequence comparisons however reveal that the chloroplastic redox regulated enzyme contains an insertion known as the 170s loop of ca. 30 amino acids that is absent in cytosolic FBPases, leading to the proposal that the redox active site is located in this loop. As four cysteines are present in this regulatory insertion (Cys_{153}, C_{173}, C_{178} and C_{190} in pea), it was necessary to use site-directed mutagenesis to discriminate which of the cysteines are involved in redox regulation. Results from three independent laboratories on three different enzymes indicate that Cys_{190} is clearly not implicated in the redox process, that Cys_{153} is clearly essential. The removal of Cys_{173} or Cys_{178} strongly deregulates the enzyme, but it does maintain some of its wild-type redox characteristics (Jacquot et al.,

1997b; Rodriguez-Suarez *et al.*, 1997; Balmer and Schürmann, 2001). From these results it has been proposed that the redox active site is probably situated between the obligatory Cys_{153} and either Cys_{173} or Cys_{178}.

1.4.4.2   3D structural properties, catalytic site and consequences on the regulatory process

The 3D structure of chloroplast FBPase has been elucidated for two enzymes, one from spinach, (presumably in a reduced form; Villeret *et al.*, 1995), and one from pea in the oxidized form (Chiadmi *et al.*, 1999). The enzyme is a tetramer comprising identical subunits of ca. 340 amino acids each. In the oxidized form, the regulatory disulfide bridge was found to be present between the sulfur atoms of Cys_{153} and Cys_{173} in the pea enzyme (Chiadmi *et al.*, 1999). In the 3D structure, the regulatory disulfides are located in the outside corners of each monomer, distant from the active sites. The regulatory redox insertion forms a loop that connects the two structural units. As this loop is believed to be flexible, Cys_{178}, which is located close to the disulfide bridge, could possibly replace (at least partially) Cys_{173} as the second partner of the disulfide bridge with Cys_{153}, which may account for the mixed results of the site directed mutagenesis experiments. No additional structural disulfide could be located in the structure, contrary to the proposal of Drescher *et al.* (1998).

Comparing the structures of the mammalian pig heart enzyme to spinach, and those of the wild-type and $Cys_{153}Ser$ mutant of the pea, identifies the active site in the chloroplastic enzymes. It is clearly similar in all structures with amino acid residues conserved for the binding of the substrate fructose-1,6-bisphosphate (FBP). One major difference is however the positioning of a critical glutamate residue which in the active or activated enzyme binds a Mg^{2+} ion in a conformation favorable for catalysis, i.e. close to the FBP molecule. In the oxidised inactive enzyme the position of the glutamate is occupied by the side chain of a valine residue. As this amino acid is not charged, it is believed that the catalytic Mg^{2+} ion cannot bind rendering the enzyme inactive. The proposed mechanism for activation of the enzyme suggests that upon reduction, the loop containing the regulatory sequence is relaxed, allowing two critical β strands, linking the regulatory loop to the active site region, to move outwards. This in turn displaces the side chain of the valine residue and gives room for the side chain of the glutamate. As a result, Mg^{2+} can now bind and the enzyme becomes catalytically active.

1.4.4.3   The thioredoxin f FBPase interaction site

As outlined previously, in order to make the enzyme catalytically active under physiological conditions, it is necessary to reduce it with thioredoxin f. Due to the high specificity of this interaction, there is indeed little doubt that thioredoxin f is the physiological partner of FBPase. However, some observations, such as the high expression of its mRNA in non-photosynthetic tissues (Pagano *et al.*,

2000), suggest that thioredoxin f could have other functions as well. In addition FBPase is certainly not the only enzyme activated by thioredoxin. The protein also activates NADP-MDH and phosphoribulokinase (PRK) quite efficiently (Schürmann *et al.*, 1981; Brandes *et al.*, 1993; Hodges *et al.*, 1994). A very recent report suggests, however, that thioredoxin f is not the physiological activator of PRK, as thioredoxin m seems to be more efficient in this process (Geck and Hartman, 2000). Since the FBPase/thioredoxin f interaction is extremely specific (unlike most other interactions involving thioredoxin), it is of interest to understand the nature of that specific interaction. In the absence of a 3D structure of the two interacting proteins this problem has been studied by site-directed mutagenesis on both protein partners. As will be detailed below, the FBPase/thioredoxin f site seems to be largely electrostatic with negative charges on the phosphatase matching corresponding positive charges on the thioredoxin f partner protein.

However, information available about the nature of the thioredoxin interaction site of FBPase is relatively scarce. It has been observed that the sequence of the regulatory insertion shows several negative charges clustered between Cys_{153} and Cys_{173}, the two residues proposed to be involved in the disulfide bond. Experiments aimed at expressing a peptide similar to this region (Hermoso *et al.*, 1996), or at deleting this region from the FBPase sequence (Sahrawy *et al.*, 1997), strongly suggest that this amino acid stretch is of primary importance for docking the thioredoxin f molecule.

On the other hand there is a lot more information about the thioredoxin f interaction site. This problem has been approached in different ways, including evaluating the capacity of thioredoxins related to thioredoxin f to activate FBPase, engineering thioredoxin molecules that have either a better or a lower reactivity with FBPase, constructing chimeric thioredoxins made of f/m fusions, and by solving the 3D structure of thioredoxin f and comparing it to the other poorly reactive thioredoxins (*E. coli*, m and h).

A first observation is that despite quite interesting homology between thioredoxin f and mammalian thioredoxin, the human thioredoxin is essentially unable to activate FBPase (Jacquot *et al.*, 1990). Site directed mutagenesis experiments have helped identify the primary nucleophile of thioredoxin f which was found to be Cys_{46} in spinach (Brandes *et al.*, 1993). Engineering the *E. coli* thioredoxin has shown that an asparagine residue of thioredoxin f, homologous to Asp_{60} in *E. coli* thioredoxin, is very important for the reactivity (de Lamotte-Guery *et al.*, 1991). It has also been observed that the addition of two positive charges to *E. coli* thioredoxin (mutations *E30K, L94K*) makes it a more efficient catalyst for FBPase reduction (Mora-Garcia *et al.*, 1998). Moreover, Geck *et al.* (1996) have identified Lys_{58}, Asn_{74}, Gln_{75}, and Asn_{77} of thioredoxin f as important determinants of its specificity. Fusions between thioredoxin f and thioredoxin m have been performed taking advantage of a restriction site present at the active site. While the f/m molecule is unstable, the m/f protein is inactive with MDH, but active with FBPase (Lopez-Jaramillo *et al.*, 1998). These authors postulate that the stability of thioredoxins resides on the N-side of the disulfide, and is associated with the acidic character of this fragment and, as a consequence, with the acidic pI of the whole molecule. In contrast, the ability of FBPase binding and enzyme catalysis depends on the structure on the C-side of the regulatory cysteines, a proposal that fits well with the effects of the mutations described above. In thioredoxin f, there is a third cysteine residue (Cys_{73} in spinach) that is also quite important for the reactivity of the protein, although it is not esential for catalysis (del Val et al., 1999).

Efforts at solving the 3D structure of thioredoxin f have recently paid off. The protein was expressed either with an N-terminal extension or, to avoid solubility problems, as an N-terminus truncated form (del Val et al., 1999). Thioredoxin f contains an additional α-helix at the N terminus compared with thioredoxin m, and an exposed third cysteine close to the active site (Cys_{73}). The overall 3D structures of the two chloroplast thioredoxins f and m are quite similar. However, the two proteins have a significantly different surface topology and charge distribution around the active site. An interesting feature that might significantly contribute to the specificity of thioredoxin f is an inherent flexibility of its active site (Capitani et al., 2000).

1.4.5   Interaction of plant thioredoxins with yeast targets

Recent progress in plant genomics has shown that there are several thioredoxin types encoded by multiple genes. One important question is to establish whether or not the products of these genes have redundant functions (i.e. if they interact with the same proteins) and eventually to discover new targets for the newly discovered thioredoxins. This can be studied biochemically with the limitation that some plant thioredoxin targets (with the notable exception of FBPase) are significantly activated in vitro by thioredoxins that are not located in the same compartment in the plant cell (e.g. thioredoxin h versus NADP-MDH). Because of the ease of manipulation, Saccharomyces cerevisiae appears as an attractive model to test the specificity of plant thioredoxins in vivo. This small eukaryote has two thioredoxin genes encoding almost redundant functions. Single mutants do not show clear phenotypes while the double mutant has numerous abnormalities including a disturbed cell cycle with a very long S phase, inability to use mineral sulfate or methonine sulfoxide as a sulfur source (but demonstrated good growth on methionine or cysteine) and a higher sensitivity to H_2O_2 or tertio butyl hydroperoxide (Muller, 1991, 1996). The protein targets implicated in this complex phenotype are not known, but the simplest hypotheses consider that in absence of thioredoxin, the ribonucleotide reductase is less efficient (because it would be reduced less efficiently by glutaredoxins) leading to a long S phase. Further, in the mutant, sulfate assimilation would be blocked because 3′-phosphoadenylylsulfate reductase (PAPS reductase) needs thioredoxin as a proton donor, methionine sulfoxide assimilation would be strongly reduced because methionine sulfoxide reductase is thioredoxin dependent, and the higher sensitivity to peroxide could be due to one or more thioredoxin dependent peroxidases. Heterologous expression of five *Arabidopsis* thioredoxins *h* in the double thioredoxin mutant of *Saccharomyces cerevisiae* leads to different phenotypes. In these, all restore a normal cell cycle and, with different efficiencies, growth on methionine sulfoxide, but only *h2* restores sulfate assimilation while only *h3* restores peroxide tolerance. This strongly suggests a functional specialization of the different members within the h family (Mouaheb *et al.*, 1998). Hybrid proteins were constructed between *h2* and *h3* showing that all determinants responsible for sulfate assimilation are located in the *C*-terminal part of the thioredoxin, while the determinants for peroxide tolerance are located in both parts. The *h3* gene has an atypical active site (TrpCys**Pro**ProCys) and mutation of this into the classic TrpCysGlyProCys site has no effect on peroxide tolerance, but restores some ability to assimilate sulfate. Thus one consequence of the atypical site TrpCysProProCys is to limit the access of thioredoxin to some targets (Brehelin *et al.*, 2000). A similar study was performed with the four members of the thioredoxin m family and with the unique member of a recently discovered new x type thioredoxin, probably located in the chloroplast (Issakidis-Bourguet *et al.*, 2001). These proteins are not able to induce sulfate assimilation in the double thioredoxin mutant, and have little effect on methionine sulfoxide assimilation. Thioredoxins m1, m2, m4 and x induce peroxide tolerance, suggesting that they could have such a function in the chloroplast. In contrast, m3 induces a hypersensitive response to hyperoxides. A tentative interpretation of this unexpected result is that m3 has no affinity for thioredoxin dependent peroxidases, but induces Yap1 reduction, the transcription factor that acts as an oxidative stress sensor in yeast and induces the cell response to oxidative stress when it is in the oxidized state. The m3 induced reduction of Yap1 would block the constitutive expression of the antioxidant defence genes previously demonstrated in the yeast mutant devoid of thioredoxins (Izawa *et al.*, 1999). It is not possible, on the basis of heterologous expression in yeast, to clearly assign a function to the multiple plant thioredoxins. Nevertheless, it is clear that even within a given type, members of gene families display different abilities to interact with target proteins, and so are implicated in processes with vastly different consequences.

1.5 New approaches for the characterization of thioredoxin targets

Analysis of plant genomes has revealed the presence of numerous plant thioredoxins compared with the small number of presently identified thioredoxin targets. Unfortunately it is not possible to deduce the presence of regulatory disulfide bridges in proteins on the basis of the amino acid sequences. Nevertheless using the sequence data provides the means to search for the presence of homologs of proteins described as thioredoxin targets in other organisms in plant genomes. Conserved cysteines at the position of the redox active cysteines strongly suggest that these proteins are redox regulated in plants. Table 1.3 shows the presence or absence of putative thioredoxin targets in the *Arabidopsis thaliana* genome, deduced from proteins already identified in bacteria, yeast or vertebrates. This survey of the *Arabidopsis* genome allows for the discovery of putative targets implicated in DNA synthesis (ribonucleotide reductase), and defense against oxidative stress (thiol peroxidases). It confirms the absence of 3'-phosphoadenylysulfate (PAPS) reductase in higher plants, sulfate assimilation taking place through 5'-adenylyl sulfate reductases (Hatzfeld *et al.*, 2000). No homolog of the redox regulated transcription factors of yeast and mammals, nor to the redox sensors, has been found, most probably because of the low conservation of the sequences in different organisms.

Table 1.3 Putative *Arabidopsis thaliana* targets characterized by homology with thioredoxin targets from other organisms: A blastn search (Altschul *et al.*, 1990) was performed using yeast, *E. coli* or human protein sequences against *Arabidopsis* genomic sequences. The sequences showing at least 40% identity and conserved cysteine positions are described by their position on *Arabidopsis* chromosomes or chromosome sections (available only for chromosome 2 and 4), and by their presence in particular bacterial cosmids used during the sequencing process. The putative localization was inferred for the deduced protein sequence using Psort (Nakai and Kanehisa, 1992)

Protein bait	Organism	*Arabidopsis* chromosome	BAC	Putative protein localization
RNR small[a]	*S. cerevisiae*	3	M0J10	Cytosolic
		3	MDB19	Cytosolic
RNR large[a]	*S. cerevisiae*	2-124	F7D8	Cytosolic
PAPS reductase	*S. cerevisiae*	none		
Tsa1 or Tsa2[b]	*S. cerevisiae*	3	T19F11	Chloroplastic
		5	MHF15	Chloroplastic
YBG4[b]	*S. cerevisiae*	1	F21D18	Cytosolic (seed)
YLR109[b]	*S. cerevisiae*	1	F12P19 (tandem)	Cytosolic
		1	F8A5	Cytosolic
		3	F8J2	Chloroplastic
PMP20[b]	*S. cerevisiae*	none		
MetSO reductase	*S. cerevisiae*	4-62	F13M23	Cytosolic
		5	K11J9	Cytosolic
Yap1[c]	*S. cerevisiae*	none		
Skn7[c]	S. cerevisiae	none		
AP1[c]	Human	none		
OxyR[d]	*E. coli*	none		

[a]RNR small and large are the subunits of ribonucleotide reductase.
[b]Peroxiredoxins.
[c]Transcription factors.
[d]Oxidative stress sensor.
Abbreviations: RNR, ribonucleotide reductase; PAPS, 3'-phosphoadenylylsulfate; MetSO, methionine sulfoxide.

Other means have been developed to uncover thioredoxin targets. Yano *et al.* (2001) set up a new method for the characterization of thioredoxin reduced proteins in peanut seeds. It is based on monobromobimane labeling of sulfhydryl groups after disulfide reduction. The untreated, dithiothreitol or thioredoxin treated proteins were separated by two-dimensional electrophoresis. Thioredoxin reduced spots were microsequenced. This allowed the characterization of ten proteins, including storage proteins and allergens previously known to form disulfide bridges, as well as one homolog to a desiccation-inducible protein in which no disulfide formation was suspected. The method can also be applied to different organs and tissues and it is expected that with the progress of proteomics, new potential thioredoxin targets will be identified this way. In order to identify the preferred interaction partner of the various thioredoxins, an interesting approach is through *in vivo* methods. The classical approach consists of the isolation of target-thioredoxin complexes. However, this is particularly difficult in the thioredoxin field due to the rapid release of the target after reduction. Nevertheless, our knowledge of the reaction mechanism helped to develop a new strategy. As indicated previously, the disulfide reduction takes place in two steps including the transient formation of a mixed disulfide-linked heterodimer. This disulfide is rapidly cleaved by the second cysteine of the wild-type thioredoxin, but can be stabilized *in vitro* when formed with a thioredoxin mutated in the second cysteine. Verdoucq *et al.* (1999), have introduced in EMY63, the thioredoxin double mutant yeast strain, an engineered thioredoxin h3 from *Arabidopsis* where a serine was substituted for the second cysteine of the active site. In addition, a polyhistidine tag was added to the N terminus to allow purification of thioredoxin-target complexes on a Ni-agarose column. A binary complex was purified in which the thioredoxin was associated with a protein encoded by the yeast Open Reading Frame YLR109 that had no known function at the time of the experiment. Low similarities with previously described thioredoxin dependent reductases suggested a possible function. The protein was produced in *E. coli* and its ability to reduce H_2O_2 in the presence of thioredoxin h and NTR was demonstrated. The protein was thus identified as a thioredoxin-dependent peroxidase. The function of this protein was demonstrated independently, and by different approaches, by two other teams (Jeong *et al.*, 1999; Lee *et al.*, 1999). Four genes encode homologous proteins in *Arabidopsis* (one chloroplastic and three cytosolic), and so can be added to the list of the potential plant thioredoxin targets. In addition, the complex was isolated only when the engineered thioredoxin h3 was introduced into the mutant yeast strain. In the wild-type, only free thioredoxin h3 was recovered, suggesting that in the wild-type the complex is directly reduced by the endogenous thioredoxin. Another possibility is that the complex is reduced by the more reduced state of the wild-type yeast (Muller, 1996). Experiments to recover other yeast thioredoxin targets by this *in vivo* approach are in progress.

One popular method to demonstrate protein–protein interaction is the two hybrid system. Here, two domains of a transcription factor are individually linked to two proteins, or protein domains. If the proteins interact, this reconstitutes the active transcription factor which induces the transcription of reporter genes. This very elegant method allows screening for cDNAs encoding proteins interacting with a bait. There is only one congress report describing a two hybrid screening for thioredoxin targets. Clearly the difficult point of this technology is the discrimination between true and false positives, particularly in the case of thioredoxin because the interaction with the target is particularly short lived. Nevertheless, a ribonucleotide reductase was among the putative positives. This protein is one of the enzymes that uses thioredoxin or a glutaredoxin as proton donor. This suggests that the two hybrid method can be successfully used for the characterisation of thioredoxin targets. In addition, the interaction could be strengthened by mutation of the second cysteine of the thioredoxin active site (described above), and by doing the experiments in a yeast devoid of thioredoxins.

1.6   Conclusions

Recently, biochemical approaches associated with structure elucidation have allowed a detailed description of the ferredoxin/FTR/thioredoxin interactions and of NADPH-MDH regulation. It is believed that with progress in recombinant protein production, in protein crystallization and NMR, our knowledge on the regulation of chloroplastic proteins will extend rapidly. At the same time some doubt persists on the respective role of thioredoxin f and thioredoxin m in the regulation of chloroplastic enzymes. This question appears now far more complex in the light of the discovery that each thioredoxin type is encoded by numerous genes and also that previously unknown thioredoxin types are present in higher plant genomes. Genetic approaches in bacteria have shown that each thioredoxin has specificity but also some overlapping functions with the other thioredoxins and even with glutaredoxins (Aslund and Beckwith, 1999). Most probably progress in the understanding of the thioredoxin/glutaredoxin cascades in higher plants will necessitate the introduction of genetics, genomics and proteomics in this research field.

References

Altschul, S.F., Gish, W., Miller, W., Myers, E.W. and Lipman, D.J. (1990) Basic local alignment search tool. *J. Mol. Biol.*, **215**, 403-410.

Ashton, A.R. and Hatch, M.D. (1983) Regulation of C4 photosynthesis: regulation of activation and inactivation of NADP-malate dehydrogenase by NADP and NADPH. *Arch. Biochem. Biophys.*, **227**, 416-424.

Aslund, F. and Beckwith, J. (1999) The thioredoxin superfamily: redudancy, specificity, and gray-area genomics. *J. Bacteriol.*, **181**, 1375-1379.

Babiychuk, E., Kushnir, S., Van Montagu, M. and Inze, D. (1994) The *Arabidopsis thaliana* apurinic endonuclease Arp reduces human transcription factors *Fos* and *Jun. Proc. Natl Acad. Sci. USA*, **91**, 3299-3303.

Balmer, Y. and Schürmann, P. (2001) Heterodimer formation between thioredoxin *f* and fructose 1,6-bisphosphatase from spinach chloroplasts. *FEBS Lett.*, **492**, 58-61.

Brandes, H.K., Larimer, F.W., Geck, M.K., Stringer, C.D., Schürmann, P. and Hartman, F.C. (1993) Direct identification of the primary nucleophile of thioredoxin *f. J. Biol. Chem.*, **268**, 18,411-18,414.

Brehelin, C., Mouaheb, N., Verdoucq, L., Lancelin, J.M. and Meyer, Y. (2000) Characterisation of determinants for the specificity of *Arabidopsis* thioredoxins *h* in yeast complementation. *J. Biol. Chem.*, **275**, 31,641-31,647.

Brugidou, C., Marty, I., Chartier, Y. and Meyer, Y. (1993) The *Nicotiana tabacum* genome encodes two cytoplasmic thioredoxine genes which are differentially expressed. *Mol. Gen. Genet.*, **238**, 285-293.

Buchanan, B.B. (1980) Role of light in the regulation of chloroplast enzymes. *Annu. Rev. Plant Physiol.*, **31**, 341-374.

Capitani, G., Markovic-Housley, Z., DelVal, G., Morris, M., Jansonius, J.N. and Schürmann, P. (2000) Crystal structures of two functionally different thioredoxins in spinach chloroplasts. *J. Mol. Biol.*, **302**, 135-154.

Carr, P.D., Verger, D., Ashton, A.R. and Ollis, D.L. (1999) Chloroplast NADP-malate dehydrogenase: structural basis of light-dependent regulation of activity by thiol oxidation and reduction. *Structure*, **7**, 461-475.

Carrasco, J.L., Chueca, A., Prado, F.E. *et al.* (1994) Cloning, structure and expression of a pea cDNA clone coding for a photosynthetic fructose-1,6-bisphosphatase with some features different from those of the leaf chloroplast enzyme. *Planta*, **193**, 494-501.

Chardot, T. and Meunier, J.C. (1991) Properties of oxidized and reduced spinach (*Spinacia oleracea*) chloroplast fructose-1,6-bisphosphatase activated by various agents. *Biochem. J.*, **278**, 787-791.

Chiadmi, M., Navaza, A., Miginiac-Maslow, M., Jacquot, J.P. and Cherfils, J. (1999) Redox signalling in the chloroplast: structure of oxidized pea fructose-1,6-bisphosphate phosphatase. *EMBO J.*, **18**, 6809-6815.

Dai, S., Saarinen, M., Ramaswamy, S., Meyer, Y., Jacquot, J.P. and Eklund, H. (1996) Crystal structure of *Arabidopsis thaliana* NADPH dependent thioredoxin reductase at 2.5 Å resolution. *J. Mol. Biol.*, **264**, 1044-1057.

Dai, S., Schwendtmayer, C., Ramaswamy, S., Eklund, H. and Schürmann, P. (1998) Crystallization and crystallographic investigations of ferredoxin:thioredoxin reductase from *Synechocystis* sp. PCC6803, in *Photosynthesis: Mechanisms and Effects* (ed. G. Garab), Kluwer Academic Publishers, Dordrecht, pp. 1931-1934.

Dai, S., Schwendtmayer, C., Johansson, K., Ramaswamy, S., Schürmann, P. and Eklund, H. (2000a) How does light regulate chloroplast enzymes? Structure–function studies of the ferredoxin/thioredoxin system. *Q. Rev. Biophys.*, **33**, 67-108.

Dai, S., Schwendtmayer, C., Schürmann, P., Ramaswamy, S. and Eklund, H. (2000b) Redox signaling in chloroplasts: cleavage of disulfides by an iron–sulfur cluster. *Science*, **287**, 655-658.

Darby, N.J. and Creighton, T.E. (1995) Characterization of the active site cysteine residues of the thioredoxin-like domains of protein disulfide isomerase. *Biochemistry*, **34**, 16,770-16,780.

de Lamotte-Guery, F., Miginiac-Maslow, M., Decottignies, P., Stein, M., Minard, P. and Jacquot, J.P. (1991) Mutation of a negatively charged amino acid in thioredoxin modifies its reactivity with chloroplastic enzymes. *Eur. J. Biochem.*, **196**, 287-294.

del Val, G., Maurer, F., Stutz, E. and Schürmann, P. (1999) Modification of the reactivity of spinach chloroplast thioredoxin *f* by site-directed mutagenesis. *Plant Sci.*, **149**, 183-190.

de Pascalis, A.R., Schürmann, P. and Bosshard, H.R. (1994) Comparison of the binding sites of plant ferredoxin for two ferredoxin-dependent enzymes. *FEBS Lett.*, **337**, 217-220.

Drescher, D.F., Follmann, H., Häberlein, I. (1998) Sulfitolysis and thioredoxin-dependent reduction reveal the presence of a structural disulfide bridge in spinach chloroplast fructose-1,6-bisphosphatase. *FEBS Lett.*, **424**, 109-112.

Droux, M., Jacquot, J.-P., Miginiac-Maslow, M., *et al.* (1987) Ferredoxin-thioredoxin reductase, an iron–sulfur enzyme linking light to enzyme regulation in oxygenic photosynthesis: purification and properties of the enzyme from C3, C4, and cyanobacterial species. *Arch. Biochem. Biophys.*, **252**, 426-439.

Faske, M., Backhausen, J.E., Sendker, M., Singer-Bayerle, M., Scheibe, R. and von Schaewen, A. (1997) Transgenic tobacco plants expressing pea chloroplast NADP-MDH cDNA in sense and antisense orientation. *Plant Physiol.*, **115**, 705-715.

Feng, J.N., Model, P. and Russel, M. (1999) A trans-envelope protein complex needed for filamentous phage assembly and export. *Mol. Microbiol.*, **34**, 745-755.

Fonolla, J., Hermoso, R., Carrasco, J.L. *et al.* (1994) Antigenic relationships between chloroplast and cytosolic fructose-1,6-bisphosphatases. *Plant Physiol.*, **104**, 381-686.

Gan, Z.R., Sardana, M.K., Jacobs, J.W. and Polokoff, M.A. (1990) Yeast thioltransferase—the active site cysteines display differential reactivity. *Arch. Biochem. Biophys.*, **282**, 110-115.

Gaymard, E. and Schürmann, P. (1995) Cloning and expression of cDNAs coding for the spinach ferredoxin:thioredoxin reductase, in *Photosynthesis: From Light to Biosphere* (ed. P. Mathis), Kluwer Academic Publishers, Dordrecht, pp. 761-764.

Gaymard, E., Franchini, L., Manieri, W., Stutz, E. and Schürmann, P. (2000) A dicistronic construct for the expression of functional spinach chloroplast ferredoxin:thioredoxin reductase in *E.coli*. *Plant Sci.*, **158**, 107-113.

Geck, M.K. and Hartman, F.C. (2000) Kinetic and mutational analyses of the regulation of phosphoribulokinase by thioredoxins. *J. Biol. Chem.*, **275**, 18,034-18,039.

Geck, M.K., Larimer, F.W. and Hartman, F.C. (1996) Identification of residues of spinach thioredoxin *f* that influence interactions with target enzymes. *J. Biol. Chem.*, **271**, 24,736-24,740.

Goyer, A., Decottignies, P., Lemaire, S. *et al.* (1999) The internal Cys 207 of sorghum leaf NADP-malate dehydrogenase can form mixed disulfides with thioredoxin. *FEBS Lett.*, **444**, 165-169.

Hatch, M.D. and Agostino, A. (1992) Bilevel disulfide group reduction in the activation of C4 leaf nicotinamide adenine dinucleotide phosphate-malate dehydrogenase. *Plant Physiol.*, **100**, 360-366.

Hatzfeld, Y., Lee, S., Lee, M., Leustek, T., and Saito, K. (2000) Functional characterization of a gene encoding a fourth ATP sulfurylase isoform from *Arabidopsis thaliana*. *Gene*, **248**, 51-58.

Hermoso, R., Castillo, M., Chueca, A., Lazaro, J.J., Sahrawy, M. and Gorge, J.L. (1996) Binding site on pea chloroplast fructose-1,6-bisphosphatase involved in the interaction with thioredoxin. *Plant Mol. Biol.*, **30**, 455-465.

Hirasawa, M., Droux, M., Gray, K.A. *et al.* (1988) Ferredoxin-thioredoxin reductase: properties of its complex with ferredoxin. *Biochim. Biophys. Acta*, **935**, 1-8.

Hirasawa, M., Schürmann, P., Jacquot, J.-P. *et al.* (1999) Oxidation–reduction properties of chloroplast thioredoxins, ferredoxin–thioredoxin reductase and thioredoxin *f*-regulated enzymes. *Biochemistry*, **38**, 5200-5205.

Hirasawa, M., Ruelland, E., Schepens, I., Issakidis-Bourguet, E., Miginiac-Maslow, M. and Knaff, D. (2000) Oxidation–reduction properties of the regulatory disulfides of sorghum chloroplast NADP-malate dehydrogenase. *Biochemistry*, **39**, 3344-3350.

Hodges, M., Miginiac-Maslow, M., Decottignies, P. *et al.* (1994) Purification and characterization of pea thioredoxin *f* expressed in *Escherichia coli*. *Plant Mol. Biol.*, **26**, 225-234.

Holmgren, A. (1968) Thioredoxin 6. The amino acid sequence of the protein from *Escherichia coli* B. *Eur. J. Biol.*, **6**, 475-484.

Holmgren, A. (1995) Thioredoxin structure and mechanism: conformational changes on oxidation of the active-site sulfhydryls to a disulfide. *Structure*, **3**, 239-243.

Huppe, H.C., Lamotte-Guéry, F.D., Jacquot, J.-P. and Buchanan, B.B. (1990) The ferredoxin–thioredoxin system of a green alga, *Chlamydomonas reinhardtii*. Identification and characterization of thioredoxins and ferredoxin–thioredoxin reductase components. *Planta*, **180**, 341-351.

Huppe, C.H., Picaud, A., Buchanan, B.B. and Miginiac-Maslow, M. (1991) Identification of an NADP/thioredoxin system in *Chlamydomonas reinhardtii*. *Planta*, **186**, 115-121.

Hur, Y. and Vasconcelos, A.C. (1998) Spinach cytosolic fructose-1,6-bisphosphatase. I. Its organ-specific and developmental expression characteristics. *Mol. Cell.*, **8**, 138-147.

Issakidis, E., Miginiac-Maslow, M., Decottignies, P., Jacquot, J.-P., Cretin, C. and Gadal, P. (1992) Site-directed mutagenesis reveals the involvement of an additional thioredoxin-dependent regulatory site in the activation of recombinant sorghum leaf NADP-malate dehydrogenase. *J. Biol. Chem.*, **267**, 21,577-21,583.

Issakidis, E., Saarinen, M., Decottignies, P. *et al.* (1994) Identification and characterization of the second regulatory disulfide bridge of recombinant sorghum leaf NADP-malate dehydrogenase. *J. Biol. Chem.*, **269**, 3511-3517.

Issakidis, E., Lemaire, M., Decottignies, P., Jacquot, J.-P. and Miginiac-Maslow, M. (1996) Direct evidence for the different roles of the *N*- and *C*-terminal regulatory disulfides of sorghum leaf NADP-malate dehydrogenase in its activation by reduced thioredoxin. *FEBS Lett.*, **392**, 121-124.

Issakidis-Bourguet, E., Mouaheb, N., Meyer, Y. and Miginiac-Maslow, M. (2001) Heterologous complementation of yeast reveals a new putative function for chloroplast *m*-type thioredoxin. *Plant J.*, **25**, 127-136.

Izawa, S., Maeda, K., Sugiyama, K.-I., Mano, J., Inoue, Y. and Kimura, A. (1999) Thioredoxin deficiency causes the constitutive activation of Yap1, an AP-1-like transcripton factor in *Saccharomyces cerevisiae*. *J. Biol. Chem.*, **274**, 28,459-28,465.

Jacquot, J.P., de Lamotte, F., Fontecave, M. *et al.* (1990) Human thioredoxin reactivity-structure/function relationship. *Biochem. Biophys. Res. Comm.*, **173**, 1375-1381.

Jacquot, J.P., Rivera-Madrid, R., Marinho, P. *et al.* (1994) *Arabidopsis thaliana* NADPH thioredoxin reductase. cDNA characterization and expression of the recombinant protein in *Escherichia coli*. *J. Mol. Biol.*, **235**, 1357-1363.

Jacquot, J.P., Stein, M., Suzuki, A., Liottet, S., Sandoz, G. and Miginiac-Maslow, M. (1997a) Residue Glu 91 of *Chlamydomonas reinhardtii* is essential for electron transfer to ferredoxin–thioredoxin reductase. *FEBS Lett.*, **400**, 293-296.

Jacquot, J.P., Lopez-Jaramillo, J., Miginiac-Maslow, M. *et al.* (1997b) Cysteine-153 is required for redox regulation of pea chloroplast fructose-1,6-bisphosphatase. *FEBS Lett.*, **401**, 143-147.

Jeong, J.S., Kwon, S.J., Kang, S.W., Rhee, S.G. and Kim, K. (1999) Purification and characterization of a second type thioredoxin peroxidase (type II TPx) from *Saccharomyces cerevisiae*. *Biochemistry*, **38**, 776-783.

Johansson, K., Ramaswamy, S., Saarinen, M., Lemaire-Chamley, M., Issakidis-Bourguet, E., Miginiac-Maslow, M. and Eklund, H. (1999) Structural basis for light activation of a chloroplast enzyme. The structure of sorghum NADP-malate dehydrogenase in its oxidized form. *Biochemistry*, **38**, 4319-4326.

Kobe, B. and Kemp, B.E. (2000) Active site-directed protein regulation. *Nature*, **402**, 373-376.

Krimm, I., Lemaire, S., Ruelland, E. *et al.* (1998) The single mutation Trp35 → Ala in the 35–40 redox site of *Chlamydomonas reinhardtii* thioredoxin h affects its biochemical activity and the pH dependence of C36-C39 1H-13C NMR. *Eur. J. Biochem.*, **255**, 185-195.

Krimm, I., Goyer, A., Issakidis-Bourguet, E., Miginiac-Maslow, M. and Lancelin, J.-M. (1999) Direct NMR observation of the thioredoxin-mediated reduction of the chloroplast NADP-malate dehydogenase provides a structural basis for the relief of auto-inhibition. *J. Biol. Chem.*, **274**, 34,539-34,542.

Laloi, C., Rayapuram, N., Chartier, Y., Grienenberger, J.M., Bonnard, G. and Meyer, Y. (2002) Identification and characterization of a mitochondrial thioredoxin system in plants. *Proc. Natl. Acad. Sci. USA* (in press).

Lancelin, J.M., Guilhaudis, L., Krimm, I., Blackledge, M.J., Marion, D. and Jacquot, J.P. (2000) NMR structures of thioredoxin *m* from green alga *Chlamydomonas reinhardtii*. *Proteins*, **41**, 334-349.

Lee, J., Spector, D., Godon, C., Labarre, J. and Toledano, M.B. (1999) A new antioxidant with alkyl hydroperoxide defense properties in yeast. *J. Biol. Chem.*, **274**, 4537-4544.

Lennon, B.W., Williams Jr, C.H. and Ludwig, M.L. (2000) Twists in catalysis: alternating conformations of *Escherichia coli* thioredoxin reductase. *Science*, **289**, 1190-1194.

Lopez-Jaramillo, J., Chueca, A., Jacquot, J.P. *et al.* (1997) High-yield expression of pea thioredoxin *m* and assessment of its efficiency in chloroplast fructose-1,6-bisphosphatase activation. *Plant Physiol.*, **114**, 1169-1175.

Lopez-Jaramillo, J., Chueca, A., Sahrawy, M. and Lopez Gorge, J. (1998) Hybrids from pea chloroplast thioredoxins *f* and *m*: physicochemical and kinetic characteristics. *Plant J.*, **15**, 155-163.

Marcus, F., Harrsch, P.B., Moberly, L., Edelstein, I. and Latshaw, S.P. (1987) Spinach chloroplast fructose-1,6-bisphosphatase: identification of the subtilisin-sensitive region and of conserved histidines. *Biochemistry*, **26**, 7029-7035.

Martin, W., Mustafa, A.Z., Henze, K. and Schnarrenberger, C. (1996) Higher-plant chloroplast and cytosolic fructose-1,6-bisphosphatase isoenzymes: origins via duplication rather than prokaryote–eukaryote divergence. *Plant Mol. Biol.*, **32(3)**, 485-491.

Mestres-Ortega, D. and Meyer, Y. (1999) The *Arabidopsis thaliana* genome encodes at least four thioredoxins *m* and a new prokaryotic-like thioredoxin. *Gene*, **240(2)**, 307-316.

Meyer, Y., Verdoucq, L. and Vignols, F. (1999) Plant thioredoxins and glutaredoxins: identity and putative roles. invited review in *Trends Plant Sci.*, **4**, 388-393.

Meyer, Y., Vignols, F. and Reicheld, J.F. (2001) Classification of plant thioredoxins by sequence similarity and intron position. *Methods Enzymol.*, Academic Press, Orlando, USA (in press).

Miginiac-Maslow, M., Johansson, K., Ruelland, E. *et al.* (2000) Light-activation of NADP-malate dehydrogenase: a highly controlled process for an optimized function. *Physiol. Plant.*, **110**, 322-329.

Mora-Garcia, S., Rodriguez-Suarez, R. and Wolosiuk, R.A. (1998) Role of electrostatic interactions on the affinity of thioredoxin for target proteins. Recognition of chloroplast fructose-1,6-bisphosphatase by mutant *Escherichia coli* thioredoxins. *J. Biol. Chem.*, **273**, 16,273-16,280.

Mouaheb, N., Thomas, D., Verdoucq, L., Monfort, P. and Meyer, Y. (1998) *In vivo* functional disrimination between plant thioredoxins by heterologous expression in the yeast *Saccharomyces cerevisiae*. *Proc. Natl Acad. Sci. USA*, **95**, 3312-3317.

Muller, E.G. (1991) Thioredoxin deficiency in yeast prolongs S phase and shortens the G1 interval of the cell cycle. *J. Biol. Chem.*, **266**, 9194-9202.

Muller, E.G. (1996) A glutathione reductase mutant of yeast accumulates high levels of oxidized glutathione and requires thioredoxin for growth. *Mol. Biol. Cell.*, **7**, 1805-1813.

Nakai, K. and Kanehisa, M. (1992) A knowledge base for predicting protein localization sites in eukaryotic cells. *Genomics*, **14**, 897-911.

Ocheretina, O., Haferkamp, I., Tellioglu, H. and Scheibe, R. (2000) Light-modulated NADP-malate dehydrogenases from mossfern and green algae: insights into evolution of the enzyme's regulation. *Gene*, **258**, 147-154.

Pagano, E.A., Chueca, A. and Lopez-Gorge, J. (2000) Expression of thioredoxins *f* and *m*, and of their targets fructose-1,6-bisphosphatase and NADP-malate dehydrogenase, in pea plants grown under normal and light/temperature stress conditions. *J. Exp. Bot.*, **51**, 1299-1307.

Raines, C.A., Lloyd, J.C., Longstaff, M, Bradley, D. and Dyer, T. (1988) Chloroplast fructose-1,6-bisphosphatase: the product of a mosaic gene. *Nucleic Acids Res.*, **16**, 7931-7942.

Riessland, R. and Jaenicke, R. (1997) Determination of the regulatory disulfide bonds of NADP-dependent malate dehydrogenase from *Pisum sativum* by site-directed mutagenesis. *Biol. Chem.*, **378**, 983-988.

Rodriguez-Suarez, R.J., Mora-Garcia, S. and Wolosiuk, R.A. (1997) Characterization of cysteine residues involved in the reductive activation and the structural stability of rapeseed (*Brassica napus*) chloroplast fructose-1,6-bisphosphatase. *Biochem. Biophys. Res. Commun.*, **232**, 388-393.

Ruelland, E. and Miginiac-Maslow, M. (1999) Regulation of chloroplast enzyme activities by thiol-disulfide interchange with reduced thioredoxin: activation or relief from inhibition? *Trends Plant Sci.*, **4**, 136-141.

Ruelland, E., Lemaire-Chamley, M., Le Maréchal, P., Issakidis-Bourguet, E., Djukic, N. and Miginiac-Maslow, M. (1997) An internal cysteine is involved in the thioredoxin-dependent activation of sorghum NADP-malate dehydrogenase. *J. Biol. Chem.*, **272**, 19,851-19,857.

Ruelland, E., Johansson, K., Decottignies, P., Djukic, N. and Miginiac-Maslow, M. (1998) The auto-inhibition of sorghum NADP-malate dehydrogenase is mediated by a *C*-terminal negative charge. *J. Biol. Chem.*, **273**, 33,482-33,488.

Sahrawy, M., Chueca, A., Hermoso, R., Lazaro J.J. and Lopez-Gorge, J. (1997) Directed mutagenesis shows that the preceding region of the chloroplast fructose-1,6-bisphosphatase regulatory sequence is the thioredoxin docking site. *J. Mol. Biol.*, **269**, 623-630.

Scheibe, R. and Jacquot, J.-P. (1983) NADP regulates the light-activation of NADP-dependent malate dehydrogenase. *Planta*, **157**, 548-553.

Schepens, I., Ruelland, E., Miginiac-Maslow, M., Le Maréchal, P. and Decottignies, P. (2000a) The role of the active-site arginines of sorghum NADP-malate dehydrogenase in thioredoxin-dependent activation and activity. *J. Biol. Chem.*, **275**, 35,792-35,798.

Schepens, I., Johansson, K., Decottignies, P. *et al.* (2000b) Inhibition of the thioredoxin-dependent activation of the NADP-malate dehydrogenase and cofactor specificity. *J. Biol. Chem.*, **275**, 20,996-21,001.

Schepens, I., Decottignies, P., Ruelland, E., Johansson, K. and Miginiac-Maslow, M. (2000c) The dimer contact area of sorghum NADP-malate dehydrogenase: role of aspartate 101 in dimer stability and catalytic activity. *FEBS Lett.*, **471**, 240-244.

Schürmann, P. (1981) The ferredoxin/thioredoxin system of spinach chloroplasts. Purification and characterization of its components, in *Photosynthesis* (ed. G. Akoyunoglou) Vol 4, Balaban Intern. Sci. Services, Philadelphia, PA, USA, pp. 273-280.

Schürmann, P. (1995) The ferredoxin/thioredoxin system, in *Methods in Enzymology*, Vol. 252, *Biothiols Part B; Glutathione and Thioredoxin: Thiols in Signal Transduction and Gene Regulation* (ed. L. Packer) Academic Press, Orlando, Florida 32887, USA, pp. 274-283.

Schürmann, P. and Buchanan, B.B. (2001) The structure and function of the ferredoxin/thioredoxin system in photosynthesis, in *Regulation of Photosynthesis. Advances in Photosynthesis, Vol. 11* (eds E.M. Aro and B. Andersson), Kluwer Academic Publishers, Dordrecht, The Netherlands, pp. 331-361.

Schürmann, P. and Jacquot, J.-P. (2000) Plant thioredoxin systems revisited. *Annu. Rev. Plant Physiol. Plant Mol. Biol.*, **51**, 371-400.

Schürmann, P., Maeda, K. and Tsugita, A. (1981) Isomers in thioredoxins of spinach chloroplasts. *Eur. J. Biochem.*, **116**, 37-45.

Schwendtmayer, C., Manieri, W., Hirasawa, M., Knaff, D.B. and Schürmann, P. (1998) Cloning, expression and characterization of ferredoxin:thioredoxin reductase from *Synechocystis* sp. PCC6803, in *Photosynthesis: Mechanisms and Effects* (ed. G. Garab), Kluwer Academic Publishers, Dordrecht, The Netherlands, pp. 1927-1930.

Staples, C.R., Ameyibor, E., Fu, W. *et al.* (1996) The nature and properties of the iron–sulfur center in spinach ferredoxin:thioredoxin reductase: a new biological role for iron-sulfur clusters. *Biochemistry*, **35**, 11,425-11,434.

Staples, C.R., Gaymard, E., Stritt-Etter, A.L. *et al.* (1998) Role of the [Fe$_4$S$_4$] cluster in mediating disulfide reduction in spinach ferredoxin:thioredoxin reductase. *Biochemistry*, **37**, 4612-4620.

Tang, G.L., Wang, Y.F., Bao, J.S. and Chen, H.B. (2000) Overexpression in *Escherichia coli* and characterization of the chloroplast fructose-1,6-bisphosphatase from wheat. *Protein Expr. Purif.*, **19**, 411-418.

Trevanion, S., Furbank, R.T. and Ashton, A.R. (1997) NADP-malate dehydrogenase in the C4 plant *Flaveria bidentis*. Cosense suppression of activity in mesophyll and bundle sheath cells and consequences for photosynthesis. *Plant Physiol.*, **113**, 1153-1165.

Udvardy, J., Godeh, M.M. and Farkas, G.L. (1982) Regulatory properties of a fructose 1,6-bisphosphatase from the cyanobacterium *Anacystis nidulans*. *J. Bacteriol.*, **151**, 203-208.

Verdoucq, L., Vignols, F., Jacquot, J.P., Chartier, Y. and Meyer, Y. (1999) *In vivo* characterization of a thioredoxin *h* target protein defines a new peroxiredoxin family. *J. Biol. Chem.*, **274**, 19,714-19,722.

Villeret, V., Huang, S., Zhang, Y., Xue, Y. and Lipscomb, W.N. (1995) Crystal structure of spinach chloroplast fructose-1,6-bisphosphatase at 2.8 Å resolution. *Biochemistry*, **34**, 4299-4306.

Xu, Z., Mayer, A., Muller, E. and Wickner, W. (1997) A heterodimer of thioredoxin and I(B)2 cooperates with Sec18p (NSF) to promote yeast vacuole inheritance. *J. Cell. Biol.*, **136**, 299-306.

Yano, H., Wong, J.H., Lee, Y.M., Cho, M.J. and Buchanan, B.B. (2001) A new strategy for the identification of proteins targeted by thioredoxin. *Proc. Natl Acad. Sci. USA*, **98**, 4794-4799.

2   The Rubisco activase–Rubisco system: an ATPase-dependent association that regulates photosynthesis

Archie R. Portis Jr

2.1   Introduction

The protein ribulose 1,5-bisphosphate carboxylase (Rubisco) activase is required for maintaining the catalytic activity of Rubisco in plants. Rubisco is the protein that captures carbon dioxide from the atmosphere and combines it with ribulose 1,5-bisphosphate (RuBP) and water to form two molecules of phosphoglyceric acid. Rubisco is arguably the most abundant protein in the biosphere (Ellis, 1979) and it is certainly one of the most important. The activity of Rubisco is a key limiting factor in the productivity of plants, and Rubisco is essentially the only carbon sequestering enzyme that directly leads to the net biosynthesis of carbohydrates, sustaining life for both autotrophic and heterotrophic organisms.

The synthesis, assembly, functioning, regulation and degradation of Rubisco involve a multitude of protein–protein interactions and many of these have served as paradigms for plant molecular biology. The focus in this chapter is on the interaction between Rubisco and the activase that is involved in regulating and maintaining the activity of Rubisco.

2.2   Structure of Rubisco and the activase

In plants and algae, Rubisco occurs as a hexadecameric protein of ca. 550 kDa, comprising a core of four large (L) subunit (ca. 52 kDa) homodimers held together at each end with four small (S) subunits (ca. 12 kDa), which can be represented by $S_4(L_2)_4S_4$. The catalytic site is at the face of an α/β barrel in the C-terminal domain of each of the large subunits and critical residues are contributed by the N-terminal domain of the other large subunit in each dimer. The large subunits are encoded in the chloroplast genome, whereas the small subunits are encoded as a multigene family in the nucleus. Thus the small subunits are targeted to the chloroplast with a transit peptide that is removed during import. Chaperones then assist with the assembly of the large and small subunits inside the chloroplast. Excellent reviews covering the many aspects of Rubisco beyond the scope of this chapter are available (see Hartman and Harpel, 1994; Portis, 1995; Spreitzer, 1999). A large number of three-dimensional (3D) structures of Rubisco are now available (mostly summarized in Duff *et al.*, 2000).

These will inform the discussion of the experiments that have investigated the nature of Rubisco's interaction with the activase.

2.2.1 Two activase isoforms from alternative splicing

In most plants, Rubisco activase comprises two isoforms of 41–43 kDa and 45–46 kDa. Because the gene(s) for activase is located in the nucleus, the activase found in the chloroplast is derived from an intermediate, cytoplasmic form containing a transit peptide. Analyses of cDNA and genomic DNA clones of *Arabidopsis* (Werneke *et al.*, 1989), spinach (Werneke *et al.*, 1989), barley (Rundle and Zielinski, 1991b) and rice (To *et al.*, 1999; Zhang and Komatsu, 2000) have demonstrated that the two isoforms arise during mRNA processing by alternative splicing of the last intron. At the mRNA level, the processing details differ in each case, but the final products are similar—two proteins that differ by about 30 amino acids at the *C*-terminus. *Arabidopsis*, spinach and rice only contain one activase gene but, curiously, barley (and perhaps other species) also contains another gene that encodes a single and divergent 41 kDa isoform. The two isoforms function in an oligomeric complex, which will be discussed later.

2.2.2 Variation in plants and algae

Protein gels (Salvucci *et al.*, 1987) and a recent survey (A.R. Portis, unpublished) of the cDNA sequences (including expressed sequence tags) available in the National Center for Biotechnology Information (NCBI) database (http://www.ncbi.nlm.nih.gov/Entrez) reveal that most species examined to date contain the two isoforms of activase. However, there are a few exceptions to this rule. The cDNAs currently available for tobacco (accession no. Z14980), *Phaseolus vulgaris* (accession no. AF041068), cucumber (accession no. X67674), maize (accession no. AF084478), and mung bean (accession no. AAD20019) provide no evidence of encoding a completely satisfactory *C*-terminal domain for a larger isoform. In common with Rubisco, the amino acid sequence of the activase in the plants surveyed thus far is highly conserved and this could reflect in part the need to maintain proper protein interactions with Rubisco (Wang *et al.*, 1992). Figure 2.1 illustrates the high degree of similarity by showing the sequences of the larger isoform in four species with documented alternative splicing and the sequences in two species that only contain the smaller isoform. The greatest variation in the sequences occurs in the transit peptide and at the extreme *N*-terminus. Otherwise, only a few regions exhibit substantial variation.

2.2.3 Member of the AAA$^+$ protein super-family

Unfortunately, a 3D structure for the activase is not yet available. In the past, standard homology searches have not provided any suggestions for distant

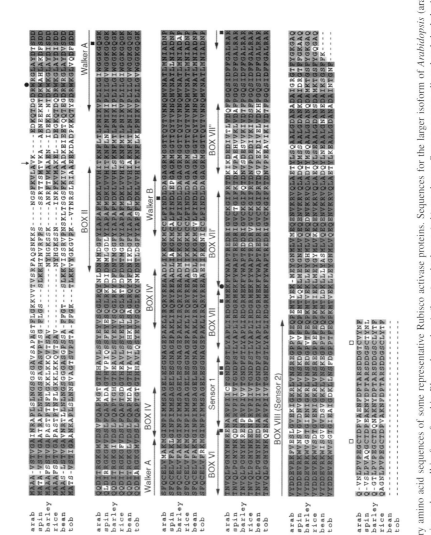

Figure 2.1 Primary amino acid sequences of some representative Rubisco activase proteins. Sequences for the larger isoform of *Arabidopsis* (arab), spinach (spin), barley, and rice and a small isoform found in *Phaseolus vulgaris* (bean) and tobacco (tob) are shown. Sequences were aligned and shaded with the MSA and BOXSHADE tools in the SDSC Biology Workbench (http://workbench.sdsc.edu). The AAA⁺ family motifs indicated above the sequences (arrows) and the closed boxes indicate some key residues, discussed by Guenther *et al.* (1997), Neuwald *et al.* (1999) and in the text. The down arrow in the first row indicates the beginning of the mature *Arabidopsis* protein. The Cys residues indicated by the open boxes in C-terminal region of the large isoforms are required for redox regulation and the Trp and Lys residues indicated by the closed circles have been identified as catalytically important for the Rubisco activation activity of the

relatives of this protein except for identification of the Walker A and B consensus sequences for nucleotide binding (Werneke *et al.*, 1988b). Recently, some biochemical evidence for a chaperone-like ability to restore activity to heat-inactivated Rubisco *in vitro* was reported (Sánchez de Jiménez *et al.*, 1995), but not confirmed in a later report (Eckardt and Portis, 1997). A breakthrough occurred with the finding of Neuwald *et al.* (1999) that Rubisco activase is a member of a sequence superset of the AAA family (ATPases associated with diverse cellular activities) (Patel and Latterich, 1998) designated the AAA$^+$ family. This family was assembled by the use of a combination of iterative database searches and multiple sequence alignment methods. The family is ancient and has undergone considerable functional divergence as witnessed by the observation that it includes: "'regulatory" components of Lon and Clp proteases, proteins involved in DNA replication, recombination and restriction (including subunits of the origin recognition complexes, replication factor C proteins, MCM DNA-licensing factors and the bacterial DnaA, RuvB, and McrB proteins), prokaryotic NtrC-related transcription regulators, the sporulation protein SpoVJ of *Bacillus*, Mg^{2+} and Co^{2+} chelatases, the GvpN gas vesicle synthesis protein of *Halobacterium*, dynein motor proteins, Torsin A' (Neuwald *et al.*, 1999) as well as Rubisco activase. The common feature of these proteins is that they contain one or more copies of the AAA$^+$ motifs and typically form supercomplexes, often including other proteins. These complexes perform chaperone-like functions that assist in the assembly, operation or disassembly of other protein complexes.

Neuwald *et al.* (1999) also provide new structural and mechanistic insights into these proteins by showing that the 3D structures of the core AAA$^+$ module in two of the proteins, *Escherichia coli* DNA polymerase III δ' subunit (1A5T) and NSF-D2 (1NSF), have a common fold formed by 11 sequence motifs. The common fold consists of a five-stranded β-sheet domain with one or more α-helices on each side (formed by the first eight motifs) that is linked to another three-helix bundle domain (last three motifs). The same fold was subsequently found in the Hs1U component of an *E. coli* ATP-dependent protease (Bochtler *et al.*, 2000) and it will probably be identified in the other proteins as more structures become available. In figure 2.1, the 11 motifs in the activase that could form this folding domain are identified using the nomenclature of Neuwald *et al.* (1999).

2.3   Activities and mechanism of activase

Characterizations of the rca^- mutant of *Arabidopsis* (Somerville *et al.*, 1982; Salvucci *et al.*, 1986) and transgenic antisense activase plants in tobacco (Mate *et al.*, 1993; Hammond *et al.*, 1998a,b) and *Arabidopsis* (Eckardt *et al.*, 1997; Mott *et al.*, 1997) have clearly demonstrated that the primary function of activase is to promote and maintain a high catalytic capacity of Rubisco, especially under otherwise limiting CO_2 concentrations. The biochemical foundation for modulating the catalytic capacity of Rubisco independent of the concentrations of its substrates, RuBP and CO_2, is provided by a process called carbamylation (Lorimer et al., 1976). Carbamylation is the addition of a CO_2 molecule to an ε-amino group of a lysine located in the catalytic site. Carbamylation of this lysine provides a binding site for Mg^{2+} and both the carbamate and the Mg^{2+} play critical roles during the catalytic cycle (Cleland et al., 1998). Carbamylation is a spontaneous process and minimal subsequent changes occur in the active site of Rubisco and the conformation of the overall protein (Taylor and Andersson, 1996). Carbamylation is facilitated by the increases in pH and Mg^{2+} that occur in the chloroplast stroma upon illumination when Rubisco activity is needed for photosynthesis.

In contrast to most enzymes, the catalytic sites of Rubisco have significant affinities for quite a variety of sugar phosphates in addition to the RuBP substrate (reviewed in Portis, 1992; Hartman and Harpel, 1994). Furthermore, the binding affinity of a particular sugar phosphate depends on whether or not the catalytic site is carbamylated and contains Mg^{2+}. Thus some sugar phosphates bind more tightly to the carbamylated form whereas others prefer the uncarbamylated form. The binding and resulting effects of many of these sugar phosphates, which are observed in vitro, are probably not physiologically significant because the RuBP level is generally in considerable excess of the available binding sites (Woodrow and Berry, 1988). But RuBP itself is of particular importance in that it binds much more tightly to the uncarbamylated form (Jordan and Chollet, 1983) and thus can severely inhibit the otherwise rapid carbamylation process that is necessary for enzyme activity (Portis et al., 1986). Indeed, under the appropriate conditions, RuBP binding to the uncarbamylated form can result in a fairly rapid decrease in the catalytic activity of Rubisco, which is initially carbamylated and highly active (Portis et al., 1995). Inhibition of Rubisco by RuBP is physiologically important because it is this inhibition that largely accounts for the reduction in Rubisco activity, which occurs at light intensities limiting for photosynthesis (Brooks and Portis, 1988; Cardon and Mott, 1989).

Another physiologically significant sugar phosphate in many plant species (e.g. Phaseolus vulgaris and tobacco) is 2-carboxy-D-arabinitol 1-phosphate (CA1P) (reviewed by Seemann et al., 1990; Hartman and Harpel, 1994). CA1P binds preferentially to carbamylated sites containing Mg^{2+}, which otherwise would have catalytic activity. CA1P accumulates in these plants during darkness or extended periods of low light and thus it also sequesters Rubisco into a form incapable of catalysis, even though the Rubisco remains carbamylated.

Yet another group of sugar phosphates that may also be physiologically significant are several that can be formed in the active site during normal catalysis (Hartman and Harpel, 1994), at least one of which may also be an oxidation product of RuBP (Kane et al., 1998). A more detailed characterization of this group of inhibitors is needed, but it is quite common for the activity of Rubisco in many plants to decline slowly *in vitro* to a much lower level as these compounds are produced (Edmondson *et al.*, 1990). There is emerging evidence that Rubisco activity *in vivo* is reduced by the formation of trace amounts of inhibitors other than RuBP and CA1P (Parry *et al.*, 1997).

2.3.1 Dissociation of Rubisco–sugar phosphate complexes

Remarkably, activase can completely restore the activity of Rubisco that has been inhibited by preincubation *in vitro* with either RuBP (Portis *et al.*, 1986), CA1P (Robinson and Portis, 1988b) or after a period of catalysis in which a reduction in its activity has occurred (Robinson and Portis, 1989b). These effects are most readily explained by the activase promoting the dissociation of the inhibitors (Robinson *et al.*, 1988). Using labeled RuBP, Wang and Portis (1992) obtained direct evidence that the activase catalyzed the release of RuBP from Rubisco. The release of the RuBP then allows spontaneous carbamylation and Mg^{2+} binding to proceed (Werneke *et al.*, 1988a) and more Rubisco activity is observed, which is what is typically measured. The rate of reactivation is a second-order process, responding to both Rubisco and activase concentrations (Robinson *et al.*, 1988; Lan and Mott, 1991). Reactivation of the RuBP-inhibited enzyme has also been analyzed extensively *in planta* using CO_2 gas-exchange measurements on leaves to follow the increase in photosynthesis (induction) following a step increase in light intensity (Mott and Woodrow, 2000). Transgenic plants with reduced levels of activase have slower induction kinetics (Mott *et al.*, 1997; Hammond *et al.*, 1998a).

The reactivation of Rubisco, which has been inhibited with CA1P or by an extended period of catalysis, has not been characterized as extensively. However, reactivation after CA1P inhibition was also confirmed *in planta* by analysis of the light-induced increase in Rubisco activity with transformed tobacco plants having reduced levels of activase (Hammond *et al.*, 1998a,b). While more direct evidence of enhanced sugar phosphate release by the activase has not been obtained, there is no reason to suspect that the same fundamental mechanism does not occur.

Activase also allows Rubisco to exhibit its maximal catalytic capacity in the presence of RuBP at otherwise suboptimal but physiological concentrations of CO_2 (Portis *et al.*, 1986). This effect of activase does not appear to be due to a direct effect on the affinity of the enzyme for the carbamylating CO_2 because no effect of the activase is observed in the absence of RuBP or another suitable sugar phosphate (Lilley and Portis, 1990). Rather a difference in the ability to promote the release of RuBP from the carbamylated and uncarbamylated forms appears to be responsible. A model (Mate *et al.*, 1996) incorporating carbamylation, RuBP binding to the various forms of the enzyme and the activase-facilitated dissociation of RuBP from these forms can fully account for the increased level of carbamylation at suboptimal CO_2 concentrations (Portis *et al.*, 1995). Based on a characterization of transformed tobacco plants expressing reduced amounts of the activase, it has even been suggested that the activase may also increase the catalytic activity of fully activated Rubisco (He *et al.*, 1997). However, obtaining further support for this idea with studies *in vitro* that can resolve an actual increase in the intrinsic maximal velocity of Rubisco from the removal of the inhibitors formed during catalysis may be difficult.

2.3.2 Hydrolysis of ATP and inhibition by ADP

ATP hydrolysis by the activase is required to restore the activity of Rubisco previously reduced by incubation with inhibitors (Streusand and Portis, 1987; Robinson and Portis, 1989a). Non-hydrolyzable analogs of ATP and other nucleotides do not support the activation of Rubisco. ADP inhibits the ATPase and Rubisco activation activities of the activase to a similar extent. Other treatments that either stimulate the ATPase activity of the activase (polyethylene glycol, Salvucci, 1992), or lower the activity (activase concentration, heat, chaotropic agents, or exposure to ultraviolet (UV) irradiation (Salvucci and Ogren, 1996) and low ionic strength media (Lilley and Portis, 1997)) cause concomitant changes in its Rubisco activation activity. Chemical modification, photoaffinity labeling and site-directed mutagenesis also reduce both activities to similar levels (Shen *et al.*, 1991; Salvucci, 1993; Salvucci *et al.*, 1994; Salvucci and Klein, 1994).

Although Rubisco activation requires the hydrolysis of ATP, ATP hydrolysis by the activase does not require the presence of Rubisco (Robinson and Portis, 1989a). Also some site-directed and *C*-terminal deletion mutants of the activase exhibit altered ratios of the two activities (Shen and Ogren, 1992; Esau *et al.*, 1996; Kallis *et al.*, 2000). The lack of coupling between the ATPase and Rubisco activation activities remains to be completely resolved, but is probably due to the self-associating property of the activase, which will be discussed more thoroughly later. The monomers and perhaps even the smaller oligomers of activase do not appear to be able to hydrolyze ATP and an analogy to the actin system has been suggested in which a dynamic equilibrium exists between monomeric and oligomeric forms, powered by ATP hydrolysis (Lilley and Portis, 1997). The lack of strict coupling may thus simply reflect the ability of the activase subunits to assemble into complexes that hydrolyze ATP even without Rubisco at the protein concentrations typical in *in vitro* studies.

For example, comparison of the maximal rates of ATP hydrolysis and Rubisco activation with high concentrations of both proteins show that more than 10 mol of ATP are hydrolyzed for every mol of Rubisco active sites affected (Salvucci and Ogren, 1996). The efficiency decreases proportionately if activase is assayed at low Rubisco concentrations due to the second order rate of the activation process. It is estimated that the rate of Rubisco activation *in vivo* is about 60 nmol sites/mg activase/min from comparisons of the induction kinetics of transgenic plants expressing different amounts of activase (Hammond *et al.*, 1998a). Using the maximal rate of the ATPase activity observed *in vitro*, the minimal efficiency is 20 mol ATP/mol sites activated. Because the activase has lower activity at physiological ATP/ADP ratios and ATPase activity could be lowered further by extensive binding to Rubisco at the protein concentrations present in the stroma, it is likely that the efficiency approaches a theoretical number of 1 to 2. Binding of most of the activase to Rubisco *in situ* is expected because the holoenzyme concentration of Rubisco is about 500 µM, the concentration of activase monomers is about 170 to 500 µM, and estimates of the dissociation constant for the complex range from 0.34 to 5 µM (Robinson *et al.*, 1988; Lan and Mott, 1991).

2.3.3 Oligomer formation and its relationship to activase activity

Consistent with its inclusion in the AAA^+ family, the activase typically self-associates to form large oligomers of variable size, depending on a number of conditions. Using gel-filtration chromatography on a Superose-6 column at room temperature, in the presence of either ATP or ATP-γ-S and Mg^{2+}, oligomers larger than Rubisco ($> 550,000$) were observed, whereas in the presence of ADP smaller oligomers (ca. 340,000) were observed (Wang *et al.*, 1993). In contrast, at 4°C in the absence of nucleotides, a molecular mass of only 58,000 was obtained with rate zonal sedimentation or chromatography on Sephacryl S-300, but 280,000 was obtained with a Superose-12 column (Salvucci, 1992). Polyethylene glycol, a solvent-excluding agent known to promote self-association of proteins, increased the apparent molecular mass of activase in both of these systems by two- to four-fold. In the other study using the Superose-6 column, trailing peaks were reported, suggesting a variety of oligomer sizes were present with smaller oligomers forming as the protein became diluted. The higher masses measured with the more rapid fast protein liquid chromatography (FPLC) columns also suggest that analysis time is probably important (Salvucci, 1992), consistent with a very dynamic system.

The specific activities of activase *in vitro* increase hyperbolically over the range of protein concentration (0–$100\,\mu g\,ml^{-1}$) typically used for studies *in vitro* (Salvucci, 1992; Wang *et al.*, 1993) and also can be increased with polyethylene glycol (Salvucci, 1992). Thus activase exhibits enhanced activity in an associated state. Oligomerization of the activase, specific to ATP or ATP-γ-S and Mg^{2+}, has been monitored in several studies by the increase in its intrinsic fluorescence (Wang *et al.*, 1993; van de Loo and Salvucci, 1996, 1998; Lilley and Portis, 1997). The fluorescence increase appears to be due to a change in the environment of one or more tryptophan residues, probably the Trp in Box VII' (figure 2.1) (van de Loo and Salvucci, 1996). Comparisons of the change in fluorescence with the other properties suggest a molecular sequence of ATP binding followed by increased oligomerization and then ATP hydrolysis. While some ATPase activity may be present in smaller oligomers of the activase, the Rubisco activation activity seems to occur only with the largest oligomeric forms exhibiting increased fluorescence. High cooperativity within the activase complex is also indicated by the inhibition of the ATPase and Rubisco activation activities by the addition of some mutant forms of activase to wild-type enzyme (Salvucci and Klein, 1994; van de Loo and Salvucci, 1998).

2.3.4 Key residues identified in activase and its AAA$^+$ domains

Several key residues in the activase important for its activities have been identified by mutagenesis experiments. The recent recognition that the activase contains the AAA$^+$ folding domain provides targets for future studies and insights about results previously published. The P-loop region (-Gly-X-X-X-Gly-Lys-Ser-) in the Walker A motif was shown to be important for the ATPase and Rubisco activation in the first site-directed mutagenesis studies of activase (Shen et al., 1991; Shen and Ogren, 1992). As indicated in figure 2.1, comparison of the activase with other AAA$^+$ proteins reveals that it has a conserved D residue in the Walker B motif, the N in the Sensor 1 motif, and the R in the Box VIII (Sensor 2) motif, all of which are in a position to interact with the bound nucleotide as discussed by Guenther et al. (1997). The Walker A–Box IV and the Sensor 1 regions have been photoaffinity-labeled with ATP γ-benzophenone (Salvucci et al., 1993) and thus implicated in binding of the nucleotide phosphates. Direct evidence for the involvement of the D residue in the Walker B region and the D residue adjacent to the N in the Sensor 1 region in the coordination of the ATP has been obtained by site-directed mutagenesis of activase (van de Loo and Salvucci, 1998). Box II has been photoaffinity labeled with 2-N$_3$ATP (Salvucci et al., 1994).

The marked E/D residue in BoxVI and R residue in BoxVII (figure 2.1) are especially interesting because they may interact with the nucleotide phosphate bound to an adjacent subunit in some AAA$^+$ proteins, as discussed by Guenther et al. (1997). This interaction could provide a critical linkage between conformational changes in the oligomers or supercomplexes of these proteins and ATP binding/hydrolysis. Indeed, changing the nearby K at the Box VII–Box VII′ boundary to Arg, Cys or Gln, by site-directed mutagenesis of the activase, resulted in proteins with a high affinity for ATP, but with very low ATPase and no Rubisco activation activity (Salvucci and Klein, 1994). The Lys$_{247}$ to Arg mutant was later shown to exhibit no increase in its intrinsic fluorescence with Mg^{2+} and ATP or ATP-γ-S, indicating that proper oligomerization into an activation complex is abolished (van de Loo and Salvucci, 1996).

Only a few critical residues not involved in nucleotide binding have been identified. A conserved tryptophan at the N-terminus (figure 2.1) has been shown to be critical for the Rubisco activation activity, but not the ATPase activity by either site-directed mutagenesis or deletion (van de Loo and Salvucci, 1996; Esau *et al.*, 1996). Indeed, deletion of the first 50 residues has little effect on the ATPase activity or the increase in intrinsic fluorescence, suggesting that this domain, and especially the conserved tryptophan, is involved only in the activation of Rubisco (van de Loo and Salvucci, 1996). This idea is consistent with the absence of AAA^+ motifs at the N-terminus as shown in figure 2.1. The two Cys residues in the C-terminal domain unique to the larger isoform of activase (see figure 2.1) are essential for the regulation of activase by reduction/oxidation (Zhang and Portis, 1999), which will be discussed later.

2.3.5   Specificity of interaction with Rubisco: identification of a Rubisco binding site

In order to facilitate the dissociation of the inhibitory sugar phosphates from Rubisco, the activase must bind somewhere on Rubisco and induce a conformational change at the active site. Given the self-associating property of the activase, one might expect that complexes including Rubisco could easily be detected, but this has proven to be experimentally challenging. In a chemical cross-linking experiment, activase could be cross-linked to the large subunit (Yokota and Tsujimoto, 1992) and co-immunoprecipitation of the two proteins has been reported (Sánchez de Jiménez *et al.*, 1995; Zhang and Komatsu, 2000), but these reports lack extensive characterization of the process. A complex of Rubisco and activase was also visualized by electron microscopy (Büchen-Osmond *et al.*, 1992) in which the activase seems to encircle the Rubisco holoenzyme.

Indirect evidence for a physical association between the proteins came from the unexpected observation that the activase exhibits dramatic species specificity (Wang *et al.*, 1992). Activase from Solanaceae species (e.g. tobacco and petunia) does not activate Rubisco from non-Solanaceae species (e.g. spinach, barley and *Chlamydomonas*) and activase from non-Solanaceae species does not activate Rubisco from Solanaceae species. Comparison of the Rubisco large subunit sequences revealed that a small set of residues clustered on the surface of the large subunit is substantially different in these two groups (Portis, 1995). Directed mutagenesis and chloroplast transformation were used to change four of these residues in *Chlamydomonas* to those found in tobacco (Larson *et al.*, 1997; Ott *et al.*, 2000). Two of these changes (Lys_{356} to Gln and Asp_{86} to Arg) had little effect on species–species interaction, but the Pro_{89} to Arg and Asp_{94} to Lys substitutions resulted in Rubisco enzymes that could no longer be activated by spinach activase very well *in vitro*, but the mutant enzymes could now be activated by tobacco activase. Thus a region on the large subunit of Rubisco was identified that interacts with the activase. In the absence of a structure for activase, the molecular basis for specificity in the interaction between these proteins remains largely speculative. However, it is obvious that complimentary amino acid differences must also be present in the activase. The relative specificity of chimeric activase proteins comprising the spinach and tobacco sequence (Esau *et al.*, 1998) suggests that the major specificity determinants are located in the *C*-terminal domain of the activase (including Box VIII), but the specific residues remain to be identified.

The location of at least part of the activase binding site on the large subunit is intriguing in considering possible mechanisms by which the activase can alter the sugar phosphate binding properties of Rubisco. The activase binding site is immediately adjacent to the active site in the large subunit and, collectively, the binding sites could enable an activase oligomer to completely encircle the Rubisco holoenzyme, consistent with the complex visualized by electron microscopy (Büchen-Osmond *et al.*, 1992).

2.3.6 *Conformational changes in Rubisco and a model for the activase mechanism*

Minimal conformational differences can be observed when the carbamylated and uncarbamylated structures of Rubisco are compared (Taylor and Andersson, 1996). In contrast, more dramatic conformational changes exist when enzymes containing bound sugar phosphates are compared (Duff *et al.*, 2000). These basically group into two alternative states corresponding to either an open or closed active site (Schreuder *et al.*, 1993; Taylor and Andersson, 1996, 1997). Figure 2.2 highlights some of the conformational differences between the open and closed states that may be most relevant to the ability of the activase to facilitate sugar phosphate release. In the closed state, the sugar phosphate is completely surrounded by residues from both large subunits in the homodimer and shielded from the outside environment because of two domains: Loop 6, comprising residues 333–338 between strand 6 and helix 6, and the *C*-terminus, comprising residues 468–475. Loop 6 is positioned in the active site so that Lys_{334} coordinates with both the sugar phosphate and Glu_{60}, a residue in the *N*-domain of the other large subunit in the homodimer. The *C*-terminus packs over Loop 6, further isolating the active site from the external medium. In the open structures, the *C*-terminus has such a high mobility that it cannot be visualized beyond residue 462. Loop 6 also cannot be seen in most of the open structures, but it is positioned out of the active site in some of them, as shown in figure 2.2. The intermolecular distance between the phosphates in the bound bisphosphate is 0.91 nm or less in closed structures and 0.94 nm or more in the open structures—the shortening of this distance after the initial binding of RuBP may trigger closure of the active site during catalysis and the formation of the tight complexes with inhibitors (Duff *et al.*, 2000).

These two states of Rubisco, open or closed, may also be relevant to the mechanism of the activase. The closed structures, particularly 1RCX (Taylor

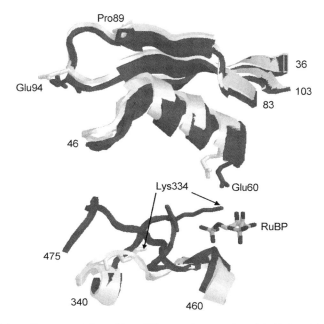

Figure 2.2 Schematic representation of key differences between 'open' (white) and 'closed' (gray) structures of Rubisco that may be relevant to the mechanism of the activase. The α-carbon backbones of the α/β-barrel domains of the 1RCX and 1RXO crystal structures, representing a 'closed' and 'open' active site, respectively (see Duff *et al.*, 2000), were aligned with Swiss-PdbViewer v3.7 and RASMOL was used to portray the α-carbon backbone of residues 36–60 and 83–103 in the *N*-domain of one subunit and residues 460–475(462) and 329–340 in the *C*-domain of the other subunit in a large subunit homodimer. Residues 60, 89, 94 and 334 are shown completely and RuBP is ribulose 1,5-bisphosphate in the 1RCX structure. RuBP is in a nearly identical position in the 1RXO structure but is not shown for clarity.

and Andersson, 1997), may correspond with the Rubisco–sugar phosphate substrates for the activase, while the open structures may correspond with the product of the activase, in which the active site is open and the dissociation of the sugar phosphate can readily proceed. Transition of the Rubisco structure from the closed to the open state would then primarily involve rotation of the *N*-domain away from the α/β barrel domain coupled with movement of the *C*-terminal strand away from Loop 6 and movement of Loop 6 out of the active site, opposite to the process described by Duff *et al.* (2000). The occasional movement of Loop 6 and the *C*-terminal stand probably occurs to some degree even when inhibitors are tightly bound because binding is not irreversible. The key process in the activase mechanism may be that after it forms a supercomplex with Rubisco, it catalyzes the movement of the *N*-domain toward the position seen in the open structures and may then form other interactions that stabilize the open conformation. As discussed previously, there is good evidence that the activase binds to the N-domain at residues 89 and 94. Movement of the N-domain would move Glu_{60} away from being able to interact with and stabilize Loop 6 (figure 2.2). Movement of the N-terminal domain may also be linked to the movement of Loop 6 and the C-terminus in more subtle ways that are discussed in detail by Duff *et al.* (2000). They also suggest that the activase may interact with a 'latch site' involving the side-chain of Asp_{473}. This is an interesting possibility but in contrast to the N-domain, there is currently no evidence that the activase interacts directly with the C-terminus.

Incorporating both new and old information, previous models for the activase mechanism (Andrews *et al.*, 1995; Portis, 1995; Salvucci and Ogren, 1996) can be elaborated further. The reported sizes of the activase complexes in the absence of Rubisco, and the equatorial location of residues 89 and 94 in the Rubisco holoenzyme, suggest that the domain movements occur in an activase–Rubisco supercomplex consisting of the Rubisco holoenzyme with its eight active sites encircled by 16 activase subunits. As noted previously, a complex consistent with such an arrangement has been visualized by electron microscopy (Büchen-Osmond *et al.*, 1992). A movement of the activase subunits relative to one another in the complex would be sufficient to catalyze the movement of the N-domains in the Rubisco large subunits to initiate opening of the active site and sugar phosphate release. A concerted movement of the subunits could be linked with the hydrolysis of ATP, consistent with the linkage between ATPase activity and oligomerization. The formation of ADP and a resulting conformational change in the activase would then favor dissociation of the complex, until the ADP is replaced by ATP to initiate another round of binding to Rubisco.

2.4   Regulation of the activase and Rubisco

The activity of Rubisco changes in response to several environmental signals including light, changes in source-sink balance, and temperature (reviewed in Portis, 1992). Changes in Rubisco activity between light and dark was recognized quite early and was readily understood by the positive effects of increased stromal pH and Mg^{2+} on the carbamylation state of the enzyme like those observed *in vitro* (Lorimer *et al.*, 1976). However, the dominant and critical role that activase has in controlling the activity of Rubisco is now apparent and a fairly detailed picture of the manner in which these environmental signals alter the activity of the activase can be presented.

2.4.1   *ADP/ATP regulation and redox regulation by thioredoxin f*

The potent inhibition by ADP causes the activity of the activase to be very sensitive to the ADP:ATP ratio at the physiological ratios of these nucleotides found in the chloroplast stroma (Robinson and Portis, 1988a, 1989a; Zhang and

Portis, 1999; Kallis *et al.*, 2000). At the ADP:ATP ratios typical of the dark (1:1), the activase appears to have minimal activity, conserving ATP when Rubisco activity is not needed. At the ratios typical of the light (1:2 to 1:3), the activase has less than half of its maximal activity. The sensitivity of activase to the ADP:ATP ratio allows Rubisco activity to be reduced when adequate sinks for reduced carbon are not available, which at the chloroplast level is expressed as a triose phosphate utilization limitation, reduced inorganic phosphate and lowered ATP:ADP ratios (Sharkey, 1990). Evidence for a triose phosphate utilization limitation can be observed at the intact leaf level by oxygen insensitive CO_2 uptake measured by gas-exchange analysis (Sharkey *et al.*, 1986). A linkage between the activation state of Rubisco and the ATP:ATP ratio was also revealed in the characterization of transgenic tobacco plants containing reduced amounts of Rubisco (Quick *et al.*, 1991). The reduced amount of Rubisco in these plants was compensated by an increase in its activation state and an increase in the ATP:ADP ratio as compared with the wild-type plants under the conditions analyzed. The importance of the structure of the P loop (in the Walker A motif) in the ADP:ATP sensitivity is indicated by the observation that the specific activity of the activase was increased and its sensitivity to the ADP:ATP ratio was diminished in recombinant activase proteins in which the Gln residue in the P loop region (figure 2.1) is replaced by Asp or Glu (Shen and Ogren, 1992; Kallis *et al.*, 2000).

The larger isoform of activase is more sensitive to ADP than the smaller isoform (Shen *et al.*, 1991). Recently it was found that the ADP:ATP sensitivity of the activity of the larger isoform of the activase is greatly diminished by reduction via thioredoxin f of a disulfide probably formed by the two Cys residues in the *C*-terminus of this isoform (Zhang and Portis, 1999). Reduction and oxidation of the native protein and mixtures of the recombinant isoforms indicate that the redox changes in the larger isoform also alter the activity of the smaller isoform, presumably thorough cooperative interactions in the activase complexes. The molecular details of how reduction/oxidation of the *C*-terminal domain changes the ADP:ATP sensitivity of the larger isoform remain to be elucidated, but it has been proposed that formation of a disulfide allows this acidic domain to be docked into the nucleotide binding site, altering its properties (Zhang and Portis, 2001).

The reduction/oxidation response of activase provides a means to fine-tune the activity of the activase and the Rubisco activation state to the prevailing light intensity. This regulatory response has been known for many years (Mächler and Nösberger, 1980; Perchorowicz *et al.*, 1981), but the biochemical mechanism was unclear. The characteristics of redox regulation of the larger isoform are consistent (Zhang and Portis, 2001) with the light/dark regulation of several key photosynthetic enzymes in the chloroplast stroma by reduction of disulfide bonds in these enzymes via thioredoxin, which is well established (Schürmann and Jacquot, 2000). A direct and quantitative link between the activation state of

Rubisco and the reduction state of either thioredoxin f or the activase remains to be established through direct measurements in leaf extracts. However, transgenic *Arabidopsis* plants expressing only the smaller isoform of activase no longer downregulate Rubisco in response to low light intensities (N. Zhang, R.P. Kallis, R.G. Ewy and A.R. Portis, unpublished).

Species that do not contain both isoforms of the activase may also regulate the activity of Rubisco in response to light, as shown by a recent study with tobacco (Ruuska *et al.*, 2000). How this is achieved is not clear, but one might envisage that a separate reduction/oxidation-sensitive protein may substitute for the regulatory function of the *C*-terminal domain in the larger isoform of the activase.

2.4.2   A role in the temperature response of photosynthesis

Evidence that the light activation of Rubisco was one of the most labile reactions associated with the inhibition of photosynthesis in spinach by high temperatures appeared several years before the role of activase in the process was discovered (Weis, 1981a,b). Subsequent work has confirmed that the inhibition of Rubisco activation at higher temperatures can be observed in wheat (Kobza and Edwards, 1987; Feller *et al.*, 1998), cotton (Feller *et al.*, 1998) and tobacco (Crafts-Brandner and Salvucci, 2000). These studies, and especially the recent and very detailed analysis of the response of photosynthesis to temperature (Crafts-Brandner and Salvucci, 2000), indicate that it is likely to be a common response in plants. A critical role for activase was initially suggested when it was found that the isolated spinach activase was very heat labile (Robinson and Portis, 1989a). A detailed investigation with the recombinant spinach isoforms (Crafts-Brandner *et al.*, 1997) revealed that stability was increased in the presence of adenine nucleotides and that the larger isoform was considerably more stable than the smaller isoform in the presence of the ATP analog, ATP-γ-S, which promotes oligomerization of the protein. The larger isoform could confer increased stability to the smaller isoform when they were combined. Finally, intrinsic fluorescence experiments indicated that subunit interactions accompanying oligomerization of the protein were disrupted by higher temperatures before denaturation of the protein itself occurs.

Activase levels increase during heat shock treatment in maize seedling leaves (Sánchez de Jiménez *et al.*, 1995). Maize appears to only have one isoform (Salvucci *et al.*, 1987) and, intriguingly, the increase is accompanied by the appearance of a larger isoform, suggesting a possible role for this isoform during heat stress. Some acclimation of photosynthesis and Rubisco activation in cotton and wheat plants was observed when heat stress is imposed gradually (Law and Crafts-Brandner, 1999) and it was suggested that new forms of activase with greater temperature stability could be involved. However, a specific role for the larger isoform in determining the temperature stability of activase may now be less likely because the primary role of the larger isoform is to allow redox regulation of the activity of the activase complex (Zhang and Portis, 1999). Also a subsequent investigation (Kallis *et al.*, 2000) with the recombinant *Arabidopsis* isoforms did not find that the larger isoform was more heat stable in this species. Whether or not reduction/oxidation alters the temperature stability of the larger isoform is not known. The role of the activase isoforms in the acclimation or resistance of photosynthesis to heat stress will continue to be an interesting area for future studies.

A series of studies suggest that the heat sensitivity of the activation of Rubisco could also be due to disrupted interactions between activase monomers and/or between activase oligomers and Rubisco (Crafts-Brandner *et al.*, 1997; Feller *et al.*, 1998; Crafts-Brandner and Law, 2000; Crafts-Brandner and Salvucci, 2000), i.e. the supercomplex itself may be very sensitive to temperature. An unusual and significant feature of the inhibition of Rubisco activation by high temperatures is that it is rapid and reversible at moderate temperatures (Weis, 1981a; Crafts-Brandner and Law, 2000). Given our current understanding of the mechanism of the activase, disruption of supercomplex formation would be consistent with rapid and reversible inhibition. However, reversibility has not yet been demonstrated in experiments with isolated activase, perhaps because of the more dilute concentrations *in vitro* that must be used.

The mechanism of the diminished Rubisco activation at higher temperatures may involve more than disrupted interactions of activase with itself or with Rubisco. Most significantly, Rubisco was shown to lose activity more rapidly at higher temperatures, which may be greater than the increase in the ability of activase to promote activation (Crafts-Brandner and Salvucci, 2000). The loss in Rubisco activity is probably due to an increased intrinsic propensity of the enzyme to form inhibitors during catalysis at higher temperature. The rapid reversibility of the inhibition at moderate temperatures would be readily explained by this mechanism. While experiments with transgenic plants expressing reduced amounts of activase indicate that activase can be in excess of that required for maintaining steady-state photosynthesis at normal temperatures (Eckardt *et al.*, 1997; Hammond *et al.*, 1998a), it is not clear to what extent this is true at higher temperatures. Experiments comparing the ability of activase to restore the activity of the various inhibited forms of Rubisco and the temperature response of plants overexpressing activase could begin to address this issue. It is also possible that the activase is downregulated in response to a limitation in some other process coming into play at higher temperatures. Some evidence (Weis, 1981b; Crafts-Brandner and Law, 2000) indicates that the known regulators of the activase, ATP:ADP ratio and stromal redox state, are unlikely to be involved. However, measurements of small changes in these factors are problematic. Sharkey (2000) has suggested that increased membrane leakiness at higher temperatures may cause a lower ATP:ADP ratio and lower activase activity, which could explain the greater heat tolerance of photosynthesis in plants with altered lipid composition (Murakami *et al.*, 2000). In any case, there may be other factors controlling the activity of the activase, which have not yet been identified.

2.4.3 Activase mRNA and protein levels

The mRNA encoding activase exhibits a diurnal rhythm (Rundle and Zielinski, 1991a), which is actually circadian and is disrupted by chilling in tomato (Martino-Catt and Ort, 1992). The promoter elements controlling activase transcription have been investigated (Orozco and Ogren, 1993; Liu *et al.*, 1996). Subsequent studies have found similar rhythms in activase gene expression in other species, including: tobacco (Klein and Salvucci, 1995), *Arabidopsis* (Pilgrim and McClung, 1993), apple (Watillon *et al.*, 1993) and rice (To *et al.*, 1999; Zhang and Komatsu, 2000). The functional significance of this rhythm for the regulation of Rubisco, if any, is unclear because a measurable diurnal change in the level of activase itself has only been observed in developing leaves grown under very low light intensities (Rundle and Zielinski, 1991a; Rundle, 1991). This is probably due to the relatively large amount (about 1–2% of total soluble protein) of activase found in plants under normal conditions. Alterations in expression of the two barley activase genes have also been observed during development and in response to illumination (Rundle and Zielinski, 1991a), but the significance of these differences are unclear. Over the long term, the abundance of activase generally follows that of Rubisco (Crafts-Brandner *et al.*, 1996; Jiang *et al.*, 1997) but the expression of Rubisco and the activase are not tightly coordinated, as demonstrated with both antisense activase tobacco transformants (Jiang *et al.*, 1994; Mate *et al.*, 1993) and a tobacco mutant defective in Rubisco synthesis (Klein and Salvucci, 1995). A limitation in the activity of Rubisco by the amount of activase has been suggested in studies of early leaf development (Fukayama *et al.*, 1996) and senescence (He *et al.*, 1997). Increases in the level of activase have also been correlated with increases in maize productivity (Martínez-Barajas *et al.*, 1997; Morales *et al.*, 1999). Thus changes in the mRNA level or the amount of activase may have long-term functional significance in the regulation of the activity of Rubisco and more work is needed to examine this question.

2.5   Future perspectives

Interest in the activase has been largely driven by the importance of Rubisco in photosynthesis. The emerging evidence that the activase is a member of the AAA^+ family, and that the activase–Rubisco interaction could be involved in determining the sensitivity of photosynthesis and plants to high temperature, should promote additional interest in the study of this system. However, progress continues to be hindered by the lack of a 3D structure for the activase and methods to study the interaction between activase and Rubisco more directly. Once these obstacles are overcome, the relative simplicity of the activase–Rubisco system may prove to be advantageous in furthering our understanding of the common molecular details of how proteins in the AAA$^+$ family carry out their functions.

Acknowledgements

I thank Mike Salvucci and Robert Ewy for critical reading of the manuscript and helpful suggestions. Support for the work from my laboratory cited in the text was provided by the US Department of Agriculture and grants from USDA-CRGO and DOE.

References

Andrews, T.J., Hudson, G.S., Mate, C.J., von Caemmerer, S., Evans, J.R. and Arvidsson, Y.B.C. (1995) Rubisco—the consequences of altering its expression and activation in transgenic plants. *J. Exp. Bot.*, **46**, 1293-1300.

Bochtler, M., Hartmann, C., Song, H.K., Bourenkov, G.P., Bartunik, H.D. and Huber, R. (2000) The structures of HsIU and the ATP-dependent protease HsIU-HsIV. *Nature*, **403**, 800-805.

Brooks, A. and Portis Jr, A.R. (1988) Protein-bound ribulose bisphosphate correlates with deactivation of ribulose bisphosphate carboxylase in leaves. *Plant Physiol.*, **87**, 244-249.

Büchen-Osmond, C., Portis Jr, A.R. and Andrews, T.J. (1992) Rubisco activase modifies the appearance of Rubisco in the electron microscope, in *Research in Photosynthesis, Vol. III; IXth International Congress On Photosynthesis, Nagoya, Japan, August 30–September 4, 1992* (ed. N. Murata), pp. 653-656. Kluwer Academic Publishers, Dordrecht, The Netherlands.

Cardon, Z.G. and Mott, K.A. (1989) Evidence that ribulose-1,5-bisphosphate (RuBP) binds to inactive sites of RuBP carboxylase *in vivo* and an estimate of the rate constant for dissociation. *Plant Physiol.*, **89**, 1253-1257.

Cleland, W.W., Andrews, T.J., Gutteridge, S., Hartman, F.C. and Lorimer, G.H. (1998) Mechanism of Rubisco: the carbamate as general base. *Chem. Rev.*, **98**, 549-561.

Crafts-Brandner, S.J. and Law, R.D. (2000) Effect of heat stress on the inhibition and recovery of the ribulose-1,5-bisphosphate carboxylase/oxygenase activation state. *Planta*, **212**, 67-74.

Crafts-Brandner, S.J. and Salvucci, M.E. (2000) Rubisco activase constrains the photosynthetic potential of leaves at high temperature and CO_2. *Proc. Natl Acad. Sci. USA*, **97**, 13,430-13,435.

Crafts-Brandner, S.J., Klein, R.R., Klein, P., Hoelzer, R. and Feller, U. (1996) Coordination of protein and mRNA abundances of stromal enzymes and mRNA abundances of the Clp protease subunits during senescence of *Phaseolus vulgaris* (L.) leaves. *Planta*, **200**, 312-318.

Crafts-Brandner, S.J., van de Loo, F.J. and Salvucci, M.E. (1997) The two forms of ribulose-1,5-bisphosphate carboxylase-oxygenase activase differ in sensitivity to elevated temperature. *Plant Physiol.*, **114**, 439-444.

Duff, A.P., Andrews, T.J. and Curmi, P.M.G. (2000) The transition between the open and closed states of Rubisco is triggered by the inter-phosphate distance of the bound bisphosphate. *J. Mol. Biol.*, **298**, 903-916.

Eckardt, N.A. and Portis Jr, A.R. (1997) Heat denaturation profiles of ribulose-1,5-bisphosphate carboxylase/oxygenase (Rubisco) and Rubisco activase and the inability of Rubisco activase to restore activity of heat-denatured Rubisco. *Plant Physiol.*, **113**, 243-248.

Eckardt, N.A., Snyder, G.W., Portis Jr, A.R. and Ogren, W.L. (1997) Growth and photosynthesis under high and low irradiance of *Arabidopsis thaliana* antisense mutants with reduced ribulose-1,5-bisphosphate carboxylase/oxygenase activase content. *Plant Physiol.*, **113**, 575-586.

Edmondson, D.L., Badger, M.R. and Andrews, T.J. (1990) Slow inactivation of ribulosebisphosphate carboxylase during catalysis is caused by accumulation of a slow, tight-binding inhibitor at the catalytic site. *Plant Physiol.*, **93**, 1390-1397.

Ellis, R.J. (1979) The most abundant protein in the world. *Trends Biochem. Sci.*, **4**, 241-244.

Esau, B.D., Snyder, G.W. and Portis Jr, A.R. (1996) Differential effects of *N*- and *C*-terminal deletions on the two activities of Rubisco activase. *Arch. Biochem. Biophys.*, **326**, 100-105.

Esau, B.D., Snyder, G.W. and Portis Jr, A.R. (1998) Activation of ribulose-1,5-bisphosphate carboxylase/oxygenase (Rubisco) with chimeric activase proteins. *Photosynthesis Res.*, **58**, 175-181.

Feller, U., Crafts-Brandner, S.J. and Salvucci, M.E. (1998) Moderately high temperatures inhibit ribulose-1,5-bisphosphate carboxylase/oxygenase (Rubisco) activase-mediated activation of Rubisco. *Plant Physiol.*, **116**, 539-546.

Fukayama, H., Uchida, N., Azuma, T. and Yasuda, T. (1996) Relationships between photosynthetic activity and the amounts of Rubisco activase and Rubisco in rice leaves from emergence through senescence. *Jpn. J. Crop Sci.*, **65**, 296-302.

Guenther, B., Onrust, R., Sali, A., O'Donnell, M. and Kuriyan, J. (1997) Crystal structure of the δ' subunit of the clamp-loader complex of *E. coli* DNA polymerase III. *Cell*, **91**, 335-345.

Hammond, E.T., Andrews, T.J., Mott, K.A. and Woodrow, I.E. (1998a) Regulation of Rubisco activation in antisense plants of tobacco containing reduced levels of Rubisco activase. *Plant J.*, **14**, 101-110.

Hammond, E.T., Andrews, T.J. and Woodrow, I.E. (1998b) Regulation of ribulose-1,5-bisphosphate carboxylase/oxygenase by carbamylation and 2-carboxyarabinitol 1-phosphate in tobacco: insights from studies of antisense plants containing reduced amounts of Rubisco activase. *Plant Physiol.*, **118**, 1463-1471.

Hartman, F.C. and Harpel, M.R. (1994) Structure, function, regulation, and assembly of D-ribulose-1,5-bisphosphate carboxylase-oxygenase, in *Annual Review of Biochemistry, Vol. 63* (ed. C.C. Richardson), pp. 197-234. Annual Reviews Inc., Palo Alto, CA, USA.

He, Z.L., von Caemmerer, S., Hudson, G.S., Price, G.D., Badger, M.R. and Andrews, T.J. (1997) Ribulose-1,5-bisphosphate carboxylase/oxygenase activase deficiency delays senescence of ribulose-1,5-bisphosphate carboxylase/oxygenase but progressively impairs its catalysis during tobacco leaf development. *Plant Physiol.*, **115**, 1569-1580.

Jiang, C.Z., Quick, W.P., Alred, R., Kliebenstein, D. and Rodermel, S.R. (1994) Antisense RNA inhibition of Rubisco activase expression. *Plant J.*, **5**, 787-798.

Jiang, C.Z., Rodermel, S.R. and Shibles, R.M. (1997) Regulation of photosynthesis in developing leaves of soybean chlorophyll-deficient mutants. *Photosynthesis Res.*, **51**, 185-192.

Jordan, D.B. and Chollet, R. (1983) Inhibition of ribulose bisphosphate carboxylase by substrate ribulose 1,5-bisphosphate. *J. Biol.Chem.*, **258**, 13,752-13,758.

Kallis, R.P., Ewy, R.G. and Portis Jr, A.R. (2000) Alteration of the adenine nucleotide response and increased Rubisco activation activity of *Arabidopsis* Rubisco activase by site-directed mutagenesis. *Plant Physiol.*, **123**, 1077-1086.

Kane, H.J., Wilkin, J.M., Portis Jr, A.R. and Andrews, T.J. (1998) Potent inhibition of ribulosebisphosphate carboxylase by an oxidized impurity in ribulose-1,5-bisphosphate. *Plant Physiol.*, **117**, 1059-1069.

Klein, R.R. and Salvucci, M.E. (1995) Rubisco, Rubisco activase and ribulose-5-phosphate kinase gene expression and polypeptide accumulation in a tobacco mutant defective in chloroplast protein synthesis. *Photosynthesis Res.*, **43**, 213-223.

Kobza, J. and Edwards, G.E. (1987) Influences of leaf temperature on photosynthetic carbon metabolism in wheat. *Plant Physiol.*, **83**, 69-74.

Lan, Y. and Mott, K.A. (1991) Determination of apparent K_m values for ribulose 1,5-bisphosphate carboxylase/oxygenase (Rubisco) activase using the spectrophotometric assay of Rubisco activity. Plant Physiol., 95, 604-609.

Larson, E.M., Obrien, C.M., Zhu, G.H., Spreitzer, R.J. and Portis Jr, A.R. (1997) Specificity for activase is changed by a Pro-89 to Arg substitution in the large subunit of ribulose-1,5-bisphosphate carboxylase/oxygenase. J. Biol. Chem., 272, 17,033-17,037.

Law, R.D. and Crafts-Brandner, S.J. (1999) Inhibition and acclimation of photosynthesis to heat stress is closely correlated with activation of ribulose-1,5-bisphosphate carboxylase/oxygenase. Plant Physiol., 120, 173-181.

Lilley, R.M. and Portis Jr, A.R. (1990) Activation of ribulose-1,5-bisphosphate carboxylase/oxygenase (Rubisco) by Rubisco activase: effects of some sugar phosphates. Plant Physiol., 94, 245-250.

Lilley, R.M. and Portis Jr, A.R. (1997) ATP hydrolysis activity and polymerization state of ribulose-1,5-bisphosphate carboxylase oxygenase activase—do the effects of Mg^{2+}, K^+, and activase concentrations indicate a functional similarity to actin? Plant Physiol., 114, 605-613.

Liu, Z.R., Taub, C.C. and McClung, C.R. (1996) Identification of an Arabidopsis thaliana ribulose-1,5-bisphosphate carboxylase oxygenase activase (RCA) minimal promoter regulated by light and the circadian clock. Plant Physiol., 112, 43-51.

Lorimer, G.H., Badger, M.R. and Andrews, T.J. (1976) The activation of ribulose-1,5-bisphosphate carboxylase by carbon dioxide and magnesium ions. Equilibria, kinetics, a suggested mechanism, and physiological implications. Biochemistry, 15, 529-536.

Mächler, F. and Nösberger, J. (1980) Regulation of ribulose bisphosphate carboxylase activity in intact wheat leaves by light, CO_2, and temperature. J. Exp. Bot., 31, 1485-1491.

Martínez-Barajas, E., Molina-Galán, J. and Sánchez de Jiménez, E. (1997) Regulation of Rubisco activity during grain-fill in maize: possible role of Rubisco activase. J. Agric. Sci., 128, 155-161.

Martino-Catt, S. and Ort, D.R. (1992) Low temperature interrupts circadian regulation of transcriptional activity in chilling-sensitive plants. Proc. Natl Acad. Sci. USA, 89, 3731-3735.

Mate, C.J., Hudson, G.S., von Caemmerer, S., Evans, J.R. and Andrews, T.J. (1993) Reduction of ribulose bisphosphate carboxylase activase levels in tobacco (Nicotiana tabacum) by antisense RNA reduces ribulose bisphosphate carboxylase carbamylation and impairs photosynthesis. Plant Physiol., 102, 1119-1128.

Mate, C.J., von Caemmerer, S., Evans, J.R., Hudson, G.S. and Andrews, T.J. (1996) The relationship between CO_2-assimilation rate, Rubisco carbamylation and Rubisco activase content in activase-deficient transgenic tobacco suggests a simple model of activase action. Planta, 198, 604-613.

Morales, A., Ortega-Delgado, M.L., Molina-Galán, J. and Sánchez de Jiménez, E. (1999) Importance of Rubisco activase in maize productivity based on mass selection procedure. J. Exp. Bot., 50, 823-829.

Mott, K.A. and Woodrow, I.E. (2000) Modelling the role of Rubisco activase in limiting non-steady-state photosynthesis. J. Exp. Bot., 51, 399-406.

Mott, K.A., Snyder, G.W. and Woodrow, I.E. (1997) Kinetics of Rubisco activation as determined from gas-exchange measurements in antisense plants of Arabidopsis thaliana containing reduced levels of Rubisco activase. Aust. J. Plant Physiol., 24, 811-818.

Murakami, Y., Tsuyama, M., Kobayashi, Y., Kodama, H. and Iba, K. (2000) Trienoic fatty acids and plant tolerance of high temperature. Science, 287, 476-479.

Neuwald, A.F., Aravind, L., Spouge, J.L. and Koonin, E.V. (1999) AAA^+: A class of chaperone-like ATPases associated with the assembly, operation, and disassembly of protein complexes. Genome Res., 9, 27-43.

Orozco, B.M. and Ogren, W.L. (1993) Localization of light-inducible and tissue-specific regions of the spinach ribulose bisphosphate carboxylase/oxygenase (rubisco) activase promoter in transgenic tobacco plants. Plant Mol. Biol., 23, 1129-1138.

Ott, C.M., Smith, B.D., Portis Jr, A.R. and Spreitzer, R.J. (2000) Activase region on chloroplast ribulose-1,5-bisphosphate carboxylase/oxygenase—nonconservative substitution in the large subunit alters species specificity of protein interaction. *J. Biol. Chem.*, **275**, 26,241-26,244.

Parry, M.A.J., Andralojc, P.J., Parmar, S. *et al.* (1997) Regulation of Rubisco by inhibitors in the light. *Plant Cell Env.*, **20**, 528-534.

Patel, S. and Latterich, M. (1998) The AAA team—related ATPases with diverse functions. *Trends Cell Biol.*, **8**, 65-71.

Perchorowicz, J.T., Raynes, D.A. and Jensen, R.G. (1981) Light limitation of photosynthesis and activation of ribulose bisphosphate carboxylase in wheat seedlings. *Proc. Natl Acad. Sci. USA*, **78**, 2985-2989.

Pilgrim, M.L. and McClung, C.R. (1993) Differential involvement of the circadian clock in the expression of genes required for ribulose-1,5-bisphosphate carboxylase–oxygenase synthesis, assembly, and activation in *Arabidopsis thaliana*. *Plant Physiol.*, **103**, 553-564.

Portis Jr, A.R. (1992) Regulation of ribulose 1,5-bisphosphate carboxylase/oxygenase activity. *Annu. Rev. Plant Physiol. Plant Mol. Biol.*, **43**, 415-437.

Portis Jr, A.R. (1995) The regulation of Rubisco by Rubisco activase. *J. Exp. Bot.*, **46**, 1285-1291.

Portis Jr, A.R., Salvucci, M.E. and Ogren, W.L. (1986) Activation of ribulosebisphosphate carboxylase/oxygenase at physiological CO_2 and ribulosebisphosphate concentrations by Rubisco activase. *Plant Physiol.*, **82**, 967-971.

Portis Jr, A.R., Lilley, R.M. and Andrews, T.J. (1995) Subsaturating ribulose-1,5-bisphosphate concentration promotes inactivation of ribulose-1,5-bisphosphate carboxylase/oxygenase (Rubisco)—studies using continuous substrate addition in the presence and absence of Rubisco activase. *Plant Physiol.*, **109**, 1441-1451.

Quick, W.P., Schurr, U., Scheibe, R. *et al.* (1991) Decreased ribulose-1,5-bisphosphate carboxylase–oxygenase in transgenic tobacco transformed with 'antisense' rbcS. I. Impact on photosynthesis in ambient growth conditions. *Planta*, **183**, 542-554.

Robinson, S.P. and Portis Jr, A.R. (1988a) Involvement of stromal ATP in the light activation of ribulose-1,5-bisphosphate carboxylase/oxygenase in intact isolated chloroplasts. *Plant Physiol.*, **86**, 293-298.

Robinson, S.P. and Portis Jr, A.R. (1988b) Release of the nocturnal inhibitor, carboxyarabinitol-1-phosphate, from ribulose bisphosphate carboxylase/oxygenase by Rubisco activase. *FEBS Lett.*, **233**, 413-416.

Robinson, S.P. and Portis Jr, A.R. (1989a) Adenosine triphosphate hydrolysis by purified Rubisco activase. *Arch. Biochem. Biophys.*, **268**, 93-99.

Robinson, S.P. and Portis Jr, A.R. (1989b) Ribulose-1,5-bisphosphate carboxylase/oxygenase activase protein prevents the *in vitro* decline in activity of ribulose-1,5-bisphosphate carboxylase/oxygenase. *Plant Physiol.*, **90**, 968-971.

Robinson, S.P., Streusand, V.J., Chatfield, J.M. and Portis Jr, A.R. (1988) Purification and assay of Rubisco activase from leaves. *Plant Physiol.*, **88**, 1008-1014.

Rundle, S.J. (1991) Isolation, characterization and expression of cDNA and genomic sequences encoding ribulosebisphosphate carboxylase/oxygenase activase from barley. *PhD thesis*, University of Illinois, Urbana, IL, USA.

Rundle, S.J. and Zielinski, R.E. (1991a) Alterations in barley ribulose-1,5-bisphosphate carboxylase/oxygenase activase gene expression during development and in response to illumination. *J. Biol. Chem.*, **266**, 14,802-14,807.

Rundle, S.J. and Zielinski, R.E. (1991b) Organization and expression of two tandemly oriented genes encoding ribulosebisphosphate carboxylase/oxygenase activase in barley. *J. Biol. Chem.*, **266**, 4677-4685.

Ruuska, S.A., Andrews, T.J., Badger, M.R., Price, G.D. and von Caemmerer, S. (2000) The role of chloroplast electron transport and metabolites in modulating Rubisco activity in tobacco. Insights from transgenic plants with reduced amounts of cytochrome *b/f* complex or glyceraldehyde 3-phosphate dehydrogenase. *Plant Physiol.*, **122**, 491-504.

Salvucci, M.E. (1992) Subunit interactions of Rubisco activase—polyethylene glycol promotes self-association, stimulates ATPase and activation activities, and enhances interactions with Rubisco. *Arch. Biochem. Biophys.*, **298**, 688-696.

Salvucci, M.E. (1993) Covalent modification of a highly reactive and essential lysine residue of ribulose-1,5-bisphosphate carboxylase–oxygenase activase. *Plant Physiol.*, **103**, 501-508.

Salvucci, M.E. and Klein, R.R. (1994) Site-directed mutagenesis of a reactive lysyl residue (Lys-247) of Rubisco activase. *Arch. Biochem. Biophys.*, **314**, 178-185.

Salvucci, M.E. and Ogren, W.L. (1996) The mechanism of Rubisco activase—insights from studies of the properties and structure of the enzyme. *Photosynthesis Res.*, **47**, 1-11.

Salvucci, M.E., Portis Jr, A.R. and Ogren, W.L. (1986) Light and CO_2 response of ribulose-1,5-bisphosphate carboxylase/oxygenase activation in *Arabidopsis* leaves. *Plant Physiol.*, **80**, 655-659.

Salvucci, M.E., Werneke, J.M., Ogren, W.L. and Portis Jr, A.R. (1987) Purification and species distribution of Rubisco activase. *Plant Physiol.*, **84**, 930-936.

Salvucci, M.E., Rajagopalan, K., Sievert, G., Haley, B.E. and Watt, D.S. (1993) Photoaffinity labeling of ribulose-1,5-bisphosphate carboxylase/oxygenase activase with ATP γ-benzophenone: identification of the ATP γ-phosphate binding domain. *J. Biol. Chem.*, **268**, 14,239-14,244.

Salvucci, M.E., Chavan, A.J., Klein, R.R., Rajagopolan, K. and Haley, B.E. (1994) Photoaffinity labeling of the ATP binding domain of Rubisco activase and a separate domain involved in the activation of ribulose-1,5-bisphosphate carboxylase/oxygenase. *Biochemistry*, **33**, 14,879-14,886.

Sánchez de Jiménez, E., Medrano, L. and Martínez-Barajas, E. (1995) Rubisco activase, a possible new member of the molecular chaperone family. *Biochemistry*, **34**, 2826-2831.

Schreuder, H.A., Knight, S., Curmi, P.M.G. *et al.* (1993) Formation of the active site of ribulose-1,5-bisphosphate carboxylase–oxygenase by a disorder–order transition from the unactivated to the activated form. *Proc. Natl Acad. Sci. USA*, **90**, 9968-9972.

Schürmann, P. and Jacquot, J.-P. (2000) Plant thioredoxin systems revisited. *Annu. Rev. Plant Physiol. Plant Mol. Biol.*, **51**, 371-400.

Seemann, J.R., Kobza, J. and Moore, B.D. (1990) Metabolism of 2-carboxyarabinitol 1-phosphate and regulation of ribulose-1,5-bisphosphate carboxylase activity. *Photosynthesis Res.*, **23**, 119-130.

Sharkey, T.D. (1990) Feedback limitation of photosynthesis and the physiological role of ribulose bisphosphate carboxylase carbamylation. *Botanical Magazine of Tokyo Special Issue*, **2**, 87-105.

Sharkey, T.D. (2000) Plant biology—some like it hot. *Science*, **287**, 435-437.

Sharkey, T.D., Stitt, M., Heineke, D., Gerhardt, R., Raschke, K. and Heldt, H.W. (1986) Limitation of photosynthesis by carbon metabolism. II. O_2 insensitive CO_2 uptake results from limitation of triose phosphate utilization. *Plant Physiol.*, **81**, 1123-1129.

Shen, J.B. and Ogren, W.L. (1992) Alteration of spinach ribulose-1,5-bisphosphate carboxylase/oxygenase activase activities by site-directed mutagenesis. *Plant Physiol.*, **99**, 1201-1207.

Shen, J.B., Orozco, E.M. and Ogren, W.L. (1991) Expression of the two isoforms of spinach ribulose 1,5-bisphosphate carboxylase activase and essentiality of the conserved lysine in the consensus nucleotide-binding domain. *J. Biol. Chem.*, **266**, 8963-8968.

Somerville, C.R., Portis Jr, A.R. and Ogren, W.L. (1982) A mutant of *Arabidopsis thaliana* which lacks activation of RuBP carboxylase *in vivo*. *Plant Physiol.*, **70**, 381-387.

Spreitzer, R.J. (1999) Questions about the complexity of chloroplast ribulose-1,5-bisphosphate carboxylase/oxygenase. *Photosynthesis Res.*, **60**, 29-42.

Streusand, V.J. and Portis Jr, A.R. (1987) Rubisco activase mediates ATP-dependent activation of ribulose bisphosphate carboxylase. *Plant Physiol.*, **85**, 152-154.

Taylor, T.C. and Andersson, I. (1996) Structural transitions during activation and ligand binding in hexadecameric Rubisco inferred from the crystal structure of the activated unliganded spinach enzyme. *Nat. Struct. Biol.*, **3**, 95-101.

Taylor, T.C. and Andersson, I. (1997) The structure of the complex between Rubisco and its natural substrate ribulose 1,5-bisphosphate. *J. Mol. Biol.*, **265**, 432-444.

To, K.Y., Suen, D.F. and Chen, S.C.G. (1999) Molecular characterization of ribulose-1,5-bisphosphate carboxylase/oxygenase activase in rice leaves. *Planta*, **209**, 66-76.

van de Loo, F.J. and Salvucci, M.E. (1996) Activation of ribulose-1,5-bisphosphate carboxylase/oxygenase (Rubisco) involves Rubisco activase Trp16. *Biochemistry*, **35**, 8143-8148.

van de Loo, F.J. and Salvucci, M.E. (1998) Involvement of two aspartate residues of Rubisco activase in coordination of the ATP γ-phosphate and subunit cooperativity. *Biochemistry*, **37**, 4621-4625.

Wang, Z.Y. and Portis Jr, A.R. (1992) Dissociation of ribulose-1,5-bisphosphate bound to ribulose-1,5-bisphosphate carboxylase/oxygenase and its enhancement by ribulose-1,5-bisphosphate carboxylase/oxygenase activase-mediated hydrolysis of ATP. *Plant Physiol.*, **99**, 1348-1353.

Wang, Z.Y., Snyder, G.W., Esau, B.D., Portis Jr, A.R. and Ogren, W.L. (1992) Species-dependent variation in the interaction of substrate-bound ribulose-1,5-bisphosphate carboxylase/oxygenase (Rubisco) and Rubisco activase. *Plant Physiol.*, **100**, 1858-1862.

Wang, Z.Y., Ramage, R.T. and Portis Jr, A.R. (1993) Mg^{2+} and ATP or adenosine 5'-[γ-thio]-triphosphate (ATPγS) enhances intrinsic fluorescence and induces aggregation which increases the activity of spinach Rubisco activase. *Biochim. Biophys. Acta*, **1202**, 47-55.

Watillon, B., Kettmann, R., Boxus, P. and Burny, A. (1993) Developmental and circadian pattern of Rubisco activase messenger RNA accumulation in apple plants. *Plant Mol. Biol.*, **23**, 501-509.

Weis, E. (1981a) Reversible heat-inactivation of the Calvin cycle: a possible mechanism of the temperature regulation of photosynthesis. *Planta*, **151**, 33-39.

Weis, E. (1981b) The temperature-sensitivity of dark-inactivation and light-activation of the ribulose-1,5-bisphosphate carboxylase in spinach chloroplasts. *FEBS Lett.*, **129**, 197-200.

Werneke, J.M., Chatfield, J.M. and Ogren, W.L. (1988a) Catalysis of ribulosebisphosphate carboxylase/oxygenase activation by the product of a Rubisco activase complementary DNA clone expressed in *Escherichia coli. Plant Physiol.*, **87**, 917-920.

Werneke, J.M., Zielinski, R.E. and Ogren, W.L. (1988b) Structure and expression of spinach leaf complementary DNA encoding ribulose bisphosphate carboxylase–oxygenase activase. *Proc. Natl Acad. Sci. USA*, **85**, 787-791.

Werneke, J.M., Chatfield, J.M. and Ogren, W.L. (1989) Alternative mRNA splicing generates the two ribulosebisphosphate carboxylase/oxygenase activase polypeptides in spinach and *Arabidopsis. Plant Cell*, **1**, 815-825.

Woodrow, I.E. and Berry, J.A. (1988) Enzymatic regulation of photosynthetic CO$_2$ fixation in C$_3$ plants. *Annu. Rev. Plant Physiol. Plant Mol. Biol.*, **39**, 533-594.

Yokota, A. and Tsujimoto, N. (1992) Characterization of ribulose-1,5-bisphosphate carboxylase/oxygenase carrying ribulose 1,5-bisphosphate at its regulatory sites and the mechanism of interaction of this form of the enzyme with ribulose-1,5-bisphosphate-carboxylase/oxygenase activase. *Eur. J. Biochem.*, **204**, 901-909.

Zhang, Z.L. and Komatsu, S. (2000) Molecular cloning and characterization of cDNAs encoding two isoforms of ribulose-1,5-bisphosphate carboxylase/oxygenase activase in rice (*Oryza sativa* L.). *J. Biochem.*, **128**, 383-389.

Zhang, N. and Portis Jr, A.R. (1999) Mechanism of light regulation of Rubisco: a specific role for the larger Rubisco activase isoform involving reductive activation by thioredoxin-f. *Proc. Natl Acad. Sci. USA*, **96**, 9438-9443.

Zhang, N., Schürmann, P. and Portis Jr, A.R. (2001) Characterization of the regulatory function of the 46-kDa isoform of Rubisco activase from *Arabidopsis. Photosynthesis Res.*, **68**, 29-37.

3 14-3-3 Protein: effectors of enzyme function

Paul C. Sehnke and Robert J. Ferl

3.1 Introduction

The ability of plants to sense and respond to their environment is a seemingly simple, yet obligatory, duty. The mechanisms by which appropriate responses are initiated are, however, quite elaborate and usually involve a multitude of protein factors and interactions. In fact, many of the proteins utilized are already present in the plant and simply require a change in the their regulatory state, thus allowing for a shortened response time. This rapid response system of post-translational regulation often involves phosphorylation of the target protein by the appropriate kinases, resulting in an altered enzymatic activity. In many cases the phosphorylation event itself is not sufficient to bring about the desired consequence, but rather provides the trigger for subsequent protein–protein interactions that then allow for activation or inhibition of the target protein. These partner proteins can simply be another subunit of the same family of proteins, such as the case with receptor kinases, or they maybe heterologous multifunctional signal transduction proteins. In the plant world, the emergence of 14-3-3 proteins as nearly ubiquitous regulators that primarily bind phosphorylated residues suggests a common yet expansive role for 14-3-3s in the environmental response cascade as well as in everyday plant functions. The presence of large 14-3-3 families, coupled with divergent extremities of the 14-3-3s supports the argument for these proteins as omnipotent multitasking factors that participate in signal transduction through direct protein-phosphoprotein interactions.

3.2 Structure and biochemical characterization of 14-3-3 proteins

While the 14-3-3s were first described by Moore and Perez in the 1960s as a class of soluble, small (ca. 30–35 kDa), mammalian brain proteins (Moore and Perez, 1967), they have since been cataloged in almost every eukaryotic organism (Rosenquist *et al.*, 2000). The 14-3-3s, named for their electrophoretic and chromatographic migration patterns, exist in their native state as homo- and heterodimers of 60–70 kDa. The different 14-3-3 monomers are a result of the different family member gene products and post-translational protein modifications that might be applied to any of the gene products (Aitken *et al.*, 1995). The conservation among the family member gene products is $>60\%$

depending upon the host system, and this degree of conservation suggests that a similar basic 14-3-3 protein architecture applies across phyla.

The three dimensional structure of 14-3-3s was determined for two mammalian isoforms, ζ and τ, and is shown in figure 3.1 (Liu *et al.*, 1995,

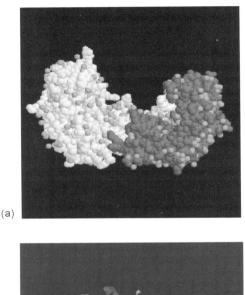

(a)

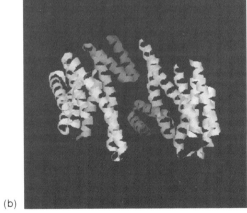

(b)

Figure 3.1 Three-dimensional structure of a 14-3-3 dimer and bound peptide. (a) A space-filling model of the human 14-3-3 ζ dimer structure with bound target peptide demonstrates the W or cupped nature of the dimer and binding channel. The 14-3-3 monomers are dark grey; a target peptide (light grey) is shown docked into the left monomer. (b) A schematic cartoon of the quaternary structure of the x-ray diffraction crystallographic model of human 14-3-3 ζ with nine antiparallel helical tertiary elements displayed. The figures were generated with the Molecular Graphics Visualization Tool RasMol 2.7.1.1 using the coordinates deposited in the Research Collaboratory for Structural Bioinformatics Protein Data Bank (file identifier 1QJB) by Rittinger *et al.* (1999).

Xiao *et al.*, 1995). The protein monomer is composed of nine antiparallel α-helices (figure 3.1a). The distal portions of both the *N*- and *C*-termini were structurally less ordered and/or took on multiple conformations, therefore rendering them discontiguous in the three dimensional (3D) electron density chain tracings. The 14-3-3 dimer forms a distinctive 'W'-shaped bowl, with a double groove of electrostatic potential responsible for stabilizing interactions with target proteins (figure 3.1b). The capacity to bind and draw two target proteins together arises from this symmetrical dimer arrangement. The contacts for the dimer interphase are predominantly in the first four helices of the 14-3-3s. The remaining portion of the molecule's cupped-like structure serves as the amphipathic channel that comprises the target protein interaction domain. Modeling of a substrate peptide co-crystallized with recombinant 14-3-3s confirmed that the charged channels were responsible for the 14-3-3 target interaction, although contact/regulation with a full-length protein is still not obvious or constant for each target (Yaffe *et al.*, 1997, Petosa *et al.*, 1998, Rittinger *et al.*, 1999). The high conservation of the amino acid sequence of 14-3-3s across phyla supports using the mammalian 3D-structures for modeling studies of non-mammalian 14-3-3s, including plants. However the length and diversity of terminal helices necessitates that constraint be exercised when examining specific features not associated with the core backbone of the 14-3-3, such as for specific isoform binding details.

Biochemical mapping of the interaction domains of 14-3-3s and other substructures provide further indications of the high degree of similarity amongst 14-3-3s. Biochemical studies by Ichimura *et al.*, determined that the C-terminal region of the 14-3-3s (Ichimura *et al.*, 1995) can regulate their interaction with the metabolic brain enzymes tyrosine and tryptophan hydroxylase. In plants, this same region was also identified as possessing a divalent cation binding domain that could alter the structure of the 14-3-3 when bound by calcium or magnesium, according to proteolytic digestion studies (figure 3.2) (Lu *et al.*, 1994b). The flexible nature of this region may allow it to act as a hinge to lock the 14-3-3 into a specific active conformation. This plastic nature may also contribute to the difficulty in determining the 3D-structure of this region. Interestingly, in virtually every plant system where 14-3-3s are involved in regulation, divalent cations are a prerequisite to the activity, suggesting that divalent binding to 14-3-3s plays a direct role in 14-3-3 binding (Athwal *et al.*, 1998a; Morsomme and Boutry, 2000). Additionally the epitope for an antibody against a plant transcription complex associated 14-3-3 mapped to this same region, suggesting that the *C*-termini of 14-3-3s is externally localized when 14-3-3s are bound to their target proteins (Lu *et al.*, 1994b).

Since the 3D-structure for 14-3-3s was determined using dimers composed of identical 14-3-3 monomers, the specifics about heterodimer formation and or dimerization preference among isoforms remains unclear. Characterization of interactions between monomers of plant 14-3-3s as both homo- and

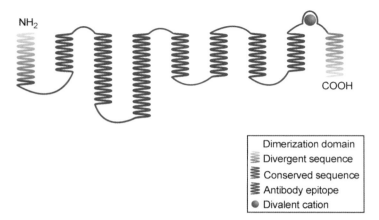

Figure 3.2 Schematic cartoon of the functional domains of plant 14-3-3s. The fundamental structure of the plant 14-3-3s is a highly conserved central core centered between distal divergent termini. The first four *N*-terminal antiparallel helices are primarily responsible for the dimerization between 14-3-3 monomers. The remaining portion of the molecule contains the target binding elements. The exposed antibody epitope of a complexed 14-3-3 is contained within the seventh, eighth and part of the ninth helix. The flexible hinge between helix eight and nine is responsible for the divalent cation binding, a necessary precursor to the 14-3-3 phosphoserine/threonine target binding.

heterodimers was examined by Wu *et al.* and the study demonstrated that preferred 14-3-3 subunit associations exist (Wu *et al.*, 1997a). Using the yeast two-hybrid system, the authors determined that the dimeric contact was supported by helices 1–4 of 14-3-3s and that homodimers of certain isoforms were preferred over heterodimerization. Using native gels and heterodimerization studies with denatured/renatured recombinant 14-3-3s, *in vitro* heterodimerization was also demonstrated between the different 14-3-3 isoforms. The inherent stability of the dimers was also authenticated, with complete denaturation required prior to re-association.

3.3   Plant 14-3-3 protein families and organization

Many large 14-3-3 gene families have been identified in plants, including *Arabidopsis*, tobacco, tomato, barley, potato, maize and rice (DeLille *et al.*, 2001; Piotrowski and Oecking, 1998; Roberts and Bowles, 1999; Testerink *et al.*, 1999; Wilczynski *et al.*, 1998; de Vetten and Ferl, 1994; Kidou *et al.*, 1993). The absolute number of isoforms per species varies. However, this may yet be attributable to the incomplete characterization of many of the genomes under study. One of the most confusing aspects of the 14-3-3 literature has been the diversity in nomenclature propagated by research groups identifying

14-3-3s as partners to a protein of particular interest. The completion of genome projects may facilitate a more cohesive understanding of isoform similarity and diversity within organisms and should allow for more direct comparisons between isoforms of different hosts.

Perhaps the best-characterized 14-3-3 plant family is the GF14s of *Arabidopsis* (DeLille *et al.*, 2001). With the entire genome of *Arabidopsis* now sequenced, it is unique in that it serves as the prototypical example for 14-3-3 family organization, with its 13 members. *Arabidopsis* was one of the original plant species that was determined to possess these hitherto mammalian-only brain proteins (Lu *et al.*, 1992), and the initial discovery of 14-3-3s was made using antibodies directed against a regulatory DNA–protein complex to screen a cDNA expression library. A total of five different isoforms were isolated by immunoscreening. Western analysis with these same antibodies recognized at least seven different molecular weight polypeptides on SDS-PAGE blots. Using the yeast two-hybrid protein–protein detection system with an *Arabidopsis* 14-3-3 as bait, an additional three members were identified. A search of expressed tag sequences (EST) of *Arabidopsis* cDNAs yielded still another two isoforms (Wu *et al.*, 1997b). In 2000 the completion of the *Arabidopsis* genome sequence provided the final source for 14-3-3 identification. Computer based genomic BLAST analysis of the *Arabidopsis* genome, using the ten *Arabidopsis* isoforms as probes, yielded three additional 14-3-3 isoforms and one severely truncated 14-3-3-like protein (Rosenquist *et al.*, 2000, DeLille *et al.*, 2001). To date this represents the largest 14-3-3 family in any organism and as such it provides insight as a model system for 14-3-3 genomic and isoform organization, evolution and function.

Phylogenetic examination of the *Arabidopsis* 14-3-3 gene family divulges several key features (figure 3.3). A phylogenetic tree based on primary sequence and gene structure of the 13 isoforms forms a bifurcated tree, with the two major branches being the ε (epsilon) and non-ε (non-epsilon) groups. This distribution and basic branching pattern is conserved with other plant and animal 14-3-3 multi-isoform gene families and as such may be indicative of an ancient and significant divergence from common ancestry. However specific homology amongst animal and plant 14-3-3 isoforms is not discernable, i.e. while the basic branching pattern is preserved, the branches themselves are clearly distinct such that there are no shared evolutionary branches between plants and animals. The *Arabidopsis* isoforms within the epsilon group are ε (epsilon), π (pi), o (omicron), ρ (rho) and μ (mu). The non-epsilon group contains the isoforms ω (omega), κ (kappa), λ (lambda), φ (phi), χ (chi), ψ (psi), υ (upsilon) and ν (nu). Some isoforms within these groupings further cluster to suggest a closer identity. Examples of this are; κ (kappa) and λ (lambda); ω (omega), φ (phi) and χ (chi); υ (upsilon), ψ (psi) and ν (nu); ε (epsilon) and π (pi); ρ (rho), μ (mu) and o (omicron).

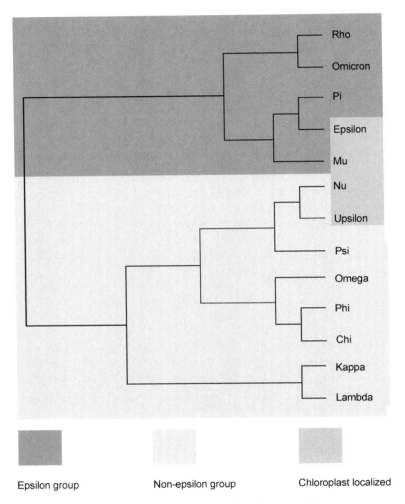

Epsilon group   Non-epsilon group   Chloroplast localized

Figure 3.3 Phylogenetic analysis of the complete family of *Arabidopsis* 14-3-3 proteins. A phylogenetic tree generated by the bootstrapped parsimony method reveals the major branching of the epsilon and non-epsilon groups of all 13 of the *Arabidopsis* 14-3-3s identified in the *Arabidopsis* genome. Sub-branching within groups is also supported, especially within the non-epsilon group. As of this writing, isoforms Rho, Omicron and Pi have yet to be demonstrated as expressed in plants.

This phylogenetic amino acid sequence grouping and branch pattern is also corroborated by the genomic organization of the isoforms (figure 3.4). Non-epsilon members fall into two categories with either four exons and three introns or five exons and four introns. The placement of the intron/exon junctions are highly conserved. Epsilon members again cluster into two groups of distinct genomic organization with either six exons and five introns, or seven exons

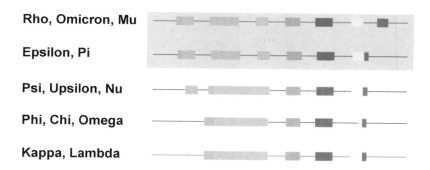

Rho, Omicron, Mu

Epsilon, Pi

Psi, Upsilon, Nu

Phi, Chi, Omega

Kappa, Lambda

Figure 3.4 Gene organization of the complete family of *Arabidopsis* 14-3-3 proteins. The genomic organization of the 13 *Arabidopsis* 14-3-3 isoforms fall into five different schemes of intron/exon organization. The exons are represented by boxes and are drawn to scale while the introns are lines and have been abbreviated.

and six introns. Interestingly two exons and one intron are conserved amongst all *Arabidopsis* isoforms, perhaps reflecting the genomic conservation of the protein structural core elements.

The presence of these distinct evolutionary groupings begs the question of whether there are functional groupings that map to the branches of the evolutionary tree. To date, only a few studies have examined the relative activities among isoforms (see below), so no conclusions are yet possible. However, the fact that different isoforms do possess different activities suggests that there is functional variation that can eventually be mapped to the tree. Further support for functional diversity comes from localization studies that indicate that certain isoforms demonstrate organellar-specific localization.

3.4 Mode of action of 14-3-3 proteins

The 14-3-3 proteins were found to bind many different target proteins that are involved in many different activities and pathways. The ability of 14-3-3s to bind to so many different targets was initially rather perplexing because of the diversity of the targets compared with the high degree of similarity amongst 14-3-3 isoforms. The identification of 14-3-3 proteins as phosphoserine/threonine binding proteins (Muslin *et al.*, 1996), explained how the wide target range was possible and also indicated the full potential of the 14-3-3s as signal transduction elements. The 14-3-3 proteins bind to target proteins that are phosphorylated at specific target sequences. It must be pointed out though that not all 14-3-3 binding partners have been shown to be phosphorylated, and several clearly interact without being phosphorylated or containing the consensus phosphorylation-binding sequence (Pan *et al.*, 1999, Masters *et al.*, 1999). Additionally, the accepted consensus 14-3-3 phospho binding target has

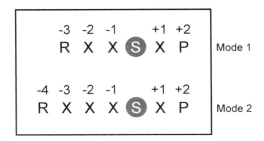

Figure 3.5 Consensus 14-3-3 binding sequences in metabolic enzyme target proteins. The classic 14-3-3-recognition site in target proteins, consisting of mainly metabolic proteins in plants to date, is composed of a central phosphorylated serine (in black circle) or threonine flanked by an upstream basic residue and a downstream proline residue. The spacing between the basic residue, either an Arg or Lys, can be two (mode 1) or three (mode 2) residues.

expanded in the past couple of years to include several variations not included in the original, mammalian-generated consensus motif, thereby rendering an exact assignment of 14-3-3 targets by simple prediction difficult at best.

In mammals, the 14-3-3 recognition site as deduced from Raf-1 kinase is considered to be Arg-X-X-p-Ser-X-Pro (where X is any amino acid, see figure 3.5) (Muslin *et al.*, 1996). Binding sites that contain this consensus sequence are known as mode 1 sites. Peptides of the specific Raf-1 site, Arg-Gln-Arg-Ser-Thr-p-Ser-Thr-Pro, bind to many 14-3-3s and have been used in competition assays to demonstrate phosphoserine dependent binding by 14-3-3s. Non-phosphorylated versions of this peptide do not interact with 14-3-3s and as such are non-functional in competition assays. Examination of the requirement for residues surrounding the phosphoserine revealed that an R at the −3 or −4 position (relative to pS at position 0) is essential in mammals. This presumably is also part of the recognition site for serine/threonine kinases essential for modification of the serine residue. The proline residue at position +2 appears to be favored, but is not critical (Rittinger *et al.*, 1999). Using peptide library screening to further identify preferred amino acids in the recognition site resulted in the refinement and expansion of the consensus site to Arg[Ser/Ar][+/Ar]pSer[Leu/Glu/Met]Pro with Ar representing an aromatic residue and + representing a basic residue. The search also identified a second possible 14-3-3 binding motif known as the mode 2 site (figure 3.5) which included an extra variable residue between the R and first Ser/Ar, thus Arg-X[Ser/Ar][+/Ar]pSer[Leu/Glu/Ala/Met]Pro.

3.5 14-3-3 proteins and the regulation of metabolic enzymes

While examples of both types of 14-3-3 recognition binding motifs exist in plants, the nitrogen assimilation enzyme nitrate reductase (NR) stands as the model for the 14-3-3 binding/regulation mechanism and contains a classic mode 1 Raf-like 14-3-3 interaction site. The 14-3-3 interaction in this system and this class of proteins (the metabolic enzymes), is to date the most prevalent 14-3-3 target in plants. In contrast, the 14-3-3 targets in animal systems seem to focus on kinases and transcriptional regulators, even though the initial discovery of mammalian 14-3-3 partners was in association with the metabolic enzymes aromatic hydroxylases (Ferl, 1996). While this dichotomy may represent a fundamental difference in the way that plants and animals utilize 14-3-3 proteins, it is more likely that this difference is a function of the range of experiments that have been explored in each system (Sehnke and Ferl, 1996).

Nitrate reductase is the rate limiting step enzyme in nitrogen assimilation by plants. Nitrate in the soil is converted into usable nitrogen by the nitrate assimilation pathway and supplies precursors to amino acid and nucleic acid production. Nitrate is taken up by the roots and converted to nitrite by NR in the cytosol. Nitrite is then further converted to ammonia by nitrite reductase in plastids. NR is a 100 kDa homodimeric enzyme with each subunit divided into three domains separated by two hinge regions. NR is light-regulated on a diurnal basis, turning off in response to loss of light. This repetitive and rapid response is accomplished by post-translational regulation via a two-step process (figure 3.6a). The first step of NR inhibition is the phosphorylation by a Ca^{2+} dependent, calmodulin-domain protein kinase (CDPK) (Douglas $et\ al.$, 1998). The major phosphorylation site is S534 in $Arabidopsis$ NR and S543 in spinach NR (Bachmann $et\ al.$, 1996a; Kanamaru $et\ al.$, 1999). The phosphorylated NR enzyme at this point is still active and requires subsequent binding of a 14-3-3 to the phosphoserine containing sequence, in the presence of Mg^{2+}, to turn the enzyme off. The binding site for the 14-3-3s on NR is Arg-Thr-Ala-p-Ser-Thr-Pro in spinach and Lys-Ser-Val-p-Ser-Thr-Pro in $Arabidopsis$. The inactivation is thought to occur as a result of the interruption of electron flow from the heme to the Mo-pterin domain of NR (Bachmann $et\ al.$, 1996b, Moorhead $et\ al.$, 1996). However there is also evidence that the 14-3-3 interaction invokes a NR conformation that is non-functional (Weiner and Kaiser, 1999, 2000). In either case, this process is reversible upon interaction of a microsystin sensitive phosphatase with NR and removal of the 14-3-3 (Douglas $et\ al.$, 1998).

While NR serves as the prime example, the number of enzymes that interact with and are regulated by 14-3-3s in plants is substantive and ever-increasing. The enzymes affected are rate limiting in many of the critical pathways in plants. For example enzymes like glutamine synthetase (Finnemann and Schjoerring, 2000) and glutamate synthase (Moorhead $et\ al.$, 1999) also play a role in the nitrogen partitioning process in plants and are regulated by 14-3-3s.

Studies of cytosolic glutamine synthetase (GS), the key enzyme for ammonia assimilation in plants, demonstrate the same two-step regulation involving 14-3-3s as observed for NR. However GS is activated by the process, instead

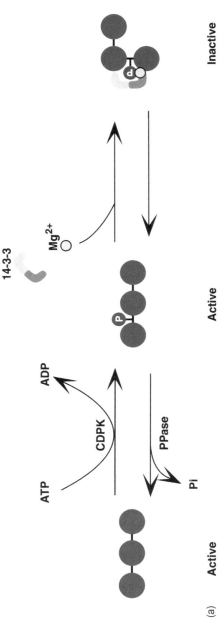

Figure 3.6 Regulation of metabolic plant enzymes by 14-3-3 proteins. The control of plant metabolic enzymes by 14-3-3s is a two-step mechanism. The target protein is phosphorylated by a kinase, which does not change the activity of the enzyme, and is then recognized by a 14-3-3 dimer. The prototypical 14-3-3 regulated plant enzyme is nitrate reductase, which, when phosphorylated and bound by 14-3-3s, is inhibited (a). However, this same mechanism has also been demonstrated to activate enzymes, such is the case with glutamine synthetase (b).

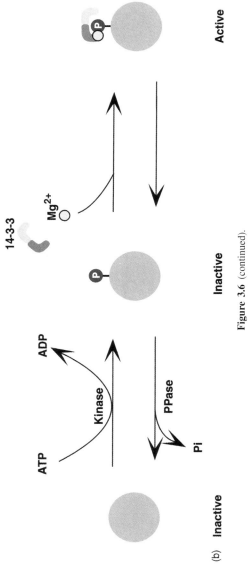

Figure 3.6 (continued).

of being inhibited as was the case with NR. Plant GS is an octameric enzyme with a native molecular mass of 320 (for the cytosolic form GS_1) or 380 kDa (for the plastid localized isozyme GS_2) (Ochs *et al.*, 1999). Historically GS activity in plants has been thought to be regulated at the transcriptional level, with GS_2 transcription sensitive to light and nitrogen (Tjaden *et al.*, 1995), while GS_1 appears to be dependent solely upon external nitrogen supplies (Finnemann and Schjoerring, 1999). However recent work has revealed that post-translational activation of GS by phosphorylation and subsequent 14-3-3 binding occurs in senescing leaves (Finnemann and Schjoerring, 2000).

Finnemann and Schoerring (2000) demonstrated that 14-3-3s bind to the only active form of GS in the senescing leaves, phosphorylated GS_1, and subsequent activation of GS_1 apparently requires 14-3-3 interaction (figure 3.6b). Pulldown assays and Western analysis indicated a direct interaction between 14-3-3s and GS_1. The activity of the phospho-GS_1/14-3-3 complex could be decreased by treatment with microsystin-sensitive serine/threonine phosphatases and rescued by protein kinases. The phosphorylated form of GS_1 was much less susceptible to proteolytic degradation than the dephosphorylated form. Also much like NR regulation, the effect of 14-3-3s upon GS was totally dependent upon the presence of millimolar levels of Mg^{2+}. Addition of ethylenediaminetetraacetic acid (EDTA) to the enzyme preparation reduced the GS activity by 75–90%, further reaffirming the role of 14-3-3s in plants as divalent cation 'charged' phospho-binding proteins. Also like NR, the interaction of 14-3-3s is light dark regulated, with increased phosphorylation/14-3-3 binding during periods of dark.

The other major metabolite partitioning in plants occurs in carbon metabolism, which is also regulated by 14-3-3s via such important enzymes as sucrose phosphate synthase (SPS) (Toroser *et al.*, 1998; MacKintosh, 1998), starch synthases (Sehnke *et al.*, 2001) and trehalose 6-phosphate synthase (Moorhead *et al.*, 1999). The exact mechanisms for 14-3-3 regulation in these systems are not well understood, but perhaps the best understood example is derived from SPS.

SPS is a relatively low-level enzyme that catalyzes the formation of sucrose and sucrose-6-phosphate from fructose-6-phosphate and UDP-glucose. Like other plant metabolic enzymes, SPS levels are responsive to environmental factors, such as light, sucrose levels, CO_2 concentrations, water availability and temperature. Additionally SPS activity is controlled by allosteric effectors and post-translational modifications. Under low light levels SPS is phosphorylated and is activated or inhibited, depending upon the residue phosphorylated (Toroser *et al.*, 1998, 1999). While it is clear that 14-3-3s affect SPS activity, currently the complete role of 14-3-3 involvement is not clear. Toroser *et al.* reported that SPS binds to 14-3-3s *in vivo* in a Mg^{2+}-dependent manner. The complex of SPS and 14-3-3s could be co-immunoprecipitated and co-eluted during gel filtration. The ratio of 14-3-3 to SPS was inversely related to the level of SPS activity observed. Moreover, the addition of 14-3-3-competing phospho-Ser$_{229}$ SPS peptide resulted in the stimulation of SPS activity. MacKintosh and colleagues, using crude extracts containing SPS and yeast 14-3-3s, reported that the addition of 14-3-3-competing Raf1 peptide decreased SPS activity (Moorhead *et al.*, 1999). With multiple sites of phosphorylation for SPS (Toroser and Huber, 1997; Toroser *et al.*, 1999), the potential for coordinated 14-3-3 regulation as a result of multi-14-3-3 binding to a single SPS subunit may attribute to the complexity of such regulatory studies.

While the precise target of 14-3-3s in starch accumulation regulation is not fully defined, the clear biological role of 14-3-3s has been demonstrated (Sehnke *et al.*, 2001). Using antisense technology to reduce the levels of specific isoforms of *Arabidopsis* 14-3-3s, μ and ε, accumulation of starch at levels that were at least twice that of wild-type plants was observed. This sort of *in vivo* experiment demonstrated in plants the direct impact that 14-3-3s have upon plant metabolism in a biological setting.

Other metabolic enzymes that have been demonstrated to interact with 14-3-3s include F1 ATP synthetase (A.H. de Boer, personal communication) (Moorhead *et al.*, 1999), 13-lipoxygenase (13-LOX) (Holtman *et al.*, 2000a,b), glyceraldehyde 3-phosphate dehyrogenase (Moorhead *et al.*, 1999) and ascorbate peroxidase (Zhang *et al.*, 1997). Similar to GS, the 14-3-3s selectively interact only with one isoform of LOX, 13-LOX, and not with 9-LOX, suggesting that 14-3-3s only regulate certain pathways of a given enzymatic activity.

3.6 14-3-3 proteins and the regulation of non-metabolic enzymes

Although the majority of 14-3-3 targets in plants discovered to date are metabolic enzymes, several other categories of proteins bind to and/or are regulated by 14-3-3s. This list includes transcription factors such as G-Box binding factors (Lu *et al.*, 1992), Em binding protein (EmBP), VIVIPAROUS1 (VP1) (Schultz *et al.*, 1998), transcription factor II B (TFIIB), tata binding protein (TBP) (Pan *et al.*, 1999) and the plant homeodomain (PHD) domain (Halbach *et al.*, 2000); kinases, such as CDPK (Camoni *et al.*, 1998) and wheat protein kinase 4 (Ikeda *et al.*, 2000); a specific class of proton pumps (Morsomme and Boutry, 2000) and endonucleases (Szopa, 1995). The 14-3-3s have also recently been implicated in chaperone activity for transport or binding of chloroplast precursors such as photosystem I N (PSI-N) subunit (Sehnke *et al.*, 2000) and others (May and Soll, 2000). A similar function of 14-3-3s first described in animals was the transit nuclear import–export activity of 14-3-3s (Muslin and Xing, 2000), which has also now been observed in *Arabidopsis* (Cutler *et al.*, 2000). No doubt the role of 14-3-3 as omnipotent regulators will expand with subsequent examinations in plants and will probably include involvement in such areas as non-enzymatic structural elements and phosphatases similar to that already observed in animal systems. The completion of the *Arabidopsis* genome should

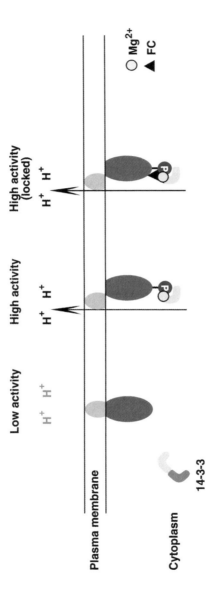

Figure 3.7 Regulation of the plasma membrane H⁺-ATPase by 14-3-3 proteins. The phosphorylation of H⁺-ATPase at a C-terminal threonine results in the creation of a 14-3-3 binding site. As with metabolic enzyme, 14-3-3 regulation divalent cations are required to broker the interaction. The result of the formation of the H⁺-ATPase/14-3-3/Mg²⁺ causes activation of the proton pump. This dynamic state allows for maintenance of the pH gradient in plants. The H⁺-ATPase/14-3-3/Mg²⁺ complex is also the target for binding of the fungal toxin fusicoccin. The toxin hyperstabilizes the association and results in unregulated ATPase activity culminating in leaf wilting in plants as a consequence of decreased turgor pressure.

allow for an active prediction/screening methodology for identifying 14-3-3 target proteins.

While NR is by far the best characterized of all of the discovered plant 14-3-3 targets and fits the classical mammalian 14-3-3 interaction, other plant specific targets have generated new interest regarding the mechanisms of 14-3-3 binding and regulation. The observed consensus binding sites and methods of activities have expanded our thinking beyond the classic animal models. Perhaps the best example of this comes from the work studying the plasma membrane H^+-ATPase proton pump (figure 3.7).

The turgor pressure of plants is maintained by plasma membrane (PM) localized H^+-ATPases (PM H^+-ATPase). The PM H^+-ATPase generates an electrochemical gradient across the plasma membrane providing a driving force for such activities as nutrient uptake, phloem loading and regulation of cytoplasmic pH, all processes important for osmoregulation (Morsomme and Boutry, 2000). The H^+-ATPase activity is thought to be regulated by an autoinhibitory domain in the C-terminus (Palmgren $et~al.$, 1990). An interesting observation regarding a fungal toxin that induces wilting in plant leaves, fusicoccin, pointed to the involvement of 14-3-3s in the regulation of H^+-ATPases. The addition of the fusicoccin allows for an almost irreversible interaction between the H^+-ATPase and 14-3-3s. This fusicoccin binding receptor is actually a heterodimer composed of a monomer of H^+-ATPase and monomer 14-3-3. When this receptor is bound by fusicoccin, the H^+-ATPase becomes uninhibited, thus inducing wilting (Piotrowski $et~al.$, 1998). The interaction between 14-3-3s and H^+-ATPase occurs in the absence of fusicoccin, albeit more transiently and subject to regulation, and is important for the normal physiological maintenance of turgor pressure. Both phosphorylation of the H^+-ATPase and magnesium was required for 14-3-3/H^+-ATPase interaction, so is therefore similar to the NR paradigm. However, a close examination of the amino acid sequence does not reveal the classic 14-3-3 phosphoserine binding consensus site. Using mutagenesis and truncation studies, the 14-3-3 binding site in H^+-ATPase was identified as Gln-Gln-X-Tyr-p-Thr-Val (Svennelid $et~al.$, 1999). In this sequence, the threonine is phosphorylated, and it is located at the extreme C-terminus end of the H^+-ATPase molecule, thus creating a binding site unlike any other 14-3-3 binding site studied to date. Interestingly however, just like the other examples of metabolic 14-3-3 regulation, the trigger for phosphorylation of H^+-ATPase is light (Kinoshita and Shimazaki, 1999).

3.7 Isoform preference for 14-3-3 target proteins

With the identification of multiple 14-3-3 target proteins and large families of 14-3-3 genes, the obvious question is whether there are specific 14-3-3 isoforms that bind specific proteins or possess other specific activities. Early investigations into the ability of 14-3-3s from different hosts to functionally substitute for endogenous 14-3-3s suggested a common capacity for all 14-3-3s. *Arabidopsis* 14-3-3 ω is able to activate the rat brain tryptophan hydroxylase, exoenzyne S of *Pseudomonas aeruginosa* and protein kinase C (Lu *et al.*, 1994a). Four *Arabidopsis* 14-3-3 isoforms are able to complement the lethal disruption of the two *Saccharomyces cerevisiae* genes encoding 14-3-3 proteins (van Heusden *et al.*, 1996). However the ability of specific 14-3-3 isoforms to bind and act differentially upon the same phyla or host target proteins was demonstrated recently for two different systems (Bachmann *et al.*, 1996a; Rosenquist *et al.*, 2000), thus suggesting that variation in affinity and/or efficacy for specific targets may exist in addition to certain fundamental interactions that may not be isoform specific.

In the first study directly designed to assay isoform specific interactions, the enzyme NR was shown to bind different 14-3-3s with different degrees of affinity and subsequent inhibition (Bachmann *et al.*, 1996a). These interactions required divalent cations under physiological conditions for both binding and activity. The specifics of the binding of 14-3-3s to NR have been further examined (Athwal *et al.*, 1998a,b, 2000). The early observation that divalent cations can bind and alter 14-3-3 structure (Lu *et al.*, 1994b) and hence binding competence was confirmed for NR regulation. The role of the divalent cation is suggested as an addition of charge that can be substituted for in binding studies by changing the pH of the buffer or addition of polyamines. It has been suggested (Athwal *et al.*, 1998a) that this modification to the charge that is presented to the target protein by the 14-3-3 may serve to regulate further the binding dynamics of 14-3-3s. It is not known, however, if this contributes to isoform specificity.

In the second study of relative 14-3-3 isoform affinities, a peptide derived from the *C*-terminus of H^+-ATPase was examined for its ability to bind different *Arabidopsis* 14-3-3 isoforms, using a Biacore plasmon resonance spectrophotometer (Rosenquist *et al.*, 2000). A distinct, reproducible preference for the isoforms was observed, with the relative affinities of $\phi > \chi > \nu > \psi > \upsilon > \varepsilon > \omega > \kappa > \lambda$. Interestingly the binding affinities show a strong correlation with the positions of the isoforms on the phylogenetic tree with the exceptions of ω and ε. These studies strongly suggests that all 14-3-3s are not equivalent in ability to function with a given target, and that the terminal residues that vary among the isoforms can direct specific interactions with targets, even though the structural core of the molecules remains the same. The conserved electrostatic pockets may serve to allow for binding to the phosphoserine motif, although additional residues are involved the specifics of the affinity of binding and or interaction.

3.8 Localization of plant 14-3-3s

Equally important to experiments *in vitro* that measure 14-3-3 affinity for a specific target protein is their biological significance based on locations of the various 14-3-3 isoforms. As with any large family of genes, 14-3-3 localization studies are complex and require assessment of both spatial and temporal distributions. The high degree of primary amino acid sequence homology further complicates such studies because of cross-reactivity when antibodies are generated from full-length proteins as antigens. One solution to overcoming these types of immunological difficulties is to generate peptide-based antibodies. Studies using such antibodies for animal 14-3-3 detection have proved successful—however, they involved only a single isoform and did not address relative distribution among isoforms (Martin *et al.*, 1993). Because of the large family sizes, a complete understanding of these variables for an entire 14-3-3 family is not available, but several recent advances in plant 14-3-3 examination have contributed to the understanding of 14-3-3 localization.

The earliest detection of 14-3-3s in plants involved immunological assays (Brandt *et al.*, 1992; Hirsch *et al.*, 1992; Lu *et al.*, 1992). The antibodies used in these systems were not specific to any one isoform but rather recognized some number of isoforms within that host, in addition to cross-reacting to 14-3-3s from other organisms. This prevented isoform specific localization studies but did provide a picture of the expansive nature of 14-3-3s within various tissues and in subcellular structures such as chloroplasts and nuclei.

The ability to address cross-reactivity of isoform-specific antibodies in plants became a real possibility with the capacity to express recombinant protein from different isoforms of one family of plant 14-3-3s, namely *Arabidopsis*. Using peptides representing the divergent *C* and *N* termini of the different 14-3-3 proteins from *Arabidopsis* as antigens, isoform specific antibodies were produced (Sehnke *et al.*, 2000). The antibodies recognized recombinant as well as the respective native plant 14-3-3 and were not cross-reactive when used within specific titres. These antibodies have been used successfully to detect plant 14-3-3s in three studies (Sehnke *et al.*, 2000; Sehnke *et al.*, 2001) and are beginning to make large-scale plant 14-3-3 localization studies possible. However, much of the information regarding plant 14-3-3 localization in the literature today is about the general presence of 14-3-3s in tissues and organelles, not specific isoforms.

Until recently 14-3-3s were thought of as mainly cytoplasmic proteins. However 14-3-3 in plants have been located inside organelles such as the nucleus (Bihn *et al.*, 1997), chloroplast (Sehnke *et al.*, 2000) and mitochondria (A.H. de Boer, personal communication).

The identity of nuclear specific 14-3-3 isoforms in *Arabidopsis* is not known, but a difference in the sizes of nuclear against cytoplasmic 14-3-3s on SDS-Poly Acrylamide Gel Electrophoresis is observed when using antibodies that recognize at least seven of the *Arabidopsis* isoforms (Bihn *et al.*, 1997). This suggests that either isoform selection is occurring for import into the nuclei or post-translational modifications to the nuclear 14-3-3s has changed their migration on SDS-PAGE. The majority of the 14-3-3s identified in the nucleus were in the insoluble fraction. This again is an incomplete data set as the antibodies used only recognize the non-epsilon 14-3-3s.

In the chloroplast, selection of certain cytoplasmic 14-3-3s is clear, with ε, μ, ν and υ present internally. These isoforms are also present in the cytoplasm, so their presence in the chloroplast is selective but not exclusive. It remains to be seen if this is reflective of an inability of certain 14-3-3s to cross the double membrane or is of functional significance. The mechanism by which 14-3-3 proteins translocate the membrane is not known, but the 14-3-3s themselves appear to be involved in shuttling proteins into the chloroplast (May and Soll, 2000). The presence of 14-3-3s in the chloroplast stroma and on thylakoid surfaces, as demonstrated by Western blotting and microscopic studies, suggest a multifunctional role for the 14-3-3s in the chloroplast. This hypothesis is supported by the identification of 14-3-3s within the starch grains from *Arabidopsis* leaves (Sehnke *et al.*, 2001) as well as associated with chloroplastic metabolic enzymes (Moorhead *et al.*, 1999). Additionally, the ability of a 14-3-3 to bind to the thylakoid precursor peptide of PSI-N subunit in a yeast two-hybrid protein–protein detection system suggests that different roles for 14-3-3s, such as a stromal chaperone, are possible inside the chloroplast.

Preliminary studies indicate that 14-3-3s are also localized within the plant mitochondria. This is in apparent contrast to their mammalian counterparts, which appear only to use cytoplasmic 14-3-3s as mitochondrial import stimulation factors for import into the mitochondria (Alam *et al.*, 1994). The 14-3-3 proteins in plant mitochondria appear to be present, at least in part, as metabolic regulators of the F1 ATP synthetases.

The presence of certain 14-3-3s in given areas within a cell may not only indicate a direct enzymatic regulatory role, but may also reflect the 14-3-3s ability to deliver or remove a target protein to a required environment. In the mammalian system, 14-3-3s have been shown to move proteins in and out of the nucleus as well as into the mitochondria.

Early studies of the export of phosphatases from the nucleus by association with 14-3-3s revealed that the 14-3-3s were acting as transient nuclear export signals. Mutagenesis experiments identified a nuclear export signal present in all 14-3-3s that was responsible for this phosphorylation-dependent gating of target proteins out of the nucleus (Dalal *et al.*, 1999). The removal of the 14-3-3 binding site from the phosphatase resulted in a decreased percentage of cytoplasmic phophatase, thereby suggesting that the 14-3-3s help to maintain cytoplasmic localization. It has also been demonstrated that 14-3-3s can affect cellular localization and thereby function by interfering with import and export signals on target proteins (Muslin and Xing, 2000). This mechanism would serve as an antagonist to the targeting signal when such movement is not warranted. Subcellular localization via 14-3-3 interaction alters the activity of histone deacetylase 4 and 5 in just such a manner (Grozinger and Schreiber, 2000; McKinsey *et al.*, 2000, Wang *et al.*, 2000). Histone deacetylases (HDACs) control transcription by dynamic acetylation and deacetylation of histones. Subclasses of HDACs in plants are themselves regulated by phosphorylation

(Kolle *et al.*, 1999) and 14-3-3 binding. Human HDAC 4 is phosphorylated at Ser_{246}, Ser_{467} and Ser_{632} by CPDK and binds 14-3-3s through these phosphoserine sites. The effect of the 14-3-3 binding is to inhibit nuclear localization by suppressing the intrinsic nuclear localization signal in HDAC 4. The regulation of HDACs is, therefore, imparted by subcellular relocalization as a consequence of 14-3-3 binding.

The question of tissue specific expression of 14-3-3 isoforms in plants is far from complete. However a study of the *Arabidopsis* isoform 14-3-3 χ demonstrated a distinct expression pattern that suggests that 14-3-3 genes have different tissue specific expression patterns (Daugherty *et al.*, 1996). Using a transgene consisting of the 14-3-3 χ promoter driving the β-glucuronidase gene as a reporter, promoter activity was tissue and developmentally regulated. Promoter activity was observed in roots of both immature and mature plants. In immature flowers, the activity was found in the buds, but as the flowers matured, staining was restricted to the stigma, anthers and pollen. In early siliques promoter activity was observed in styles and abscission zones but later expanded to the entire organ. *In situ* hybridization illustrated that the *Arabidopsis* χ mRNA was present in predominantly epidermal tissue of roots, petals and sepals of flower buds, papillae cells of the flowers, siliques and the endosperm of immature seeds. This type of tissue and cell-specific expression is reminiscent of that for certain 14-3-3s in the mammalian systems, indicating a specific pattern of expression that impacts specific isoform involvement in physiological processes.

3.9   Potential 14-3-3 binding partners in plants

With the advent of the completion of various genomic sequencing projects, the scientific focus now lies towards functional characterization of the genome products. One progressive approach to this problem is to find commonalities amongst diverse gene groups such that overlapping regulatory entities allow for accelerated discovery of functional classes of proteins. Historically this approach has served to decrease the time spent characterizing individual proteins, which in reality function in a similar manner on albeit diverse pathways. This process has been the basis for determination of homologues and protein class nomenclature. It is now possible to identify potential 14-3-3 binding partners based upon the presence of a 14-3-3 phosphoserine/threonine consensus sequence in the protein. All proteins regulated by 14-3-3s through such a consensus sequence would involve kinase and phosphatase regulation as well, suggesting pathways of involvement. An example of such a search can be demonstrated using the recently completed *Arabidopsis* genome. Applying the three classes of recognized plant consensus 14-3-3 binding sequences namely Arg/Lys-X-X-p-Ser/p-Thr-X-Pro, Arg/Lys-X-X-X-p-Ser/p-Thr-X-Pro and Glu-Glu-X-Tyr-p-Thr-Val, as probes to search the potential 25,498 gene products results in the finding that over half of the *Arabidopsis* genes might be regulated by association with 14-3-3s. This search of course correctly identifies the *bona fide* 14-3-3 binding partners such as NR and H^+-ATPase. This further suggests that the roles identified to date for pathways utilizing 14-3-3s are only a small fraction of the potential targets and, therefore, that current understanding reveals only a small portion of the physiological impact of 14-3-3s. Verification of these additional partners will require large-scale proteomic studies in addition to the passive single target studies that have characterized 14-3-3 discoveries to date.

3.10 Conclusion

While the specifics of 14-3-3 binding to different targets have not been fully illuminated, the overall scheme of a common mechanism for enzyme regulation by 14-3-3 proteins is becoming clear. The two-step process involving both kinases and divalent cation-activated 14-3-3s to impose reversible regulation is highly controllable. For the examples of both light-regulated NR and blue-light regulated H^+-ATPase, the diurnal application of this mechanism is highly successful. So successful is the mechanism that an invading fungal organism has tapped into the machinery to promote increased susceptibility by producing toxin-induced, hyper-responsive, unregulated H^+-ATPase activity. The list of enzymes that bind 14-3-3s is already large and represents many major pathways in the plants, so to envision a truly large-scale regulatory role for 14-3-3s in plants is not far fetched. The unveiling of the concerted efforts and interactions between the 14-3-3s in different pathways has not yet been explored and may provide the real picture to the complex feedback regulation that exists in 14-3-3-controlled pathways that are involved in environmental stress response in plants.

Acknowledgements

We thank Kathy Sehnke for preparing the graphics used in this chapter. Our work on 14-3-3 proteins is supported by grants from the USDA and the NSF 2010 Project.

References

Aitken, A., Howell, S., Jones, D. *et al.* (1995) Post-translationally modified 14-3-3 isoforms and inhibition of protein kinase C. *Mol. Cell Biochem.*, **149-150**, 41-49.
Alam, R., Hachiya, N., Sakaguchi, M., Kawabata, S. *et al.* (1994) cDNA cloning and characterization of mitochondrial import stimulation factor (MSF) purified from rat liver cytosol. *J. Biochem. (Tokyo)*, **116**, 416-425.

Athwal, G.S., Huber, J.L. and Huber, S.C. (1998a) Biological significance of divalent metal ion binding to 14-3-3 proteins in relationship to nitrate reductase inactivation. *Plant Cell Physiol.*, **39**, 1065-1072.

Athwal, G.S., Huber, J.L. and Huber, S.C. (1998b) Phosphorylated nitrate reductase and 14-3-3 proteins. Site of interaction, effects of ions, and evidence for an AMP-binding site on 14-3-3 proteins. *Plant Physiol.*, **118**, 1041-1048.

Athwal, G.S., Lombardo, C.R., Huber, J.L., Masters, S.C., Fu, H. and Huber, S.C. (2000) Modulation of 14-3-3 protein interactions with target polypeptides by physical and metabolic effectors. *Plant Cell Physiol.*, **41**, 523-533.

Bachmann, M., Huber, J.L., Athwal, G.S., Wu, K., Ferl, R.J. and Huber, S.C. (1996a) 14-3-3 proteins associate with the regulatory phosphorylation site of spinach leaf nitrate reductase in an isoform-specific manner and reduce dephosphorylation of Ser-543 by endogenous protein phosphatases. *FEBS Lett.*, **398**, 26-30.

Bachmann, M., Huber, J.L., Liao, P.C., Gage, D.A. and Huber, S.C. (1996b) The inhibitor protein of phosphorylated nitrate reductase from spinach (*Spinacia oleracea*) leaves is a 14-3-3 protein. *FEBS Lett.*, **387**, 127-131.

Bihn, E.A., Paul, A.L., Wang, S.W., Erdos, G.W. and Ferl, R.J. (1997) Localization of 14-3-3 proteins in the nuclei of arabidopsis and maize. *Plant J.*, **12**, 1439-1445.

Brandt, J., Thordal-Christensen, H., Vad, K., Gregersen, P.L. and Collinge, D.B. (1992) A pathogen-induced gene of barley encodes a protein showing high similarity to a protein kinase regulator. *Plant J.*, **2**, 815-820.

Camoni, L., Harper, J.F. and Palmgren, M.G. (1998) 14-3-3 proteins activate a plant calcium-dependent protein kinase (CDPK). *FEBS Lett.*, **430**, 381-384.

Cutler, S.R., Ehrhardt, D.W., Griffitts, J.S. and Somerville, C.R. (2000) Random GFP::cDNA fusions enable visualization of subcellular structures in cells of arabidopsis at a high frequency. *Proc. Natl Acad. Sci. USA*, **97**, 3718-3723.

Dalal, S.N., Schweitzer, C.M., Gan, J. and DeCaprio, J.A. (1999) Cytoplasmic localization of human cdc25C during interphase requires an intact 14-3-3 binding site. *Mol. Cell Biol.*, **19**, 4465-4479.

Daugherty, C.J., Rooney, M.F., Miller, P.W. and Ferl, R.J. (1996) Molecular organization and tissue-specific expression of an *Arabidopsis* 14-3-3 gene. *Plant Cell*, **8**, 1239-1248.

DeLille, J., Sehnke, P.C. and Ferl, R.J. (2001) The *Arabidopsis thaliana* 14-3-3 family of signaling regulators. *Plant Physiol.*, **126**, 35-38.

de Vetten, N.C. and Ferl, R.J. (1994) Two genes encoding GF14 (14-3-3) proteins in *Zea mays*. Structure, expression, and potential regulation by the G-box binding complex. *Plant Physiol.*, **106**, 1593-1604.

Douglas, P., Moorhead, G., Hong, Y., Morrice, N. and MacKintosh, C. (1998) Purification of a nitrate reductase kinase from *Spinacea oleracea* leaves, and its identification as a calmodulin-domain protein kinase. *Planta*, **206**, 435-442.

Ferl, R.J. (1996) 14-3-3 Proteins and signal transduction. *Annu. Rev. Plant Physiol. Plant Mol. Biol.*, **47**, 49-73.

Finnemann, J. and Schjoerring, J.K. (1999) Translocation of NH_4^+ in oilseed rape plants in relation to glutamine synthetase isogene expression and activity. *Physiol. Plant*, **105**, 469-477.

Finnemann, J. and Schjoerring, J.K. (2000) Post-translational regulation of cytosolic glutamine synthetase by reversible phosphorylation and 14-3-3 protein interaction. *Plant J.*, **24**, 171-181.

Grozinger, C.M. and Schreiber, S.L. (2000) Regulation of histone deacetylase 4 and 5 and transcriptional activity by 14-3-3-dependent cellular localization. *Proc. Natl Acad. Sci. USA*, **97**, 7835-7840.

Halbach, T., Scheer, N. and Werr, W. (2000) Transcriptional activation by the PHD finger is inhibited through an adjacent leucine zipper that binds 14-3-3 proteins. *Nucleic Acids Res.*, **28**, 3542-3550.

Hirsch, S., Aitken, A., Bertsch, U. and Soll, J. (1992) A plant homologue to mammalian brain 14-3-3 protein and protein kinase C inhibitor. *FEBS Lett.*, **296**, 222-224.

Holtman, W.L., Roberts, M.R., Oppedijk, B.J., Testerink, C., van Zeijl, M.J. and Wang, M. (2000a) 14-3-3 proteins interact with a 13-lipoxygenase, but not with a 9-lipoxygenase. *FEBS Lett.*, **474**, 48-52.

Holtman, W.L., Roberts, M.R. and Wang, M. (2000b) 14-3-3 Proteins and a 13-lipoxygenase form associations in a phosphorylation-dependent manner. *Biochem. Soc. Trans.*, **28**, 834-836.

Ichimura, T., Uchiyama, J., Kunihiro, O. *et al.* (1995) Identification of the site of interaction of the 14-3-3 protein with phosphorylated tryptophan hydroxylase. *J. Biol. Chem.*, **270**, 28,515-28,518.

Ikeda, Y., Koizumi, N., Kusano, T. and Sano, H. (2000) Specific binding of a 14-3-3 protein to autophosphorylated WPK4, an SNF1-related wheat protein kinase, and to WPK4-phosphorylated nitrate reductase. *J. Biol. Chem.*, **275**, 31,695-31,700.

Kanamaru, K., Wang, R., Su, W. and Crawford, N.M. (1999) Ser-534 in the hinge 1 region of *Arabidopsis* nitrate reductase is conditionally required for binding of 14-3-3 proteins and *in vitro* inhibition. *J. Biol. Chem.*, **274**, 4160-4165.

Kidou, S., Umeda, M., Kato, A. and Uchimiya, H. (1993) Isolation and characterization of a rice cDNA similar to the bovine brain-specific 14-3-3 protein gene. *Plant Mol. Biol.*, **21**, 191-194.

Kinoshita, T. and Shimazaki, K. (1999) Blue light activates the plasma membrane H(+)-ATPase by phosphorylation of the *C*-terminus in stomatal guard cells. *EMBO J.*, **18**, 5548-5558.

Kolle, D., Brosch, G., Lechner, T. *et al.* (1999) Different types of maize histone deacetylases are distinguished by a highly complex substrate and site specificity. *Biochemistry*, **38**, 6769-6773.

Liu, D., Bienkowska, J., Petosa, C., Collier, R.J., Fu, H. and Liddington, R. (1995) Crystal structure of the zeta isoform of the 14-3-3 protein. *Nature*, **376**, 191-194.

Lu, G., DeLisle, A.J., de Vetten, N.C. and Ferl, R.J. (1992) Brain proteins in plants: an *Arabidopsis* homolog to neurotransmitter pathway activators is part of a DNA binding complex. *Proc. Natl Acad. Sci. USA*, **89**, 11,490-11,494.

Lu, G., de Vetten, N.C., Sehnke, P.C., Isobe, T. *et al.* (1994a) A single *Arabidopsis* GF14 isoform possesses biochemical characteristics of diverse 14-3-3 homologues. *Plant Mol. Biol.*, **25**, 659-667.

Lu, G., Sehnke, P.C. and Ferl, R.J. (1994b) Phosphorylation and calcium binding properties of an *Arabidopsis* GF14 brain protein homolog. *Plant Cell*, **6**, 501-510.

McKinsey, T.A., Zhang, C.L. and Olson, E.N. (2000) Activation of the myocyte enhancer factor-2 transcription factor by calcium/calmodulin-dependent protein kinase-stimulated binding of 14-3-3 to histone deacetylase 5. *Proc. Natl Acad. Sci. USA*, **97**, 14,400-14,405.

MacKintosh, C. (1998) Regulation of cytosolic enzymes in primary metabolism by reversible protein phosphorylation. *Curr. Opin. Plant Biol.*, **1**, 224-229.

Martin, H., Patel, Y., Jones, D., Howell, S., Robinson, K. and Aitken, A. (1993) Antibodies against the major brain isoforms of 14-3-3 protein: an antibody specific for the *N*-acetylated amino-terminus of a protein. *FEBS Lett.*, **336**, 189.

Masters, S.C., Pederson, K.J., Zhang, L., Barbieri, J.T. and Fu, H. (1999) Interaction of 14-3-3 with a nonphosphorylated protein ligand, exoenzyme S of *Pseudomonas aeruginosa*. *Biochemistry*, **38**, 5216-5221.

May, T. and Soll, J. (2000) 14-3-3 proteins form a guidance complex with chloroplast precursor proteins in plants. *Plant Cell*, **12**, 53-64.

Moore, B.W. and Perez, V.J. (1967) Specific acidic proteins of the nervous system, in *Physiological and Biochemical Aspects of Nervous Integration* (ed. F. Carlson), Prentice Hall, Woods Hole, MA, USA, pp. 343-359.

Moorhead, G., Douglas, P., Morrice, N., Scarabel, M., Aitken, A. and MacKintosh, C. (1996) Phosphorylated nitrate reductase from spinach leaves is inhibited by 14-3-3 proteins and activated by fusicoccin. *Curr. Biol.*, **6**, 1104-1113.

Moorhead, G., Douglas, P., Cotelle, V. *et al.* (1999) Phosphorylation-dependent interactions between enzymes of plant metabolism and 14-3-3 proteins. *Plant J.*, **18**, 1-12.

Morsomme, P. and Boutry, M. (2000) The plant plasma membrane H(+)-ATPase: structure, function and regulation. *Biochim. Biophys. Acta*, **1465**, 1-16.

Muslin, A.J. and Xing, H. (2000) 14-3-3 proteins: regulation of subcellular localization by molecular interference. *Cell Signal.*, **12**, 703-709.

Muslin, A.J., Tanner, J.W., Allen, P.M. and Shaw, A.S. (1996) Interaction of 14-3-3 with signaling proteins is mediated by the recognition of phosphoserine. *Cell*, **84**, 889-897.

Ochs, G., Schock, G., Trischler, M., Kosemund, K. and Wild, A. (1999) Complexity and expression of the glutamine synthetase multigene family in the amphidiploid crop *Brassica napus*. *Plant Mol. Biol.*, **39**, 395-405.

Palmgren, M.G., Larsson, C. and Sommarin, M. (1990) Proteolytic activation of the plant plasma membrane H(+)-ATPase by removal of a terminal segment. *J. Biol. Chem.*, **265**, 13,423-13,426.

Pan, S., Sehnke, P.C., Ferl, R.J. and Gurley, W.B. (1999) Specific interactions with TBP and TFIIB *in vitro* suggest that 14-3-3 proteins may participate in the regulation of transcription when part of a DNA binding complex. *Plant Cell*, **11**, 1591-1602.

Petosa, C., Masters, S.C., Bankston, L.A., Pohl, J., Wang, B., Fu, H. and Liddington, R.C. (1998) 14-3-3zeta binds a phosphorylated Raf peptide and an unphosphorylated peptide via its conserved amphipathic groove. *J. Biol. Chem.*, **273**, 16,305-16,310.

Piotrowski, M. and Oecking, C. (1998) Five new 14-3-3 isoforms from *Nicotiana tabacum* L.: implications for the phylogeny of plant 14-3-3 proteins. *Planta*, **204**, 127-130.

Piotrowski, M., Morsomme, P., Boutry, M. and Oecking, C. (1998) Complementation of the *Saccharomyces cerevisiae* plasma membrane H^+-ATPase by a plant H^+-ATPase generates a highly abundant fusicoccin binding site. *J. Biol. Chem.*, **273**, 30,018-30,023.

Rittinger, K., Budman, J., Xu, J., Volinia, S. *et al.* (1999) Structural analysis of 14-3-3 phosphopeptide complexes identifies a dual role for the nuclear export signal of 14-3-3 in ligand binding. *Mol. Cell*, **4**, 153-166.

Roberts, M.R. and Bowles, D.J. (1999) Fusicoccin, 14-3-3 proteins, and defense responses in tomato plants. *Plant Physiol.*, **119**, 1243-1250.

Rosenquist, M., Sehnke, P., Ferl, R.J., Sommarin, M. and Larsson, C. (2000) Evolution of the 14-3-3 protein family: does the large number of isoforms in multicellular organisms reflect functional specificity? *J. Mol. Evol.*, **51**, 446-458.

Schultz, T.F., Medina, J., Hill, A. and Quatrano, R.S. (1998) 14-3-3 proteins are part of an abscisic acid-VIVIPAROUS1 (VP1) response complex in the Em promoter and interact with VP1 and EmBP1. *Plant Cell.*, **10**, 837-847.

Sehnke, P.C. and Ferl, R.J. (1996) Plant metabolism: enzyme regulation by 14-3-3 proteins. *Curr. Biol.*, **6**, 1403-1405.

Sehnke, P.C., Henry, R., Cline, K. and Ferl, R.J. (2000) Interaction of a plant 14-3-3 protein with the signal peptide of a thylakoid-targeted chloroplast precursor protein and the presence of 14-3-3 isoforms in the chloroplast stroma. *Plant Physiol.*, **122**, 235-242.

Sehnke, P.C., Chung, H.J., Wu, K. and Ferl, R.J. (2001) Regulation of starch accumulation by granule-associated plant 14-3-3 proteins. *Proc. Natl Acad. Sci. USA*, **98**, 765-770.

Svennelid, F., Olsson, A., Piotrowski, M. *et al.* (1999) Phosphorylation of thr-948 at the *C* terminus of the plasma membrane H(+)-ATPase creates a binding site for the regulatory 14-3-3 protein. *Plant Cell.*, **11**, 2379-2392.

Szopa, J. (1995) Expression analysis of a *Cucurbita* cDNA encoding endonuclease. *Acta Biochim. Pol.*, **42**, 183-189.

Testerink, C., van der Meulen, R.M., Oppedijk, B.J. *et al.* (1999) Differences in spatial expression between 14-3-3 isoforms in germinating barley embryos. *Plant Physiol.*, **121**, 81-88.

Tjaden, G., Edwards, J.W. and Coruzzi, G.M. (1995) *Cis* elements and *trans*-acting factors affecting regulation of a nonphotosynthetic light-regulated gene for chloroplast glutamine synthetase. *Plant Physiol.*, **108**, 1109-1117.

Toroser, D. and Huber, S.C. (1997) Protein phosphorylation as a mechanism for osmotic-stress activation of sucrose-phosphate synthase in spinach leaves. *Plant Physiol.*, **114**, 947-955.

Toroser, D., Athwal, G.S. and Huber, S.C. (1998) Site-specific regulatory interaction between spinach leaf sucrose-phosphate synthase and 14-3-3 proteins. *FEBS Lett.*, **435**, 110-114.

Toroser, D., McMichael, R., Krause, K.P. *et al.* (1999) Site-directed mutagenesis of serine 158 demonstrates its role in spinach leaf sucrose-phosphate synthase modulation. *Plant J.*, **17**, 407-413.

van Heusden, G.P., van der Zanden, A.L., Ferl, R.J. and Steensma, H.Y. (1996) Four *Arabidopsis thaliana* 14-3-3 protein isoforms can complement the lethal yeast bmh1 bmh2 double disruption. *FEBS Lett.*, **391**, 252-256.

Wang, A.H., Kruhlak, M.J., Wu, J. *et al.* (2000) Regulation of histone deacetylase 4 by binding of 14-3-3 proteins. *Mol. Cell Biol.*, **20**, 6904-6912.

Weiner, H. and Kaiser, W.M. (1999) 14-3-3 Proteins control proteolysis of nitrate reductase in spinach leaves. *FEBS Lett.*, **455**, 75-78.

Weiner, H. and Kaiser, W.M. (2000) Binding to 14-3-3 proteins is not sufficient to inhibit nitrate reductase in spinach leaves. *FEBS Lett.*, **480**, 217-220.

Wilczynski, G., Kulma, A. and Szopa, J. (1998) The expression of 14-3-3 isoforms in potato is developmentaly regulated. *J. Plant Phys.*, **153**, 118-126.

Wu, K., Lu, G., Sehnke, P. and Ferl, R.J. (1997a) The heterologous interactions among plant 14-3-3 proteins and identification of regions that are important for dimerization. *Arch. Biochem. Biophys.*, **339**, 2-8.

Wu, K., Rooney, M.F. and Ferl, R.J. (1997b) The *Arabidopsis* 14-3-3 multigene family. *Plant Physiol.*, **114**, 1421-1431.

Xiao, B., Smerdon, S.J., Jones, D.H. *et al.* (1995) Structure of a 14-3-3 protein and implications for coordination of multiple signalling pathways. *Nature*, **376**, 188-191.

Yaffe, M.B., Rittinger, K., Volinia, S. *et al.* (1997) The structural basis for 14-3-3: phosphopeptide binding specificity. *Cell*, **91**, 961-971.

Zhang, H., Wang, J., Nickel, U., Allen, R.D. and Goodman, H.M. (1997) Cloning and expression of an *Arabidopsis* gene encoding a putative peroxisomal ascorbate peroxidase. *Plant Mol. Biol.*, **34**, 967-971.

4 Proteinase inhibitors

William Laing and Michael T. McManus

4.1 Introduction

Proteases, proteinases, peptidases and proteolytic enzymes are names all used interchangeably to describe hydrolytic enzymes that cleave a peptide bond in a protein (Barrett *et al.*, 1999). Thus proteinases, by definition, interact with other proteins. A typical proteinase recognizes target amino acid residues on its substrate protein, binds to the protein and then hydrolyzes a peptide bond at a specific point in the sequence. Proteinases generally show a degree of specificity for their cleavage site, although the occurrence of these target sequences is not always sufficient to ensure cleavage will occur. The conformation of the protein will also determine accessibility of the proteinase to the target residues. Plant proteinases have been reviewed extensively recently (Vodkin and Scandalios, 1981; Nielsen, 1988; Huffaker, 1990; Vierstra, 1993; Callis, 1995; Kervinen *et al.*, 1995; Adam, 1996; Vierstra, 1996; Andersson and Aro, 1997; Callis and Setlow, 1997; Turk *et al.*, 1997; Cordeiro *et al.*, 1998; Faro *et al.*, 1998; Galleschi, 1998; Glaser *et al.*, 1998; Verissimo *et al.*, 1998; Bogacheva, 1999; Fu *et al.*, 1999; Keegstra and Froehlich, 1999; Mutul and Gal, 1999; Adam, 2000; Chrispeels and Herman, 2000), and will only be discussed here in terms of interactions with specific proteinase inhibitors.

Proteinaceous proteinase inhibitors (PIs) inhibit the action of proteinases most commonly because the target (or bait) residues on the PI occupy the active site of the proteinase. While the PI may be cleaved at a particular bait or target residue, it is generally difficult to dislodge the inhibitory protein from the active site. The PI acts as a tight binding competitive inhibitor (Bieth, 1995) and the PIs target residues are frequently in an exposed and stabilized binding loop. By occupying the active site, the PI prevents other proteins from being cleaved. There are also many examples of non-proteinaceous PIs, both natural and chemically synthesized. Unless otherwise stated, in this chapter we will use PI to mean proteinaceous PI.

Proteinaceous PIs of plants have also been the subject of many previous reviews (Laskowski and Kato, 1980; Garcia-Olmedo *et al.*, 1987; Ryan, 1989; Otlewski, 1990; Casaretto and Corcuera, 1995; Birk, 1996; Otlewski and Krowarsch, 1996; Brown and Dziegielewska, 1997; Bowles, 1998; Margis *et al.*, 1998; Mosolov, 1998) and so in this review, we take the approach of analyzing the full complement of PIs that we can discover in the recently released *Arabidopsis* genome (Initiative, 2000; Munich Information Center for Protein Sequences, MIPS at http://www.mips.biochem.mpg.de/proj/thal). Where PIs are present in other species but not in *Arabidopsis*, we also refer to these. We also use the proteinase MEROPS database (http://www.merops.co.uk/merops) as well as other Internet based resources.

A potential problem is the ability of protein finder algorithms to correctly recognize protein sequence start codons, open reading frames (ORFs) and intron/exon boundaries, especially in short proteins (approximately 100 residues or less in length). Thus significant numbers of proteins may be missing from this analysis. It should also be recognized that the presence of an identified ORF in *Arabidopsis* does not necessarily prove that the gene is transcribed or expressed as a protein. Also, even though an identified putative PI gene may show high homology to a known active PI, including the occurrence of sequences characteristic of a PI, this does not prove that the translated protein functions as a PI; it may serve in another role.

One approach that MIPS uses to identify protein functionality is to use a combined approach from the PROSITE, PRINTS, Pfam and ProDom databases called InterPro (Apweiler *et al.*, 2000, 2001; http://www.ebi.ac.uk/interpro). However, this approach searches for diagnostic sequences of a particular protein, which does not necessarily prove functionality. The structures of many of these PIs, alone and in interaction with their target proteinases, can be retrieved from the Research Collaboration for Structural Bioinformatics Protein Data Bank (http://www.rcsb.org/pdb) and viewed using the Swiss-Pdb viewer (http://www.expasy.ch/spdbv/mainpage.html).

This approach to classify plant PIs also assumes that all plant PIs have been discovered, the genes sequenced and the corresponding protein been shown to be functional as a PI. This is by no means certain. For example, squash aspartic proteinase inhibitor (SQAPI) (Christeller *et al.*, 1998), a proteinaceous inhibitor of pepsin-like enzymes, showed no homology to any other sequences in the database when discovered. This PI was discovered as a protein inhibitory activity, the purified protein *N*-terminal sequenced and the gene subsequently cloned and sequenced. Another example of an unexpected PI is a recently described weak inhibitor of thrombin and chymotrypsin from mice which was identified as the phosphatidylethanolamine-binding protein, a phospholipid binding protein found in a wide range of species from plants to mammals (Hengst *et al.*, 2001). There are six predicted protein sequences in the *Arabidopsis* MIPS database with good homology to these phosphatidylethanolamine-binding proteins from mice, all with similar molecular weights (around 20,000 Da). We have no idea whether these predicted proteins are also PIs. Ultimately, a gene can only be shown to have the proposed function based on homology through either purification of the functional protein and sequencing, or by expressing the protein in a heterologous background and then characterizing the protein product.

4.2   The functions of PIs

Proteinase inhibitors are ubiquitous in nature, being found in all groups of organisms including viruses. They function to inhibit proteinases, and a major unanswered question in plants is: what are their target proteinases? Two main roles are accepted: PIs act as protection agents and/or as storage proteins (McManus *et al.*, 2000). In addition, PIs may function to protect the cell against its own proteinases and, in analogy with animal systems, to participate in the regulation of controlled biochemical switches.

Proteinase inhibitors appear to have a very significant role of protection of the cell against the proteinases of pests and pathogens (Ryan, 1990; McManus *et al.*, 2000). In this protective role they are often induced locally by wounding or by systemic signals (McManus *et al.*, 1994a; Botella *et al.*, 1996; Ryan, 2000; Leon *et al.*, 2001) or are present in storage organs at high levels (Rackis and Anderson, 1964; Gatehouse *et al.*, 1979). For a PI to be effective in targeting an insect proteinase, it should have a tight affinity for its target proteinase and thus inactivate the proteinase. This prevents the target insect from digesting protein. Further evidence that is often cited for the role of PIs as protectants in plants is the demonstration of the efficiency of these proteins when expressed in transgenic plants (Hilder *et al.*, 1987; Johnson *et al.*, 1990; McManus *et al.*, 1994b, 1999). While such studies do show that accumulation of these proteins can confer insect resistance to the transformants, as evidence for a similar role for these proteins *in vivo*, this is circumstantial. Perhaps more insight can be gained from examining the mechanisms of insect resistance to PIs.

Insect resistance to PIs develops through three main mechanisms (Jongsma *et al.*, 1996; Jongsma and Bolter, 1997). One mechanism is through overproduction of insect target proteinases, in order to overcome the levels of PI provided through the diet (Broadway *et al.*, 1986; Dymock *et al.*, 1989; Markwick *et al.*, 1998). This is thought to increase the metabolic demands on the organism and may lead to slow growth and possibly death. A second mechanism is for the insect to develop a degree of immunity to the PI by evolving a proteinase with a low affinity to the PI. For example, potato tuber moth (*Phthorimaea operculella*), a successful pest on potato, was less inhibited by PIs 1 and 2 from potato when compared with a range of other lepidopteran pests of hosts other than potato (Christeller *et al.*, 1992). It might be suggested that any insect that is capable of successfully living off a plant must be at least partially resistant to the PIs produced by that plant. A third proposed mechanism is the short-term induction of PI resistant proteinases by the insect (Jongsma *et al.*, 1995; Gruden *et al.*, 1998), although this has not been observed in *Helicoverpa armigera* larvae feeding on an artificial diet (Gatehouse *et al.*, 1997). Together, these studies lead to the suggestion that a key adaptation for herbivory is for the insect to be at least partially resistant to the PIs produced by that plant, and may point to the significance of PIs as protectants in higher plants.

As storage proteins, PIs often constitute a high proportion of the seed protein (McManus *et al.*, 2000). For this reason, it has been suggested that PIs may serve as both a protective role against pests and diseases before germination, and then be degraded during germination and serve as a source of amino acids for seedling growth. In some cases, there are several subgroups within the PI class, some of which may be inhibitors and others which may be highly homologous but not inhibitory. For example plant serpins come in several groups, including the true serpin PIs, and albumins. The albumins may serve as storage proteins.

PIs have also been suggested to regulate endogenous proteinases in seed dormancy, protein mobilization during germination, protein mobilization during senescence and other similar metabolic processes (McManus *et al.*, 2000). However, many of the earlier studies are limited because of their use of crude proteinase preparations and general (caseinolytic) substrates that do not allow characterization of the specificity of the proteinase–PI relationship. In addition, no genes with significant homology to bovine trypsin have been identified in plants, and yet many of the inhibitors tested were active against bovine trypsin.

More recent studies support the concept that PIs may serve to regulate endogenous proteinase. For example, rice has three cystatin species, which have been shown to have different specifities for endogenous rice cysteine proteinases and for insect pathogen proteinases (Arai *et al.*, 1998), and the authors suggest that different rice cystatins target either endogenous proteinases or serve in defense. A wheat cystatin that inhibits an endogenous wheat cysteine proteinase has been described (Kuroda *et al.*, 2001). Further, in transgenic tobacco expressing a rice cystatin gene, pleiotropic effects of the inhibitor were observed, suggesting that endogenous cysteine proteinases were also effected by the inhibitor (Gutierrez Campos *et al.*, 2001). On the other hand, wheat serpins did not inhibit plant proteinases tested, only exogenous proteinases (Ostergaard *et al.*, 2000).

Some PIs are only found at very low concentrations in plant tissues (Ryan *et al.*, 1998; McManus *et al.*, 2000). These levels are much lower than the serine PIs, and probably too low to be functional in plant protection. However, to the best of our knowledge, this role of regulating endogenous proteinases has not been established beyond correlative observations (McManus *et al.*, 2000). More sophisticated roles for PIs such as control of proteolytic switches (e.g. serpins in blood clotting) or inhibitors of apoptosis have not been shown in plants.

One implication of this type of endogenous proteinase regulation is that in many cases there must be a mechanism to relieve inhibition when proteinase activity is required. A possible mechanism would be the degradation of the inhibitor, either when it is free or when it is bound to the proteinase. Degradation of the free inhibitor would cause the proteinase–PI complex to slowly dissociate, releasing active proteinase. We would speculate that degradation of the PI would require its own proteinase or targeting system. Little is known about proteinases that might target the PI, except perhaps the ubiquitin/proteasome system (see chapter 6). We can also envisage systems where the cell needs permanent irreversible inhibition of an endogenous proteinase; for example, if a PI serves to protect a cell from proteinase activity in unwanted locations, then the action of the PI should be irreversible. However, this implies that the proteinase can cross membranes or is accidentally expressed or transported to the incorrect compartment. As proteinases often have pre-protein sequences that serve to inactivate the proteinase until it can be removed, this sort of protective mechanism would not usually be necessary.

4.3   The classification of PIs

Proteinases are classified in terms of broad catalytic types of active site mechanism that are based on the active site residues involved in hydrolysis. Thus aspartate, cysteine, serine (including threonine) and metallo proteinase have been identified (Barrett *et al.*, 1999). These types include endo- and exoproteinases, proteinases that are tightly regulated (e.g. the proteasome, Lon and Clp, see chapter 6), as well as highly specialized proteinases such as those involved in processing signal pre-sequences on newly synthesized proteins. They also include some retrovirus proteinases identified in the *Arabidopsis* genome that presumably only process the virus poly-protein as it is made in the infected cell.

The classification of PIs generally follows the proteinase catalytic type (Laskowski and Kato, 1980). Thus there are many families of PIs that inhibit serine proteinases, several families that inhibit cysteine proteinases, plus others that inhibit metallo- and aspartate proteinases. In addition, there are multiheaded inhibitors that have more than one active site on one PI peptide chain (Lenarcic and Turk, 1999; Bode and Huber, 2000). The probable extreme is a predicted human protein with PI activities against a range of proteinases (Trexler *et al.*, 2001). In addition, there are PIs that are composed of multiple peptide chains with different specificities, as well as multifunctional PIs that have only one active site and inhibit more than one PI class (Bode and Huber, 1993, 2000). These include the α-2-macroglobulin family inhibitors that inhibit all four catalytic types of proteinases (Sottrup-Jensen, 1989).

However, there is some confusion in the naming of PIs. The pancreatic trypsin inhibitor family and the soybean trypsin inhibitors are both known as 'Kunitz inhibitors' (Laskowski and Kato, 1980). The PI families 1 and 2 are also known as the 'potato inhibitors' and examples can be found where a chymotrypsin inhibitor 2 is a member of the PI 1 family (Williamson *et al.*, 1987). Often inhibitors are named after the species in which they were first discovered, and later they are discovered in a range of other species. We contend that PI nomenclature needs an overall system to ensure consistency and rigor in the naming of PIs. Such a system would need to take into account the observation that inhibitors from one homology/structural class may inhibit proteinases of different families, and that often a PI has only been categorized based on its nucleotide sequence; the actual specifity of inhibition has not been measured experimentally.

We propose a numbering system where inhibitor clans are assigned numbers instead of existing names. A 'clan contains all the modern-day peptidases that have arisen from a single evolutionary origin of peptidases. It represents one or more families that show evidence of their evolutionary relationship' (see MEROPS), and we use a similar definition for PIs. In plants there appear to be 13 clans of inhibitors (table 4.1). If details of the inhibited proteinase are known, we follow this assigned number by a code (S, A, M and C) for the

Table 4.1 Classification of the proteinase inhibitors

Proteinase inhibited	#	Name	Active unit molecular weight range (mature chain)[a]	Known distribution and tissue[b] (and other information)
Serine	1S	Bowman–Birk	6000 to 9500 Da (except rice, wheat and barley which are 14,000 to 15,000 Da)	Legume seeds, *Zea mays*, rice, wheat, barley. Related sequences in potato have not been reported to have been tested for inhibitory activity. Structures available.
	2S	Cereal trypsin/ amylase inhibitors	11,500 to 14,000 Da	Cereals: barley, wheat, sorghum, *Eleusine coracana*, rice, rye. Structures available.
	3S	Kunitz	19,000 to 24,000 Da	Legumes. Also reported in potato, cereals and *Arabidopsis* and in other species[c]. Usually found in seed. Structures available.
	4S	Mustard seed inhibitors	6600 to 7100 Da	Brassicaceae, including mustard, rape and *Arabidopsis*.
	5S	Proteinase inhibitor 1	7200 to 9100 synthesized as a larger precursor. Can be tetramer.	Solanaceous species, cereals, squash (seed and phloem), *Arabidopsis*, legumes. Structures available.
	6S	Proteinase inhibitor 2	Base unit is 5000 to 6000 Da, but it is synthesized as a much larger protein, which is often cleaved. Preprotein can be anything from 12,000 to 40,300 Da.	Solanaceous species including potato, tomato, tobacco, aubergine and capsicum, in tubers, fruits, seeds, leaves and flowers. Structures available.
	7S	Serpins	42,000 to 44,000 Da (*Arabidopsis* more variable)	Cereals: wheat, rice, barley. *Arabidopsis*, squash, soybean[d] cotton[d]. Structures available.
	8S	Squash	3000 to 3500 Da	Cucurbit seeds. Structures available.

Table 4.1 (continued)

Proteinase inhibited	#	Name	Active unit molecular weight range (mature chain)[a]	Known distribution and tissue[b] (and other information)
Cysteine	9C	Cystatins	10,000 to 16,000 Da with occasional even smaller or greater than this.	Wide variety of species, including rice, maize, wheat, potatoes, soybean and *Arabidopsis*. Seeds and tubers as well as pollen, flowers and leaves. Structures available.
		Multicystatin	6000 to 87,000 Da	A multiple active site multicystatin has been reported from potato (8 active sites), sunflower (3) and other species.
	3C	Kunitz	20,089	Potato.
	1C	Pineapple bromellain inhibitor[e]	5800 to 5900 Da	Pineapple.
Aspartic	3A	Kunitz	20,000 to 21,000 Da	Potato.
	11A	SQAPI	10,500 Da	Squash.
	12A	Wheat inhibitor[e]	58,000 Da	Wheat.
Metallo	13M	Carboxy-peptidase inhibitor	4100 to 4300 Da	*Solanum*. Structures available.

[a] Based on predicted gene sequences and the correct identification of signal peptides and prosequences.
[b] Distribution based mainly on protein purification and sequences and sometimes from mRNA expression. For identification based on sequences, further reassurance that the claimed specified PI activity actually occurred was required, except in the case of *Arabidopsis* where everything was based on sequence homology.
[c] *Theobroma cacao, Acacia confusa, Adenanthera pavonina, Solanum tuberosum, Lycopersicon esculentum, Populus balsamifera, Salix viminalis.* While Kunitz inhibitors occur in the Solanaceae, the sequences reported from these other species may not necessarily be trypsin inhibitors. The Kunitz family includes albumins and inhibitors of other proteinases.
[d] Based on expressed sequence tag (EST) sequences.
[e] Not cloned or sequenced at the DNA level.
For references, see the text.

catalytic type (Barrett *et al.*, 1999) of proteinase they inhibit. Where members of the same inhibitor clan inhibit proteinases of a different catalytic type, they are distinguished by the code for the different catalytic type they inhibit. For example, a Kunitz inhibitor that inhibits serine proteinases is called 3S, while the Kunitz inhibitor that inhibits cysteine proteinases is called 3C. If greater specificity as to the family (Barrett *et al.*, 1999) of proteinase is known, then the proteinase code is extended e.g. a Kunitz inhibitor that inhibits subtilisin is called 3S8, while an inhibitor that inhibits chymotrypsin is called 3S1. Where a single inhibitor inhibits more than one class, then it would be designated by the letters/numbers of each enzyme class that is inhibited. As in MEROPS, different PIs belonging to the same clan/family, but encoded by a different gene, would be sequentially numbered (e.g. 3S1.023).

4.4 Classes of serine inhibitors from plants

Plant serine PIs can be classified into at least eight groups; serpins, Kunitz, Bowman–Birk, squash, PI 1 and 2, cereal trypsin/amylase inhibitors and mustard seed inhibitors. Other serine PIs are found in other kingdoms, but apparently not in plants (e.g. the Kazal inhibitors, pancreatic Kunitz inhibitors and smapins). Plant inhibitors from storage organs are the best studied and often found in the highest amounts, where it is likely they function as both storage proteins and as protectants against pathogens (Ryan, 1990). This is most commonly observed for the PI 1 and 2 gene families in potato, and the Kunitz inhibitors in legume seeds.

Serine PIs generally are very tight binding inhibitors that behave as substrates, often being cleaved in the process. This is referred to as the canonical model of proteinase inhibitor action (Bode and Huber, 1993, 2000). The cleavage is typically on the C-side of the P1 or diagnostic residue for the serine proteinase specifity (e.g. Arg or Lys in trypsin like proteinases). Further specificity may also be determined by other residues to either side of P1, although these usually are less important than the P1 residue (Bode and Huber, 1993). The process is often reversible, with an equilibrium forming between the cleaved and intact forms of the PI while bound to the proteinase.

4.4.1 Serpins

Serpins are a group of related proteins, found in both animals and plants and some viruses, that inhibit a wide range of serine proteinases including trypsin, chymotrypsin and elastase and, in some cases, cysteine proteinases (Wright, 1996; Schick *et al.*, 1998; Irving *et al.*, 2000). In animals, they generally have a range of molecular masses from 50,000 to 65,000 Da, and a C-terminal reactive region that acts as a 'bait' for an appropriate serine protease. Typically, native animal serpins have three β-sheets and nine α-helices with an exposed reactive center loop (Irving *et al.*, 2000). In the native state, the reactive loop is exposed and ready to interact with target proteinases. After binding a proteinase, the reactive loop is cleaved at the scissile bond and the loop forms a stand inserted into a β-sheet, creating a more stable state. The serpin forms a stable covalent complex with the target proteinase through the active site serine (Nair and Cooperman, 1998) and the serpin is almost always irreversibly cleaved during the inhibition process (Wright, 1996). The proteinase–PI complex may slowly dissociate into active proteinase and inactive PI. This contrasts with the reversible mechanisms (which may involve cleavage of the peptide chain) employed by most other serine PIs. In animal serpins, the irreversible nature of the cleavage is due to the massive conformational change that occurs upon active site cleavage (Wright, 1996), locking the proteinase onto the inhibitor and separating the cleaved ends of the serpin. In addition, animal serpins may form a latent state where the active loop is inaccessible to the target proteinase, and the serpin is non-inhibitory, and a polymeric form (Irving *et al.*, 2000).

Plant serpins form a distinct clade separate from other animal serpins suggesting that at the time of plant from animal separation there was only one serpin gene (Irving *et al.*, 2000). We have been unable to determine whether plant serpins function in the same way as animal serpins since little information is available either on the structure of plant serpins or on detailed biochemical analysis of their mechanism of action. Inhibition of chymotrypsin and elastase, and cleavage of the reactive loop, has been reported (Rosenkrands *et al.*, 1994; Rasmussen *et al.*, 1996a; Yoo *et al.*, 2000) as well as some evidence for conformational changes (Dahl *et al.*, 1996b). The formation of a stable chymotrypsin–serpin complex (including after heating in the presence of sodium dodecyl sulfate (SDS); Rosenkrands *et al.*, 1994) suggests that a covalent linkage between the serpin and the proteinase does form, in a similar manner to that found in animal serpins (Nair and Cooperman, 1998).

A prominent barley protein called protein Z was shown to have a peptide sequence very similar to human α1-antitrypsin, but no proteinase inhibitory activity could be associated with this protein (Hejgaard *et al.*, 1985). Subsequently, proteins have been identified in several plant species that have been shown to have serpin-like proteinase inhibitory activity, some with similar suicide inhibition characteristics, including proteins from wheat (Ostergaard *et al.*, 2000), barley (Rasmussen *et al.*, 1996b), rye (Hejgaard, 2001) and squash (Yoo *et al.*, 2000). The cereal inhibitors are found in the seed at high concentrations (Ostergaard *et al.*, 2000), but their physiological function remains unclear. They may serve to protect the grain from insect predation (Yoo *et al.*, 2000; Ostergaard *et al.*, 2000) (although no growth inhibition of a sap sucking aphid was observed), or to prevent breakdown of storage protein by endogenous proteinases (Ostergaard *et al.*, 2000). These proteins have been shown to inhibit a wide range of mammalian serine proteinases including trypsin, chymotrypsin and cathepsin G, in some cases all by the same inhibitor (Dahl *et al.*, 1996a; Ostergaard *et al.*, 2000). The target proteinase is determined by the bait sequence (e.g. Hejgaard, 2001) and the ability of one inhibitor to inhibit several different proteinases has been ascribed to overlapping reactive centres along the bait loop (Dahl *et al.*, 1996a, b). However, it appears that these serpins are unable to significantly inhibit either plant serine (e.g. cucumism) or cysteine peptidases tested (Ostergaard *et al.*, 2000), although it has been reported that this is because the serpin is rapidly cleaved as a substrate (Dahl *et al.*, 1996b). On the basis of sequence similarities, some proteins with no known inhibitory activity are thought to belong to this family, including some of the predicted *Arabidopsis* serpin-like proteins.

We have searched the *Arabidopsis* predicted protein database for serpins by several means including searching for PROSITE and prints signatures, as well as their combination in InterPro. In addition, we took typical squash (Yoo *et al.*, 2000) and barley (Rasmussen, 1993) serpins and did BLAST searches on the MIPS database. At least 11 serpin genes from *Arabidopsis* have the 'bait' region at the C-terminal end of the protein, suggesting they are active proteinase inhibitors (listed in table 4.2).

These proteins, while showing homology to serpins as well as a C-terminal serpin motif, vary in molecular weight from 13,000 to 46,000 Da. Three predicted serpins in this list (AT1g62160, AT1g64010 and AT2g35560), are considerably smaller than known animal and plant serpins, and thus are possibly too small to be active serpins. These shorter genes are effectively truncated at the N-terminal end and include the full C-terminal sequence and serpin motif. In the predicted evolutionary tree (figure 4.1), they cluster with other more likely larger *Arabidopsis* serpins, suggesting they may have originated from these putative serpins. In this tree, the cereal and squash serpins fall in a separate grouping from the *Arabidopsis* serpins, which, in turn, appear to fall in three groups.

The homology between the sequences from *Arabidopsis* and other serpin sequences occurs mainly in the C-terminal half of the molecule, where the serpin reactive center loop determines the inhibitory specificity of the serpin. In barley and wheat, the cleaved sequence has been well defined, and many cases uniquely include a Glu (Dahl *et al.*, 1996a; Ostergaard *et al.*, 2000). None of the predicted *Arabidopsis* sequences at the reactive center loop appears similar to these cereal sequences. However, in animals (Goodwin *et al.*, 1996) and plants (Ostergaard *et al.*, 2000), this region is particularly variable. It must be emphasized that we know of no work that has been done characterizing

Table 4.2 Serpins in *Arabidopsis* that show sequence homology to serpins in the GenBank database

Protein entry code	MIPS title	Molecular weight (Da)
AT1g47710	serpin	42639.57
AT1g62160	serpin	30091.00
AT1g62170	hypothetical protein	48174.14
AT1g64010	serpin	20658.53
AT2g14540	serpin	45887.07
AT2g25240	serpin	42723.89
AT2g26390	serpin	43221.66
AT2g35560	serpin	13329.48
AT2g35580	serpin	41507.49
AT2g35590	serpin	37134.74
AT3g45220	serpin	44170.24

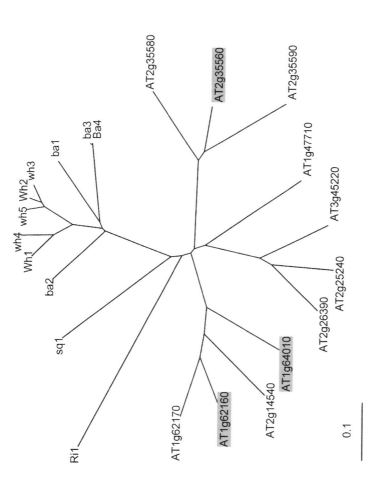

Figure 4.1 ClustalX analysis (Thompson *et al.*, 1997) of serpin genes from *Arabidopsis* and other species. The three shaded *Arabidopsis* sequences are the three listed in table 4.2 as having a low molecular mass. Gene codes are (code, accession number, species): *ba1*, T06183, *Hordeum vulgare*; *ba2*, S29819, *Hordeum vulgare*; *ba3*, S13822, *Hordeum vulgare*; *Ba4*, CAA06232, *Hordeum vulgare*; *Ri1*, BAA88536, *Oryza sativa*; *sq1*, AAG02411, *Cucurbita maxima*; *Wh1*, T06597, *Triticum aestivum*; *Wh2*, T06488, *Triticum aestivum*; *wh3*, S65782, *Triticum aestivum*; *wh4*, CAB52710, *Triticum aestivum*; *wh5*, CAB52709, *Triticum aestivum*. *Arabidopsis* codes are from Munich Information Center for Protein Sequences (MIPS).

Arabidopsis serpins at a protein level similar to that done in wheat, barley and squash.

Another 12 sequences in the *Arabidopsis* MIPS database show serpin motives, but little else that would classify them as serpins. Most have been either left unidentified or classified as particular proteins. Only one has been classified by MIPS as a serpin.

4.4.2   The Kunitz family

The Kunitz or soybean trypsin inhibitor (STI) family is a widespread family best defined in legume seeds, including soybean (Kim *et al.*, 1985) winged bean (Habu *et al.*, 1992) and *Erythrina variegata* (Kimura *et al.*, 1993). Inhibitors related to these legume proteins are found in many other plant families including cereals (Svendsen *et al.*, 1986; Leah and Mundy, 1989) and potatoes (Ishikawa *et al.*, 1994). Many of the Kunitz family members, but not all, have proteinase inhibitory activity. The family includes PIs, a taste-modifying protein, wound responsive proteins, storage proteins and amylase inhibitors (Odani *et al.*, 1996). There is also one study that reports that a soybean trypsin inhibitor can reduce dehydroascorbate when in the reduced form, but is a trypsin inhibitor in the oxidized form (Trumper *et al.*, 1994). Although Kunitz inhibitors are often found at high levels in seeds, these proteins are also found in other organs, including nodules (Manen *et al.*, 1991).

Kunitz inhibitors are usually effective against trypsin or chymotrypsin (Iwanaga *et al.*, 1999), and sometimes subtilisin (Terada *et al.*, 1994) and tissue plasmogen activator (Heussen *et al.*, 1984). However, there are examples of homologs of STI from potato that inhibit the aspartic proteinase cathepsin D (Barrett *et al.*, 1990; Strukelj *et al.*, 1990; Maganja *et al.*, 1992; Strukelj *et al.*, 1995) and cysteine proteinases (Brzin *et al.*, 1988; Krizaj *et al.*, 1993; Turk *et al.*, 1993). Another inhibitor has been reported that inhibits α-amylase as well as subtilisin (Leah and Mundy, 1989; Ohtsubo and Richardson, 1992). This inhibitor is probably a Kunitz inhibitor as determined by BLAST searches, and shows no homology to the cereal amylase/trypsin inhibitor family (see later). The papaya latex Kunitz family PI is a glycoprotein (Odani *et al.*, 1996) with a newly identified oligosaccharin chain (Shimazaki *et al.*, 1999). These authors do not comment on whether the carbohydrate effects inhibitory activity.

The soybean STI is found in minor variants with 181 amino acid residues and molecular weight of ca. 20,000 Da (Kim *et al.*, 1985). The mechanism of action follows a standard model for PIs; the STI binds rapidly with the target proteinase, with very slow dissociation (Iwanaga *et al.*, 1999). The dissociated STI is a mixture of cleaved (between Arg_{63} and Ile_{64}) and uncleaved molecules (Blow *et al.*, 1974). STI consists of 12 antiparallel β strands joined by long loops and these active site residues lie on a protruding outer loop of the roughly spherical STI, readily accessible to the target proteinase. Unlike many other PIs, is not constrained by secondary structural elements or disulfide bridges (Song and Suh, 1998). However, this active site loop does not change significantly in conformation upon binding to trypsin (Song and Suh, 1998), typical of non-serpin substrate type serine PIs. Up to 12 residues are in contact with trypsin during inhibition, suggesting that more than the target Arg residue is essential for inhibition. This is supported by site specific mutagenesis studies using the chymotrypsin Kunitz inhibitor from *Erythrina*, which was still inhibitory after four residues in the active site loop, including the target cleaved Leu, had been changed to Ala (Iwanaga *et al.*, 1999).

We detected eight sequences in the *Arabidopsis* MIPS database using the InterPro STI Kunitz signature. However, closer analysis revealed that only six appeared to be likely to be Kunitz inhibitors with the diagnostic Kunitz signature near the *N*-terminal end (table 4.3). To determine this, we blasted the *Arabidopsis* sequences back into the GenBank database to find out what other sequences have highest identity. Four of the eight closely matched known Kunitz inhibitors. For example, At1g17860 matched a range of PIs in the database including *LeMir* (Brenner *et al.*, 1998), a tobacco hypersensitive response (HR) related gene (Karrer *et al.*, 1998) and another PI (Tai *et al.*, 1991). In other cases, other Kunitz-type inhibitors were matched (Downing *et al.*, 1992; Annamalai and Yanagihara, 1999). However, two other sequences (AT3g04320 and AT3g04330), although identified as putative Kunitz type PIs in MIPS, showed little homology to proteins other than *Arabidopsis* sequences in the GenBank database. They showed high homology to each other and some homology to the other four sequences.

Several other sequences had Kunitz InterPro signatures, but for other reasons (e.g. a very high molecular weight) were considered unlikely to be Kunitz inhibitors. There appears no obvious signature to separate the *Arabidopsis* Kunitz inhibitors by the type of proteinase they inhibit.

4.4.3   The Bowman–Birk family

Bowman–Birk inhibitors are small serine proteinase inhibitors, usually double-headed with a molecular weight of 6000 to 9000 Da and seven disulfide bonds.

Table 4.3 Kunitz inhibitors (STI) in *Arabidopsis* that show sequence homology to Kunitz inhibitors in the GenBank database

Protein entry code	MIPS title	Comment	Molecular weight (Da)
AT1g17860	Hypothetical protein	Homology to Kunitz	22081.78
AT1g72290	Drought induced protein	Homology to Kunitz	23095.89
AT1g73260	Putative trypsin inhibitor	Homology to Kunitz	23001.37
AT1g73330	No title	Homology to Kunitz	23112.78
AT3g04330	Putative trypsin inhibitor	Low homology	22914.13
AT3g04320	Putative trypsin inhibitor	Low homology	27788.99

Typically, they are found in legume seeds and cereals (Laskowski and Kato, 1980; Odani *et al.*, 1986; Tashiro *et al.*, 1987; Nagasue *et al.*, 1988), although there are reports that these inhibitors have been identified in potato (Mitsumori *et al.*, 1994). These potato inhibitors only show low homology to soybean Bowman–Birk inhibitors, and it is likely that the potato inhibitor is widely divergent. A BLAST search of this inhibitor against the National Center for Biotechnology Information (NCBI) protein database results in no matches to Bowman–Birk inhibitors. An inhibitor protein from rice is a duplicated double-headed Bowman–Birk inhibitor (Tashiro *et al.*, 1987), with a much higher calculated molecular weight (15,074 Da) than is typically found. Similarly, the inhibitor from barley rootlets has a molecular weight of 13,826 Da and appears to be a duplicated double-headed inhibitor (Nagasue *et al.*, 1988) while the inhibitor from wheat germ appears in both the 7,000 and 14,000 Da forms (Odani *et al.*, 1986). These authors suggested that the 7000 Da single-headed inhibitor was a relic of an ancestral inhibitor before gene duplication led to the formation of the larger inhibitor. It has been suggested that cystatins have significant homology to the Bowman–Birk inhibitors, perhaps suggesting an evolutionary relationship (Saitoh *et al.*, 1991). In maize and alfalfa, a Bowman–Birk inhibitor gene has been reported to be induced by wounding (Brown *et al.*, 1985; Eckelkamp *et al.*, 1993; Rohrmeier and Lehle, 1993), in contrast to legume Bowman–Birk inhibitors.

Bowman–Birk inhibitors typically have seven conserved disulfide bonds giving structural rigidity, with the dual active sites on the loops of antiparallel β-strands for the two proteinases they inhibit (Chen *et al.*, 1992; Werner and Wemmer, 1992). The inhibitors are made up of two tandem repeats, with a bait residue on each repeat for trypsin (Lys or Arg), chymotrypsin (Leu, Phe or Tyr) or elastase (Ala). These active sites are on opposite sides of the molecule allowing access by two proteinases at the same time (Odani and Ikenaka, 1973; Chen *et al.*, 1992). It has been suggested that the active site loop is similar in structure to the active site loop of the Kazal family inhibitor (Chen *et al.*, 1992).

The Bowman–Birk inhibitors are another example of substrate type inhibitors, with bait residues matching the specifity of the proteinase. However, interaction between the PI and the proteinase is considered to be broader than just the bait residues in the active site, extending further into the loops. In the case of the soybean trypsin (Lys) loop, polar residues, hydrophilic bridges and weak hydrophobic contacts stabilize the interaction of trypsin with the PI. In the case of the chymotrypsin (Leu) loop, hydrophobic contacts could stabilize both chymotrypsin and trypsin in this position (Koepke *et al.*, 2000). In common with other inhibitors, the Bowman–Birk inhibitor is slowly cleaved by the proteinase at the active site, but both cleaved and uncleaved forms bind tightly to the proteinase, causing inhibition (Jensen *et al.*, 1996). The structure of the duplicated double headed 16 kDa barley seed inhibitor has been determined and, like the half-sized double-headed inhibitor, consists only of β-strands and connecting loops (Song *et al.*, 1999). The two halves of the inhibitor form equivalent domains, each with two apparent active site loops per half. Instead of seven disulfide bridges, each half only has five. However, alignment of a soybean trypsin/trypsin double-headed inhibitor against the barley duplicated double-headed trypsin inhibitor results in the first domain of the double-headed PI aligning to the first domain of the duplicated inhibitor, and the second domain of the double headed inhibitor aligning to the third domain of the duplicated PI. Whether the second and fourth domains of the duplicated inhibitor function as PI domains does not appear to be established. The stoichiometry of trypsin inhibition is reported to be only two trypsins per PI (Odani *et al.*, 1986), suggesting that steric hindrances may stop more trypsin enzymes binding. However, we are unable to determine whether more than two proteinases can bind to the duplicated double-headed inhibitor at the same time, and it has not been reported whether this inhibitor inhibited proteinases other than trypsin.

Recently, a novel 14 amino acid cyclic Bowman–Birk inhibitor from sunflower seeds was described with a very low (100 pM) dissociation constant for trypsin (Luckett *et al.*, 1999). This inhibitor is only able to inhibit one trypsin at a time. The protein is made up of one of the reactive loops of a typical Bowman–Birk inhibitor and can be satisfactorily superimposed over this loop in some other Bowman–Birk inhibitors (Luckett *et al.*, 1999). The rigid structure of the inhibitor is thought to contribute significantly to its inhibitory properties.

Although there is one gene which MIPS has identified as being similar to the Bowman–Birk inhibitors in *Arabidopsis* (AT4g14270, molecular weight 28069.92), this does not match other Bowman–Birk inhibitor genes in the GenBank database or any other protein. Thus it appears not to be a Bowman–Birk inhibitor.

4.4.4 The PI 1 family

This family is found in a wide range of plants including the Solanaceae (Richardson, 1974) where it is found in high levels in the tuber and fruit, is likely to be a defense protein against insect predation (Ryan, 1990), and is induced by wounding (Ryan, 1984). Generally, these inhibitors target chymotrypsin and elastase type proteinases, but specific inhibitors of trypsin and subtilisin have been identified (Plunkett *et al.*, 1982; Katayama *et al.*, 1994). It is also found in the phloem exudate of squash (Murray and Christeller, 1995) and seeds of squash (Krishnamoorthi *et al.*, 1990), barley (Svendsen *et al.*, 1980, 1981, 1982; Peterson *et al.*, 1991), legumes (Katayama *et al.*, 1994) and maize (Cordero *et al.*, 1994), and in the leech *Hirudo medicinalis* (Seemuller *et al.*, 1980). Inhibitors in this family are small proteins (molecular weight around 8000 Da) either lacking a disulfide bond (squash seeds, Krishnamoorthi *et al.*, 1990; squash phloem, Murray and Christeller, 1995) or containing one disulfide bond (Beuning *et al.*, 1994). The PI 1 proteins are often synthesized as 12,600 Da precursors with a signal peptide to direct export to the vacuole and a charged 19 amino acid residue pro-sequence (Graham *et al.*, 1985a), although the squash phloem inhibitor has no signal sequence (Murray and Christeller, 1995). In potato, they behave as tetramers or pentamers on gel filtration (Melville and Ryan, 1970, 1972; Plunkett *et al.*, 1982) with an effective molecular weight of 40,000 Da and, when in complex with chymotrypsin, four proteinases are bound per multimer, demonstrating the independence of the active sites. However, they can also inhibit as separate monomeric subunits.

Proteinase inhibitor 1 interacts with its target proteinase through one active site on the unit molecule in the typical serine PI canonical model with cleavage of the active site bait residues (Richardson and Cossins, 1974). The bait loop containing the active site residue lies on the sharp side of the wedge-shaped inhibitor (McPhalen and James, 1988). On complexing with its target proteinase, the conformation of the active site loop becomes restrained by a network of hydrogen bonds and electrostatic interactions when compared with the inhibitor free in solution (McPhalen and James, 1987). On the blunt side of the wedge is an α-helix, with the diagnostic consensus pattern, far removed from the active site loop.

The PI 1 family is represented by at least six examples in the *Arabidopsis* database (table 4.4). They vary in molecular weight from 7895 to 10,444 Da, and clearly show the PROSITE PI 1 family signature. In an alignment using ClustalX, the *Arabidopsis* sequences appear to align close to the non-solanaceae PI members (data not shown).

4.4.5   The PI 2 family

The PI 2 family is a well-studied group of inhibitors of the Solanaceae that are capable of inhibiting trypsin, chymotrypsin, subtilisin and pronase (Bryant *et al.*, 1976; Ryan, 1989, 1990). The base proteinaceous inhibitory unit of PI 2 has a molecular weight from 5000 to 6000 Da. However, the gene coding this inhibitor codes for two, three, four or six of these units (Schirra *et al.*, 2001) and this is translated as a single multidomain chain with a molecular weight ranging up to 43,000 Da (Atkinson *et al.*, 1993; Balandin *et al.*, 1995). The multiple inhibitors are processed by proteolysis after translation into the

Table 4.4 Proteinase 1 inhibitors in *Arabidopsis* that show sequence homology to PI 1 in the GenBank database

Protein entry code	MIPS title	Molecular weight (Da)
AT2g38870	Putative protease inhibitor	8015.14
AT2g38900	Putative protease inhibitor	7894.88
AT3g46860	Proteinase inhibitor-like protein	9573.16
AT3g50020	Putative protein	10444.39
AT5g43570	Unknown protein	10366.38
AT5g43580	Unknown protein	7871.31

smaller single domain inhibitor of 6000 Da (Hass *et al.*, 1982; Heath *et al.*, 1995; Lee *et al.*, 1999), or remain as double domain proteins of about 10,000 to 11,000 Da (Bryant *et al.*, 1976). The double domain proteins are often able to inhibit both chymotrypsin and trypsin, consistent with the presence of an Arg at the P1 site of the first domain and a Phe in the P1 site of the second domain (Graham *et al.*, 1985b). These double-domain proteins can then interact as 20,000 Da dimers (Bryant *et al.*, 1976) while the single domain inhibitors have generally been reported to remain as monomers (Hass *et al.*, 1982). Other examples of processing to generate smaller inhibitors have also been described (McManus *et al.*, 1994c). The vacuolar located protein is induced by wounding in tomato leaves (Graham *et al.*, 1985b) and in tobacco leaves (Pearce *et al.*, 1993; McManus *et al.*, 1994a).

Structurally, the single domain inhibitor shows a three-stranded antiparallel β-sheet on the side of the molecule removed from the active site loop (Greenblatt *et al.*, 1989). Generally single domain members of the PI 2 family have four disulfide bonds, two linking from close to the active site residue (Arg in the case of trypsin inhibitors) to other sites more central in the protein, giving a high degree of structural rigidity to the active site loop. While the reactive site loop of the inhibitor is still the most disordered part of the structure, the general shape of this region is always maintained (Nielsen *et al.*, 1994). In addition, both the *N*-terminal and *C*-terminal ends of the inhibitor are anchored by disulfide bonds.

The proteinase–inhibitor complex is non-covalent and the active site for the trypsin inhibitor was uncleaved when the structure was determined (Greenblatt *et al.*, 1989), although in the case of the aubergine inhibitor cleavage does occur after prolonged incubation with trypsin (Richardson, 1979). This suggests that the inhibitor behaves in the canonical manner typical of other serine PIs. The homologous inhibitors from tobacco style tissue show a similar structure (Nielsen *et al.*, 1994, 1995), with the nature of the side-chain on the primary binding residue being the main determinant of the proteinase specificity of the inhibitor, rather than the overall backbone fold (Nielsen *et al.*, 1995).

The 43,000 Da precursor from tobacco styles appears typical in its proteolytic processing. This protein is formed as six repeats and it is then processed into six inhibitors (Heath *et al.*, 1995). The six processing sites are within each homologous repeat, so effectively one *C*-terminal segment from one domain repeat forms a contiguous inhibitor with the *N*-terminal of the next domain (figure 4.2). What is particularly interesting is that the propeptide *N*-terminal and *C*-terminal domains associate to form a sixth inhibitor linked as two domains by disulfide bonds (figure 4.2, C2) (Lee *et al.*, 1999). The original six domain protein has been modeled as a circularized protein, joined by disulfide bridges linking the *N*-terminal and *C*-terminal domains of the 43,000 Da protein (Lee *et al.*, 1999). The large circularized protein only shows partial PI activity and must be cleaved to express full activity. Expression of a single repeat of the

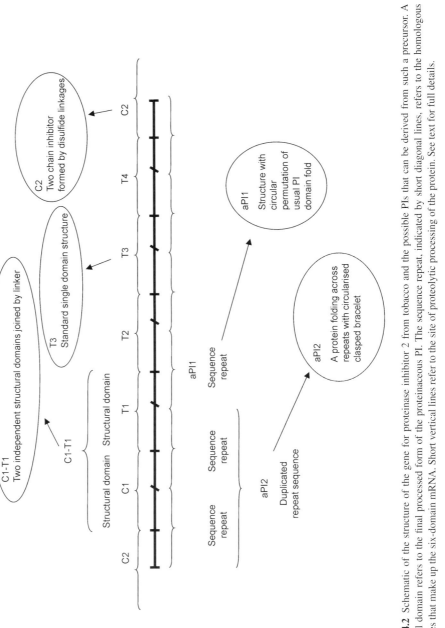

Figure 4.2 Schematic of the structure of the gene for proteinase inhibitor 2 from tobacco and the possible PIs that can be derived from such a precursor. A structural domain refers to the final processed form of the proteinaceous PI. The sequence repeat, indicated by short diagonal lines, refers to the homologous sequences that make up the six-domain mRNA. Short vertical lines refer to the site of proteolytic processing of the protein. See text for full details.

43,000 Da protein in *Escherichia coli* results in a functional inhibitor (figure 4.2, aPI1; Lee *et al.*, 1999; Scanlon *et al.*, 1999), showing that the inhibitor does not need to span between *C*-and *N*-terminal domains. In addition, when only the first two repeat domains of the 43,000 Da protein are synthesized in *E. coli*, they fold in a manner analogous to the parent six domain PI. That is, the *N*-terminal segment of the first domain associates through disulfide bonds with the *C*-terminal segment of the second domain and the second inhibitor is formed from the two continuous internal segments in apparent reverse order (figure 4.2, aPI2) (Lee *et al.*, 1999). However, when the inhibitors C1 and T1, as released by proteolysis from the 43,000 Da parent protein, are synthesized as a single protein in *E. coli*, they fold into two consecutive PIs, joined by the linker region which is normally cleaved (figure 4.2, C1–T1) (Schirra *et al.*, 2001). These authors also found no interactions between the domains and concluded that cross-repeat folds were thermodynamically more stable than folds along repeats (Schirra *et al.*, 2001).

The protein sequence of the 5000 to 6000 Da potato PI (Hass *et al.*, 1982) is compatible in sequence with having been formed by the proteolytic release from the two-domain parent protein in a manner similar to that which occurs for the tobacco six-domain protein (i.e. in the order *C*- then *N*-terminal segments). Thus the single domain inhibitors are also cross-inhibitor domains similar to the tobacco inhibitor (Greenblatt *et al.*, 1989; Scanlon *et al.*, 1999). We assume the main difference between these single-domain and the two-domain potato inhibitors is that they do not undergo proteolytic cleavage but still fold in the classical circularized clasped bracelet fold (Lee *et al.*, 1999). As far as we can determine, the potential inhibitor formed from the *C*- and *N*-terminal regions of the duplicated potato gene has not been found in potato. This may be due to the fact that the strong reducing conditions used to purify PIs from potato would reduce the disulfide bridges holding the two halves of the molecule together.

Four proteins in the *Arabidopsis* database are identified as being members of the proteinase inhibitor 2 family (AT2G02100, AT2g02120, AT2g02130 and AT2G02140). A fifth unidentified gene (*AT1g72060*) shows some homology to a potato II inhibitor from potato (Dammann *et al.*, 1997) and to a tobacco style PI 2 family member (Scanlon *et al.*, 1999), but shows much weaker homology with the other identified PI 2 inhibitors from *Arabidopsis*. When the identified proteinase inhibitor genes in the *Arabidopsis* database are BLASTed back into the GenBank database, they do not show homology to any of the Solanaceae PI 2 family members but only to thionins and to a Bowman–Birk-like PI from potato. When a range of protein sequences from known PI 2 family members are aligned using ClustalX, the four identified *Arabidopsis* sequences form their own grouping separate from the unidentified *Arabidopsis* sequence and from the rest of the known PI 2 sequences (figure 4.3). These four sequences show less than 10% identity with any of the Solanaceae PI 2 protein sequences,

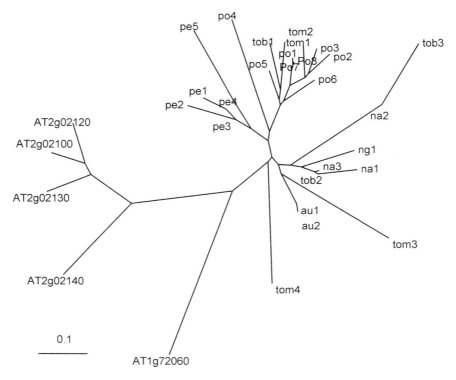

Figure 4.3 ClustalX analysis of the putative proteinase inhibitor 2 protein sequences from *Arabidopsis* and known sequences from other species, showing how the *Arabidopsis* sequences are very different from other known PI 2 proteins. Gene codes are (code, accession number, species): *po1*, XKPOC1, *Solanum tuberosum*; *po2*, AAD09849, *Solanum tuberosum*; *po3*, CAA55082, *Solanum tuberosum*; *po4*, XKPO2B, *Solanum tuberosum*; *po5*, XKPOT, *Solanum tuberosum*; *po6*, CAA78277, *Solanum tuberosum*; *Po7*, AAA53278, *Solanum tuberosum*; *Po8*, CAA27730, *Solanum tuberosum*; *tom1*, B24048, *Lycopersicon esculentum*; *tom2*, CAA64416, *Lycopersicon esculentum*; *tom3*, AAC37397, *Lycopersicon esculentum*; *tom4*, S72492, *Lycopersicon esculentum*; *tob1*, CAA82652, *Nicotiana tabacum*; *tob2*, JQ2269, *Nicotiana tabacum*; *tob3*, BAA95792, *Nicotiana tabacum*; *na1*, JQ2153, *Nicotiana alata*; *na2*, 1CE3, *Nicotiana alata*; *na3*, 1TIH, *Nicotiana alata*; *ng1*, AAF18451, *Nicotiana glutinosa*; *pe1*, AAF25496, *Capsicum annuum*; *pe2*, AAB94771, *Capsicum annuum*; *pe3*, AAB49533, *Capsicum annuum*; *pe4*, P56615, *Capsicum annuum*; *pe5*, AAF63518, *Capsicum annuum*; *au1*, P01078, *Solanum melongena*; *au2*, TIEO1, *Solanum melongena*.

while the unidentified *Arabidopsis* sequence shows a closer relationship to these unidentified sequences.

4.4.6   The squash family inhibitors

The squash family inhibitors are small proteins (about 30 residues) found in cucurbit seeds in several variants (e.g. Hayashi *et al.*, 1994; Hamato *et al.*,

1995). In addition to inhibiting trypsin and elastase (Hamato *et al.*, 1995), they have been shown to inhibit several proteinases involved in blood clotting (Hayashi *et al.*, 1994). These proteins have been well studied structurally (Bode *et al.*, 1989; Holak *et al.*, 1989; Huang *et al.*, 1993) and have three disulfide bonds, giving rigidity to the structure. In fact, the PROSITE motif is based mainly upon the cysteine residues found in the inhibitor. Their structure has also been studied in complexes with trypsin, which clearly shows the interaction of the inhibitor with the active site of trypsin (Helland *et al.*, 1999; Zhu *et al.*, 1999). This small inhibitor consists of β-turns in a simple flat shape with a protruding binding loop. During inhibition, the inhibitor is cleaved at a typical trypsin cleavage site in this binding loop after an Arg (or Lys) near the *N*-terminus, but the two fragments are held together by a disulfide bridge (Krishnamoorthi *et al.*, 1992). Kinetic analysis has supported the concept that these inhibitors follow the canonical model of PI action and that both the cleaved and uncleaved form of the inhibitor are effective in inhibiting trypsin (Otlewski and Zbyryt, 1994). It has been suggested that this inhibitor topologically resembled the potato carboxypeptidase inhibitor (Bode *et al.*, 1989) (see below).

Recent reports (Hernandez *et al.*, 2000; Felizmenio-Quimio *et al.*, 2001) have described a circular form of this family of PIs from *Momordica cochinchinensis* with a head-to-tail cyclized peptide backbone. The structure of this molecule was very similar to the structure of other homologous non-cyclic squash PIs (Felizmenio-Quimio *et al.*, 2001). They suggest the cyclization would not stabilize the active site loop, nor play a significant role in determining its conformation. Instead these authors propose that the cyclization may protect the PI from protease degradation, especially seed exoproteinases.

As these inhibitors appear confined to the cucurbits, it is very unlikely that any are found in the *Arabidopsis* genome. This was confirmed by a BLAST search of GenBank using sequences from typical family members that matched sequences found only in the cucurbits (Wilusz *et al.*, 1983; Joubert, 1984; Wieczorek *et al.*, 1985; Otlewski *et al.*, 1987; Favel *et al.*, 1989; Hara *et al.*, 1989; Holak *et al.*, 1989; Hatakeyama *et al.*, 1991; Ling *et al.*, 1993; Hayashi *et al.*, 1994; Hernandez *et al.*, 2000). In the *Arabidopsis* MIPS database, three sequences (AT1g04310, AT1g21700, and AT1g48570) were found to have squash PI family motifs, but these were all large proteins with no other homology to these very small inhibitors.

4.4.7 The mustard seed inhibitors

A low molecular weight (7000 Da) inhibitor from mustard and rape seeds has been isolated and characterized as a trypsin inhibitor (Menegatti *et al.*, 1992; Ceciliani *et al.*, 1994; Ceci *et al.*, 1995; Volpicella *et al.*, 2000). BLAST searches in the GenBank database using these sequences identified only sequences from mustard, rape and *Arabidopsis*, all closely related species, as having any similarity. These inhibitors all show a common pattern of cysteine residues, involved in disulfide bridges.

This inhibitor protein inhibits trypsin and to a lesser extent chymotrypsin (Ceciliani *et al.*, 1994). It is cleaved by trypsin between arginine and isoleucine, has four disulfide bridges and binds trypsin with a 1:1 stoichiometry (Ceciliani *et al.*, 1994), suggesting that this inhibitor also follows the standard substrate model for serine PIs. The inhibitor is expressed in seeds during the later stages of maturation and after mechanical wounding (Ceci *et al.*, 1995), suggesting it may play a role in defense.

ClustalX analysis using the *Arabidopsis* sequences and the published other sequences showed that five *Arabidopsis* sequences (table 4.5) showed homology to the other species, while a sixth *Arabidopsis* sequence, although showing some homology, would code for a protein with a significantly greater molecular mass (26,188 Da).

4.4.8   The cereal trypsin/α-amylase inhibitor family

α-Amylase/trypsin inhibitors (Odani *et al.*, 1983; Garcia-Olmedo *et al.*, 1987) are found in cereals, including rice (Ohtsubo and Richardson, 1992), wheat (Gautier *et al.*, 1990; Sanchez de la Hoz *et al.*, 1994), ragi (Srinivasan *et al.*, 1991), forage grasses (Tasneem *et al.*, 1994) and barley (Barber *et al.*, 1986). The inhibitor family includes members that inhibit proteinases and members that inhibit α-amylases (Oda *et al.*, 1997), both rather different classes of enzymes (Odani *et al.*, 1983). Some of these family members are also seed allergenic proteins (Adachi *et al.*, 1993; Alvarez *et al.*, 1995). The ragi inhibitor appears unique, in that it inhibits trypsin and α-amylase independently and simultaneously, suggesting that their binding sites are well separated (Alagiri and Singh, 1993; Maskos *et al.*, 1996). In addition to these inhibitors inhibiting trypsin, versions also inhibit the Hageman factor (Wen *et al.*, 1992) or chymotrypsin (Tasneem *et al.*, 1996) and subtilisin (Ohtsubo and Richardson, 1992).

The structure of the inhibitor from ragi has been determined by NMR and X-ray crystallography (Strobl *et al.*, 1995; Gourinath *et al.*, 1999, 2000). The structure of the 13,000 Da protein contains four α-helices and two short

Table 4.5 Mustard inhibitors in *Arabidopsis* that show sequence homology to mustard inhibitors in the GenBank database

Protein entry code	MIPS title	Molecular weight (Da)
AT1g47540	Hypothetical protein	10992.7
AT2g43510	Putative trypsin inhibitor	9885.53
AT2g43520	Putative trypsin inhibitor	9660.5
AT2g43530	Putative trypsin inhibitor	9596.43
AT2g43550	Putative trypsin inhibitor	10192.2

antiparallel β-strands with five disulfide bridges in a roughly globular structure. The trypsin binding loop of the ragi inhibitor forms the canonical substrate-like conformation, similar to other serine PIs and lies between and is stabilized by helices one and two (Strobl et al., 1995). In the case of the ragi inhibitor, the inhibitor is reversibly cleaved by trypsin after an arginine residue in the active site (Maskos et al., 1996).

It has been suggested that the amylase-binding site is via a Lys on the other side of the molecule from the trypsin binding loop (Gourinath et al., 1999). Despite the ragi inhibitor only sharing 26% identity with the wheat α-amylase inhibitor, their α-helices can be reasonably superimposed (Oda et al., 1997). Differences between the two lie in the loops, especially the trypsin binding loop, which might be expected as they inhibit such different enzymes.

BLAST and InterPro searches of the Arabidopsis MIPS database uncovered no proteinase with homology to these inhibitors. It appears restricted to cereal species.

4.4.9   Other serine proteinase inhibitors

Is thaumatin a PI? The amino acid sequence of a purified α-amylase/trypsin inhibitor from maize has been reported to be similar to thaumatin (Richardson et al., 1987; Rebmann et al., 1991). BLAST searches show that this published PI sequence is almost identical to other antifungal zeamatin-like PR5 proteins (Vigers et al., 1991; Huynh et al., 1992) and not to any known PI. In addition, it was suggested that thaumatin is more likely to be a protease rather than a PI (Skern et al., 1990), although this was later shown not to be the case (Beynon and Cusack, 1990). Further characterization of this maize thaumatin protein has shown that the signal peptide of thaumatin was very similar to the signal peptide of a PI (Lazaro et al., 1988). Zeamatin has been reported to have no trypsin inhibitory activity (Vigers et al., 1991) without any details given, nor was a barley homolog of thaumatin (Hejgaard et al., 1991). Thus we are left in a confused state as to the proteinase inhibitory activity of thaumatin, zeamatin and other PR5 proteins. In many studies the antifungal nature of thaumatin-like PR5 proteins has been well established (Vigers et al., 1991; Huynh et al., 1992).

More recently zeamatin has been reported to have a weak antitrypsin activity (Schimoler-O'Rourke et al., 2001). We calculated a K_i of 0.5 µM from the data presented in this paper, although this may be lower if the substrate concentration used affected the K_i. These authors also stated that thaumatin had no antitrypsin activity at a 100 to 1 ratio of thaumatin to trypsin. Given the state of published information, and as no PI activity of thaumatin was described in the original Nature paper (Richardson et al., 1987), we must conclude that thaumatin is not a trypsin inhibitor and that it has been misidentified as a PI in the literature. We also conclude that zeamatin is a very weak trypsin inhibitor. Two obvious alternatives exist: that these antifungal proteins are not PIs or that their target proteinase has yet to be identified. As suggested by Skern *et al.* (1990), this 'emphasizes the care with which the results of computer searches must be treated'.

4.5   Classes of cysteine PIs

The main clan of cysteine PIs is the cystatins (Turk *et al.*, 1997), found in both plants and animals. In addition, there are several other groups of cysteine PIs including thyropins in animals (Lenarcic and Bevec, 1998; Lenarcic and Turk, 1999), Kunitz or STI-like cysteine PIs from potato (Gruden *et al.*, 1997), a cysteine PI known only as a protein sequence from pineapple stem (Lenarcic *et al.*, 1992), a cysteine PI from the silkworm *Bombyx mori* that is homologous to the pro-region of cysteine proteinases (Yamamoto *et al.*, 1999) and clitocypins known only from a fungus (Brzin *et al.*, 2000). Only the cystatins and perhaps the Kunitz/STI cysteine PIs are found in *Arabidopsis* genome sequences.

4.5.1   The cystatins

The main group of cysteine proteinase inhibitors are the cystatins that, in animals, are divided into three groups: type 1 cystatins (or stefins), type 2 cystatins, and the kininogens. Plant cystatins or phytocystatins have been assigned to belong to group 1, as judged by size, homology and lack of disulfide bonds. However in amino acid sequence, the plant cystatins are closer to type 2 cystatins and it has therefore been proposed that plant cystatins should be put in a family of their own (Abe *et al.*, 1991; Turk *et al.*, 1997; Arai *et al.*, 1998; Margis *et al.*, 1998). They can be abundant proteins in many different plants with a molecular weight around 11,000 to 13,000 Da. The cystatins from rice have been divided into two groups sharing 55% identity based on sequence homology, and these show some differences in inhibitory specificity (Kondo *et al.*, 1990). Oryzacystatin I inhibits papain more strongly than cathepsin H, while oryzacystatin II is the reverse.

The cystatin family may include proteins that are not PIs. The sweet tasting protein from the African Berry (*Dioscoreophyllum cumminsii*) shows some homology to cystatins, suggesting a common origin (Murzin, 1993). Cysteine proteinases have also been isolated in complexes with cystatins (Yamada *et al.*, 2000). The presence of a cystatin in phloem sap from *Ricinus communis* has also been reported (Balachandran *et al.*, 1997). An insoluble cystatin from carrot cell walls was reported to inhibit a carrot proteinase (Ojima *et al.*, 1997). These authors drew an analogy to extracellular animal cystatins and observed that this was the first reported plant extracellular cystatin.

The 3D solution structure of a rice cystatin or oryzacystatin (Nagata *et al.*, 2000) shows that the 3D structures of plant cystatins follow the same general outline as the animal type 1 and type 2 cystatins. There appear to be three points of contact with the cysteine proteinase active site, that are all present in the outer parts of the protein and have been identified in animal cystatins by site-specific mutagenesis (Nagata *et al.*, 2000). One is at the *N*-terminal, the other two at the ends of hairpin loops that connect β-sheets (Turk *et al.*, 1997). Truncation of the *N*-terminal results in major loss of inhibition in chicken cystatin (Machleidt *et al.*, 1993). The first hairpin loop on cystatins has a signature motif of -Gln-X-Val-X-Gly- where X represents any amino acid and the second hairpin loop frequently has a Trp (Margis *et al.*, 1998). However, inhibition appears to be fundamentally different from that observed with typical serine PIs (Stubbs *et al.*, 1990; Machleidt *et al.*, 1993) in that interactions between the inhibitor and the proteinase are non-covalent, involving hydrophobic interactions (Bode and Huber, 1993; Machleidt *et al.*, 1993). The active loops of cystatin are thought to interact with subsites on papain to either side of the catalytic cysteine and thus cystatin is not cleaved while inhibiting papain. This multiple contact ensures that cystatin is a tight binding reversible inhibitor of papain-like proteinases usually with a K_i in the sub-nM regions. Even if the cysteine proteinase is inactivated by modification of the active site cysteine, it is still able to bind cystatins. In addition, kinetic analysis of the interaction of chicken cystatin with a range of cysteine proteinases suggested that no conformational change is involved in the inhibition (Bjork *et al.*, 1989; Bjork and Ylinenjarvi, 1989, 1990).

In the *Arabidopsis* database, ten predicted proteins are identified as cysteine proteinase inhibitors or have cystatin signatures (figure 4.4). These fall into two groups, one group of seven showing reasonable homology to each other and to other cystatins (table 4.6, figure 4.4). Three fall isolated from the *Arabidopsis* sequences and from other cystain sequences. The seven *Arabidopsis* cystatin sequences cluster with cystatin sequences from other species.

While some of these predicted proteins seem too large for cystatins, a cystatin from soybean has been isolated that is even larger (Misaka *et al.*, 1996), a cystatin from corn has a ca. 40 amino acid signal presequence (Abe *et al.*, 1992), while potato multicystatin has eight active domains with a molecular weight of 85 kDa (Waldron *et al.*, 1993; Walsh and Strickland, 1993). The activity of the soybean cystatin was unaffected by removal of sequences outside what might be regarded as the core ca. 12 kDa rice sequence (Misaka *et al.*, 1996).

4.5.2   The Kunitz cysteine PIs

As mentioned previously (section 4.4.2), some Kunitz inhibitors have the ability to inhibit cysteine proteinases (Brzin *et al.*, 1988; Krizaj *et al.*, 1993; Gruden *et al.*, 1997). These inhibitors are found in species related to potato, e.g. tobacco (Choi *et al.*, 2000) and can be induced by pathogens in potato (Valueva *et al.*, 1998). However, they have been little studied at the biochemical level and it can only be assumed that they would follow the canonical serine proteinase inhibitor substrate model for Kunitz inhibitors. Their active site remains to be published.

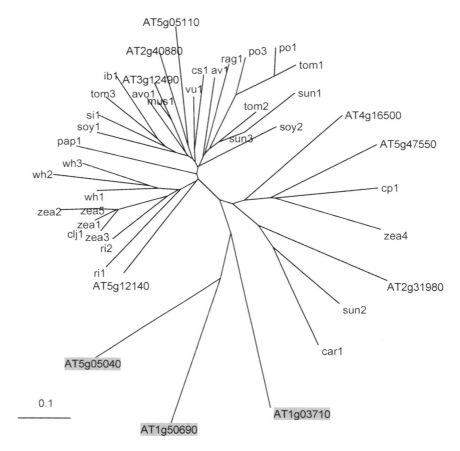

Figure 4.4 ClustalX analysis of cystatins from *Arabidopsis* and other species. The three shaded *Arabidopsis* sequences are the three identified in table 4.6 as being unlikely to be cystatins. In general, probable cystain sequences from *Arabidopsis* cluster with cystatin sequences from other species. Gene codes are (code, accession number, species): *av1*, AAD33907, *Artemisia vulgaris*; *avo1*, JH0269, *Persea americana*; *car1*, T14323, *Daucus carota*; *cs1*, CAA11899, *Castanea sativa*; *ib1*, AAD13812, *Ipomoea batatas*; *mus1*, S65071, *Brassica rapa*; *pap1*, JC4259, *Carica papaya*; *po3*, Q03196, *Solanum tuberosum*; *rag1*, JN0906, *Ambrosia artemisiifolia*; *ri1*, P20907, *Oryza sativa*; *ri2*, P09229, *Oryza sativa*; *soy1*, BAA19610, *Glycine max*; *soy2*, P25973, *Glycine max*; *sun1*, JC7333, *Helianthus annuus*; *sun2*, Q10993, *Helianthus annuus*; *sun3*, Q10992, *Helianthus annuus*; *tom1*, AAF23128, *Lycopersicon esculentum*; *tom2*, AAF23127, *Lycopersicon esculentum*; *tom3*, AAF23126, *Lycopersicon esculentum*; *vu1*, CAA79954, *Vigna unguiculata*; *wh1*, BAB18769, *Triticum aestivum*; *wh2*, BAB18767, *Triticum aestivum*; *wh3*, BAB18766, *Triticum aestivum*; *zea1*, JC4882, *Zea mays*; *zea2*, BAA07327, *Zea mays*; *zea3*, P31726, *Zea mays*; *zea4*, P19864, *Zea mays*; *zea5*, BAA09666, *Zea mays*.

Table 4.6 Cystatins in *Arabidopsis* that show sequence homology to cystatins in the GenBank database

Protein entry code	MIPS title	Molecular weight (Da)
AT2g31980	Putative cysteine proteinase inhibitor B (cystatin B)	16089.86
AT2G40880	Putative cysteine proteinase inhibitor B (cystatin B)	14421.76
AT3g12490	Hypothetical protein	26294.1
AT4g16500	Cysteine proteinase inhibitor-like protein	12555.97
AT5g05110	Cysteine proteinase inhibitor-like protein	26582.11
AT5g12140	Cystatin (emb|CAA03929.1)	11255.63
AT5g47550	Putative protein	13372.39
The following three predicted proteins show weak cysteine PI motifs, but otherwise do not match with the other seven proteins		
AT1g50690	Hypothetical protein	20251.44
AT5g05040	Putative protein	19471.48
AT1g03710	Hypothetical protein	23205.69

4.5.3   The bromelain inhibitor from pineapple

This inhibitor is small, with a molecular weight of around 5900 Da and comprises two chains joined by disulfide bonds (Reddy *et al.*, 1975; Heinrikson and Kezdy, 1976). One chain consists of 11 residues, and is thought to have the active site arginine, and the other has 41 residues and is thought to confer structural rigidity (Hatano *et al.*, 1995). There is no reported DNA sequence for this inhibitor, and it is thought to be synthesized as a precursor that is cleaved to form the two chains (Hatano *et al.*, 1995).

The reported dissociation constants (around 100 nM, Heinrikson and Kezdy, 1976) are about 100 times higher than those generally found for cystatins, and show strong pH dependency and a specificity towards bromelain rather than papain and ficin. This inhibitor shares some common structural elements and homology with the double-headed chymotrypsin and trypsin Bowman–Birk inhibitors (BBI) (Hatano *et al.*, 1996) in that the arrangement of disulfide bonds corresponds with the bromelain inhibitor and BBI and there is some homology between the bromelain inhibitor and the BBI from cowpea (Hilder *et al.*, 1989), with 16 residues out of 52 identical. The bromelain inhibitor also has a weak capacity to inhibit trypsin and chymotrypsin. These inhibitors are not found in the MIPS *Arabidopsis* database.

4.5.4   Miscellaneous cysteine PIs

Other cysteine proteinase inhibitors have been reported from plants. The pro-region from papain and other plant cysteine proteinases acts as an inhibitor of the native enzyme (Taylor *et al.*, 1995). In another example, the pro-region of papaya proteinase IV inhibits cysteine proteinases from insects (Visal *et al.*, 1998). A cysteine proteinase inhibitor from pearl millet had some characteristics (Joshi *et al.*, 1998, 1999) that would distinguish it from typical cystatins, except no DNA sequence information is available. Recently, proteins have been identified that belong to a new class of PIs related to the miraculin protein (Brenner *et al.*, 1998; Odani *et al.*, 1996).

4.6   Proteinaceous inhibitors of metalloproteinases

Proteinaceous metalloproteinase inhibitors are not common in nature (Bode and Huber, 1993; Vendrell *et al.*, 2000), being known only from *Ascaris suum*, the leech (*Hirudo medicinalis*) and rat brain, as well as in solanaceous plants (Vendrell *et al.*, 2000). In plants, metalloproteinase inhibitor sequences are only reported from tomato and potato. A small carboxypeptidase inhibitor was discovered in potato tubers (Ryan *et al.*, 1974) and shown to consist of four isoforms, some of which may be encoded by the same gene (Hass *et al.*, 1979). It has 38 to 39 amino acid residues with a molecular weight of about 4200 Da (Hass *et al.*, 1975). All isoforms inhibit carboxypeptidase A and B in a similar manner with dissociation constants in the low nM region. The mRNA and protein are expressed early in the development of the tomato fruit and at low levels in leaves, with a strong rise in mRNA and protein after wounding (Martineau *et al.*, 1991).

The structure of the inhibitor has been determined in complex with carboxypeptidase A (Rees *et al.*, 1983). The globular molecule is stabilized by three disulfide bridges but shows little other structure. The *C*-terminal of the inhibitor inserts into the active site cleft of carboxypeptidase A and the terminal P1′ residue (Gly) is cleaved but remains trapped in the cleft by the rest of the inhibitor. A mutant without the terminal Gly showed a slight reduction in binding energy but was still an effective inhibitor of carboxypeptidase A (Marino-Buslje *et al.*, 2000). The hydrophobic nature of the penultimate Val, and the preceding Tyr are also critical for inhibition (Vendrell *et al.*, 2000). Residues closer to the core on the *C*-terminal tail also affect inhibition, probably through tail mobility and positioning and a synthetic peptide consisting of the five *C*-terminal residues was only a weak inhibitor (Vendrell *et al.*, 2000). There is also a secondary interaction between the inhibitor and the proteinase through a Trp at residue 28 close to the core of the inhibitor (Vendrell *et al.*, 2000). Thus like the canonical Ser PI, the proteinase–carboxypeptidase inhibitor complex represents a reaction intermediate in the inhibition of carboxypeptidase A. In free solution NMR structures show the *C*-terminal region has considerable flexibility (Clore *et al.*, 1987) which becomes ordered upon binding to carboxypeptidase A (Bode and Huber, 1993). Interestingly, the squash seed inhibitor shows topographical similarity with the carboxypeptidase inhibitor (Bode *et al.*, 1989).

BLAST searches of the MIPS *Arabidopsis* database using protein sequences from potato (Villanueva *et al.*, 1998) did not reveal any *Arabidopsis* sequences with significant homology to the sequence of the potato carboxypeptidase inhibitor. Again the issue is raised as to whether carboxypeptidase inhibitors are considered rare in plants because they are present in low abundance, or because they have not yet been searched for and show little homology to known inhibitors.

4.7 Proteinaceous inhibitors of the aspartic proteinases

Proteinaceous aspartic proteinase inhibitors are rare in nature, known from only six different sources, three of which are from plants. The three non-plant examples are from *Ascaris* (Abu-Erreish and Peanasky, 1974), sea anemone (Lenarcic and Turk, 1999) and yeast (Schu *et al.*, 1991).

As discussed earlier, the potato cathepsin D inhibitors (Keilova and Tomasek, 1976; Mares *et al.*, 1989; Ritonja *et al.*, 1990; Kreft *et al.*, 1997) belong to a multigene family (Strukelj *et al.*, 1992) and appear to have been derived from a Kunitz/STI precursor, presumably by active site modification. This inhibitor apparently inhibits cathepsin D (but not pepsin), as well as trypsin and this was not separable from the trypsin inhibitory activity by affinity chromatography on a cathepsin D affinity column (Mares *et al.*, 1989).

The wheat inhibitor (Galleschi *et al.*, 1993) has not been characterized extensively, and no peptide or nucleic acid sequence is available. It has been reported to have a molecular weight of 58 kDa, very high for proteinase inhibitors from plants. The squash inhibitor (Christeller *et al.*, 1998) has been characterized at the protein level and its mRNA cloned. It also shows no homology to other PIs, and so information about its mechanism of interaction with target proteinases (pepsin and other gastric aspartate peptidases) as inferred by homology is limited.

4.8 Concluding remarks

Many proteinases are synthesized as inactive precursors or zymogens. They often have pro-sequences that control access to the active site of the enzyme until the pro-sequence is removed by proteolysis to release an active enzyme (Pak *et al.*, 1997; Vendrell *et al.*, 2000). In one sense, the pro-sequence can be regarded as a specific inhibitor of the proteinase. In addition, the separate pro-region may also inhibit the proteinase. For example, the pro-region of papain is a potent inhibitor of papain and several other related cysteine proteinases (Taylor *et al.*, 1995), although it also probably serves a function in folding of the proteinases (Wiederanders, 2000).

While some inhibitors appear to be very restricted in their distribution (e.g. mustard seed inhibitors and the squash family), many are found in a few very different branches of the angiosperms. For example, the Bowman–Birk inhibitors are found in legumes and in cereals; the PI 1 family is found in solanaceous species and in barley and squash. This suggests that these inhibitors may actually be more widely distributed, but no one has systematically detected their presence either at the gene or at the protein level.

Whether new inhibitors remain to be found in plants is unanswered. There are many unidentified proteins in the genome of *Arabidopsis*, and it is likely that these may represent PIs with specifity to proteinases that have not been tested for. Further, the genome of *Arabidopsis* does not contain homologs of some of the existing PIs and so a wider variety of plant species need to be tested to determine the true spectrum of the PIs in the plant kingdom.

References

Abe, K., Kondo, H., Watanabe, H., Emori, Y. and Arai, S. (1991) Oryzacystatins as the first well-defined cystatins of plant origin and their target proteinases in rice seeds. *Biomed. Biochim. Acta*, **50**, 637-641.

Abe, M., Abe, K., Kuroda, M. and Arai, S. (1992) Corn kernel cysteine proteinase inhibitor as a novel cystatin superfamily member of plant origin. Molecular cloning and expression studies. *Eur. J. Biochem.*, **209**, 933-937.

Abu-Erreish, G.M. and Peanasky, R.J. (1974) Pepsin inhibitors from *Ascaris lumbricoides*. Pepsininhibitor complex: stoichiometry of formation, dissociation, and stability of the complex. *J. Biol. Chem.*, **249**, 1566-1571.

Adachi, T., Izumi, H., Yamada, T. *et al.* (1993) Gene structure and expression of rice seed allergenic proteins belonging to the alpha-amylase/trypsin inhibitor family. *Plant Mol. Biol.*, **21**, 239-248.

Adam, Z. (1996) Protein stability and degradation in chloroplasts. *Plant Mol. Biol.*, **32**, 773-783.

Adam, Z. (2000) Chloroplast proteases: possible regulators of gene expression? *Biochimie*, **82**, 647-654.

Alagiri, S. and Singh, T.P. (1993) Stability and kinetics of a bifunctional amylase/trypsin inhibitor. *Biochim. Biophys. Acta*, **1203**, 77-84.

Alvarez, A.M., Adachi, T., Nakase, M., Aoki, N., Nakamura, R. and Matsuda, T. (1995) Classification of rice allergenic protein cDNAs belonging to the alpha-amylase/trypsin inhibitor gene family. *Biochim. Biophys. Acta*, **1251**, 201-204.

Andersson, B. and Aro, E.M. (1997) Proteolytic activities and proteases of plant chloroplasts. *Physiologia Plantarum*, **100**, 780-793.

Annamalai, P. and Yanagihara, S. (1999) Identification and characterization of a heat-stress induced gene in cabbage encodes a Kunitz type protease inhibitor. *J. Plant Physiol.*, **155**, 226-233.

Apweiler, R., Attwood, T.K., Bairoch, A. *et al.* (2000) InterPro—an integrated documentation resource for protein families, domains and functional sites. *Bioinformatics*, **16**, 1145-1150.

Apweiler, R., Attwood, T.K., Bairoch, A. *et al.* (2001) The InterPro database, an integrated documentation resource for protein families, domains and functional sites. *Nucleic Acids Res.*, **29**, 37-40.

Arai, S., Matsumoto, I. and Abe, K. (1998) Phytocystatins and their target enzymes: from molecular biology to practical application: a review. *J. Food Biochem.*, **22**, 287-299.

Atkinson, A.H., Heath, R.L., Simpson, R.J., Clarke, A.E. and Anderson, M.A. (1993) Proteinase inhibitors in *Nicotiana alata* stigmas are derived from a precursor protein which is processed into five homologous inhibitors. *Plant Cell*, **5**, 203-213.

Balachandran, S., Xiang, Y., Schobert, C., Thompson, G.A. and Lucas, W.J. (1997) Phloem sap proteins from *Cucurbita maxima* and *Ricinus communis* have the capacity to traffic cell to cell through plasmodesmata. *Proc. Natl Acad. Sci. USA*, **94**, 14,150-14,155.

Balandin, T., van der Does, C., Albert, J.M., Bol, J.F. and Linthorst, H.J. (1995) Structure and induction pattern of a novel proteinase inhibitor class II gene of tobacco. *Plant Mol. Bio.*, **27**, 1197-1204.

Barber, D., Sanchez-Monge, R., Mendez, E., Lazaro, A., Garcia-Olmedo, F. and Salcedo, G. (1986) New alpha-amylase and trypsin inhibitors among the CM-proteins of barley (*Hordeum vulgare*). *Biochim. Biophys. Acta*, **869**, 115-118.

Barrett, A.J., Ritonja, A., Buttle, D.J. *et al.* (1990) The amino acid sequence of a novel inhibitor of cathepsin D from potato. *FEBS Lett.*, **267**, 13-15.

Barrett, A.J., Rawlings, N.D. and Woessner, J.F. (1999) Introduction, in *Handbook of Proteolytic Enzymes* (eds A.J. Barrett, N.D. Rawlings and J.F. Woessner) Academic Press, pp. xxv-xxix.

Beuning, L.L., Spriggs, T.W. and Christeller, J.T. (1994) Evolution of the proteinase inhibitor I family and apparent lack of hypervariability in the proteinase contact loop. *J. Mol. Evol.*, **39**, 644-654.

Beynon, R. and Cusack, M. (1990) Thaumatin not proteolytic. *Nature*, **344**, 498.

Bieth, J.G. (1995) Theoretical and practical aspects of proteinase inhibition kinetics. *Methods Enzymol.*, **248**, 59-84.

Birk, Y. (1996) Protein proteinase inhibitors in legume seeds—overview. *Arch Latinoam Nutr.*, **44**, 26S-30S.

Bjork, I. and Ylinenjarvi, K. (1989) Interaction of chicken cystatin with inactivated papains. *Biochem. J.*, **260**, 61-68.

Bjork, I. and Ylinenjarvi, K. (1990) Interaction between chicken cystatin and the cysteine proteinases actinidin, chymopapain A, and ficin. *Biochemistry*, **29**, 1770-1776.

Bjork, I., Alriksson, E. and Ylinenjarvi, K. (1989) Kinetics of binding of chicken cystatin to papain. *Biochemistry*, **28**, 1568-1573.

Blow, D.M., Janin, J. and Sweet, R.M. (1974) Mode of action of soybean trypsin inhibitor (Kunitz) as a model for specific protein-protein interactions. *Nature*, **249**, 54-57.

Bode, W. and Huber, R. (1993) Structural basis of the proteinase–protein inhibitor interaction, in *Innovations in proteases and their inhibitors* (ed. F.X. Aviles) Walter de Gruyter & Co, Berlin, pp. 81-121.

Bode, W. and Huber, R. (2000) Structural basis of the endoproteinase-protein inhibitor interaction. *Biochim. Biophys. Acta*, **1477**, 241-252.

Bode, W., Greyling, H.J., Huber, R., Otlewski, J. and Wilusz, T. (1989) The refined 2.0 A X-ray crystal structure of the complex formed between bovine beta-trypsin and CMTI-I, a trypsin inhibitor from squash seeds (*Cucurbita maxima*). Topological similarity of the squash seed inhibitors with the carboxypeptidase A inhibitor from potatoes. *FEBS Lett.*, **242**, 285-292.

Bogacheva, A.M. (1999) Plant subtilisins. *Biochemistry*, **64**, 287-293.

Botella, M.A., Xu, Y., Prabha, T.N. *et al.* (1996) Differential expression of soybean cysteine proteinase inhibitor genes during development and in response to wounding and methyl jasmonate. *Plant Physiol.*, **112**, 1201-1210.

Bowles, D. (1998) Signal transduction in the wound response of tomato plants. *Phil. Trans R. Soc. Lond. B*, **353**, 1495-1510.

Brenner, E.D., Lambert, K.N., Kaloshian, I. and Williamson, V.M. (1998) Characterization of *LeMir*, a root-knot nematode-induced gene in tomato with an encoded product secreted from the root. *Plant Physiol.*, **118**, 237-247.

Broadway, R.M., Duffey, S.S., Pearce, G. and Ryan, C.A. (1986) Plant proteinase inhibitors: mechanism of action and effect on the growth and digestive physiology of larval *Heliothis zea* and *Spodoptera exiqua.* Plant proteinase inhibitors: a defense against herbivorous insects? *J. Insect Physiol.*, **41**, 827-833.

Brown, W.M. and Dziegielewska, K.M. (1997) Friends and relations of the cystatin superfamily—new members and their evolution. *Protein Sci.*, **6**, 5-12.

Brown, W.E., Takio, K., Titani, K. and Ryan, C.A. (1985) Wound-induced trypsin inhibitor in alfalfa leaves: identity as a member of the Bowman-Birk inhibitor family. *Biochemistry*, **24**, 2105-2108.

Bryant, J., Green, T.R., Gurusaddaiah, T. and Ryan, C.A. (1976) Proteinase inhibitor II from potatoes: isolation and characterization of its protomer components. *Biochemistry*, **15**, 3418-3424.

Brzin, J., Popovic, T., Drobnic-Kosorok, M., Kotnik, M. and Turk, V. (1988) Inhibitors of cysteine proteinases from potato. *Biol. Chem. Hoppe-Seyler*, **369**, 233-238.

Brzin, J., Rogelj, B., Popovic, T., Strukelj, B. and Ritonja, A. (2000) Clitocypin, a new type of cysteine proteinase inhibitor from fruit bodies of mushroom clitocybe nebularis. *J. Biol. Chem.*, **275**, 20,104-20,109.

Callis, J. (1995) Regulation of protein degradation. *Plant Cell*, **7**, 845-857.

Callis, J. and Setlow, J.K. (1997) Regulation of protein degradation in plants. *Genet. Eng.*, **19**, 121-148.

Casaretto, J.A. and Corcuera, L.J. (1995) Plant proteinase inhibitors: a defensive response against insects. *Biol. Res.*, **28**, 239-249.

Ceci, L.R., Spoto, N., de Virgilio, M. and Gallerani, R. (1995) The gene coding for the mustard trypsin inhibitor-2 is discontinuous and wound-inducible. *FEBS Lett.*, **364**, 179-181.

Ceciliani, F., Bortolotti, F., Menegatti, E., Ronchi, S., Ascenzi, P. and Palmieri, S. (1994) Purification, inhibitory properties, amino acid sequence and identification of the reactive site of a new serine proteinase inhibitor from oil-rape (*Brassica napus*) seed. *FEBS Lett.*, **342**, 221-224.

Chen, P., Rose, J., Love, R., Wei, C.H. and Wang, B.C. (1992) Reactive sites of an anticarcinogenic Bowman-Birk proteinase inhibitor are similar to other trypsin inhibitors. *J. Biol. Chem.*, **267**, 1990-1994.

Choi, D., Park, J.A., Seo, Y.S., Chun, Y.J. and Kim, W.T. (2000) Structure and stress-related expression of two cDNAs encoding proteinase inhibitor II of *Nicotiana glutinosa* L. *Biochim. Biophys. Acta*, **1492**, 211-215.

Chrispeels, M.J. and Herman, E.M. (2000) Endoplasmic reticulum-derived compartments function in storage and as mediators of vacuolar remodeling via a new type of organelle, precursor protease vesicles. *Plant Physiol.*, **123**, 1227-1234.

Christeller, J.T., Laing, W.A., Markwick, N.P. and Burgess, E.P.J. (1992) Midgut protease activities in 12 phytophagous lepidopteran larvae: dietary and protease inhibitor interactions. *Insect Biochem. Mol. Biol.*, **22**, 735-746.

Christeller, J.T., Farley, P.C., Ramsay, R.J., Sullivan, P.A. and Laing, W.A. (1998) Purification, characterization and cloning of an aspartic proteinase inhibitor from squash phloem exudate. *Eur. J. Biochem.*, **254**, 160-167.

Clore, G.M., Gronenborn, A.M., Nilges, M. and Ryan, C.A. (1987) Three-dimensional structure of potato carboxypeptidase inhibitor in solution. A study using nuclear magnetic resonance, distance geometry, and restrained molecular dynamics. *Biochemistry*, **26**, 8012-8023.

Cordeiro, M., Lowther, T., Dunn, B.M. *et al.* (1998) Substrate specificity and molecular modelling of aspartic proteinases (cyprosins) from flowers of *Cynara cardunculus* subsp. *Flavescens* cv. cardoon. *Adv. Exp. Med. Biol.*, **436**, 473-479.

Cordero, M.J., Raventos, D. and San Segundo, B. (1994) Expression of a maize proteinase inhibitor gene is induced in response to wounding and fungal infection: systemic wound-response of a monocot gene. *Plant J.*, **6**, 141-150.

Dahl, S.W., Rasmussen, S.K. and Hejgaard, J. (1996a) Heterologous expression of three plant serpins with distinct inhibitory specificities. *J. Biol. Chem.*, **271**, 25,083-25,088.

Dahl, S.W., Rasmussen, S.K., Petersen, L.C. and Hejgaard, J. (1996b) Inhibition of coagulation factors by recombinant barley serpin BSZx. *FEBS Lett.*, **394**, 165-168.

Dammann, C., Rojo, E. and Sanchez-Serrano, J.J. (1997) Abscisic acid and jasmonic acid activate wound-inducible genes in potato through separate organ-specific signal transduction pathways. *Plant J.*, **11**, 101-110.

Downing, W.L., Mauxion, F., Fauvarque, M.O. *et al.* (1992) A *Brassica napus* transcript encoding a protein related to the Kunitz protease inhibitor family accumulates upon water stress in leaves, not in seeds. *Plant J.*, **2**, 685-693.

Dymock, J.J., Laing, W.A., Christeller, J.T. and Shaw, B.D. (1989) The effect of trypsin inhibitors on grass grub (*Costelytra zealandica* (White)) larval growth and trypsin activity, in *Proceedings of the Forty Second New Zealand Weed and Pest Control Conference, New Plymouth, 8-10 August, 1989*, pp. 67-70.

Eckelkamp, C., Ehmann, B. and Schopfer, P. (1993) Wound-induced systemic accumulation of a transcript coding for a Bowman–Birk trypsin inhibitor-related protein in maize (*Zea mays L.*) seedlings. *FEBS Lett.*, **323**, 73-76.

Faro, C., Ramalho-Santos, M., Verissimo, P. *et al.* (1998) Structural and functional aspects of cardosins. *Adv. Exp. Med. Biol.*, **436**, 423-433.

Favel, A., Mattras, H., Coletti-Previero, M.A., Zwilling, R., Robinson, E.A. and Castro, B. (1989) Protease inhibitors from *Ecballium elaterium* seeds. *Int. J. Pept. Protein Res.*, **33**, 202-208.

Felizmenio-Quimio, M.E., Daly, N.L. and Craik, D.J. (2001) Circular proteins in plants. solution structure of a novel macrocyclic trypsin inhibitor from *Momordica cochinchinensis. J. Biol. Chem.*, **276**, 22,875-22,882.

Fu, H., Girod, P.A., Doelling, J.H. *et al.* (1999) Structure and functional analysis of the 26S proteasome subunits from plants. *Mol. Biol. Rep.*, **26**, 137-146.

Galleschi, L. (1998) Proteolysis in germinating cereal grains. *Recent Research Developments in Phytochemistry*, **2**, 95-106.

Galleschi, L., Friggeri, M., Repiccioli, R. and Come, D. (1993) Aspartic proteinase inhibitor from wheat: some properties, in *Proceedings of the Fourth International Workshop of Seeds: Basic and applied aspects of seed biology* (ed. F. Corbineau) Angers, France, pp. 207-211.

Garcia-Olmedo, F., Salcedo, G., Sanchez-Monge, R., Gomez, L., Royo, J. and Carbonero, P. (1987) Plant proteinaceous inhibitors of proteinases and alpha-amylases. *Oxf. Surv. Plant Mol. Cell. Biol.*, **4**, 275-334.

Gatehouse, A.M.R., Gatehouse, J.A., Dobie, P., Kilminster, A.M., Boulter, D. and Baker, A.M.R. (1979) Biochemical basis of insect resistance in *Vigna unguiculata. J. Sci. Food Agri.*, **30**, 948-958.

Gatehouse, L.M., Shannon, A.L., Burgess, E.P.J. and Christeller, J.T. (1997) Characterization of major midgut proteinase cDNAs from *Helicoverpa armigera* larvae and changes in gene expression in response to four proteinase inhibitors in the diet. *Insect Biochem. Mol. Biol.*, **27**, 929-944.

Gautier, M.F., Alary, R. and Joudrier, P. (1990) Cloning and characterization of a cDNA encoding the wheat (*Triticum durum* Desf.) CM16 protein. *Plant Mol. Biol.*, **14**, 313-322.

Glaser, E., Sjoling, S., Tanudji, M. and Whelan, J. (1998) Mitochondrial protein import in plants. Signals, sorting, targeting, processing and regulation. *Plant Mol. Biol.*, **38**, 311-338.

Goodwin, R.L., Baumann, H. and Berger, F.G. (1996) Patterns of divergence during evolution of alpha 1-proteinase inhibitors in mammals. *Mol. Biol. Evol.*, **13**, 346-358.

Gourinath, S., Srinivasan, A. and Singh, T.P. (1999) Structure of the bifunctional inhibitor of trypsin and alpha-amylase from ragi seeds at 2.9 Å resolution. *Acta Crystallogr. D Biol. Crystallogr.*, **55**, 25-30.

Gourinath, S., Alam, N., Srinivasan, A., Betzel, C. and Singh, T.P. (2000) Structure of the bifunctional inhibitor of trypsin and alpha-amylase from ragi seeds at 2.2 Å resolution. *Acta Crystallogr. D Biol. Crystallogr.*, **56**, 287-293.

Graham, J.S., Pearce, G., Merryweather, J., Titani, K., Ericsson, L. and Ryan, C.A. (1985a) Wound-induced proteinase inhibitors from tomato leaves. I. The cDNA-deduced primary structure of pre-inhibitor I and its post-translational processing. *J. Biol. Chem.*, **260**, 6555-6560.

Graham, J.S., Pearce, G., Merryweather, J., Titani, K., Ericsson, L.H. and Ryan, C.A. (1985b) Wound-induced proteinase inhibitors from tomato leaves. II. The cDNA-deduced primary structure of pre-inhibitor II. *J. Biol. Chem.*, **260**, 6561-6564.

Greenblatt, H.M., Ryan, C.A. and James, M.N. (1989) Structure of the complex of *Streptomyces griseus* proteinase B and polypeptide chymotrypsin inhibitor-1 from Russet Burbank potato tubers at 2.1 Å resolution. *J. Mol. Biol.*, **205**, 201-228.

Gruden, K., Strukelj, B., Ravnikar, M. *et al.* (1997) Potato cysteine proteinase inhibitor gene family: molecular cloning, characterisation and immunocytochemical localisation studies. *Plant Mol. Biol.*, **34**, 317-323.

Gruden, K., Strukelj, B., Popovic, T. *et al.* (1998) The cysteine protease activity of Colorado potato beetle (*Leptinotarsa decemlineata* Say) guts, which is insensitive to potato protease inhibitors, is inhibited by thyroglobulin type-1 domain inhibitors. *Insect Biochem. Mol. Biol.*, **28**, 549-560.

Gutierrez Campos, R., Torres Acosta, J., Perez Martinez, J.D. and Gomez Lim, M.A. (2001) Pleiotropic effects in transgenic tobacco plants expressing the oryzacystatin I gene. *Hortscience*, **36**, 118-119.

Habu, Y., Peyachoknagul, S., Umemoto, K., Sakata, Y. and Ohno, T. (1992) Structure and regulated expression of Kunitz chymotrypsin inhibitor genes in winged bean [*Psophocarpus tetragonolobus* (L.) DC.]. *J. Biochem. (Tokyo)*, **111**, 249-258.

Hamato, N., Koshiba, T., Pham, T.N. *et al.* (1995) Trypsin and elastase inhibitors from bitter gourd (*Momordica charantia* LINN.) seeds: purification, amino acid sequences, and inhibitory activities of four new inhibitors. *J. Biochem.*, **117**, 432-437.

Hara, S., Makino, J. and Ikenaka, T. (1989) Amino acid sequences and disulfide bridges of serine proteinase inhibitors from bitter gourd (*Momordica charantia* LINN.) seeds. *J. Biochem.*, **105**, 88-91.

Hass, G.M., Nau, H., Biemann, K., Grahn, D.T., Ericsson, L.H. and Neurath, H. (1975) The amino acid sequence of a carboxypeptidase inhibitor from potatoes. *Biochemistry*, **14**, 1334-1342.

Hass, G.M., Derr, J.E., Makus, D.J. and Ryan, C.A. (1979) Purification and characterization of the carboxypeptidase isoinhibitors from potatoes. *Plant Physiol.*, **64**, 1022-1028.

Hass, G.M., Hermodson, M.A., Ryan, C.A. and Gentry, L. (1982) Primary structures of two low molecular weight proteinase inhibitors from potatoes. *Biochemistry*, **21**, 752-756.

Hatakeyama, T., Hiraoka, M. and Funatsu, G. (1991) Amino acid sequences of the two smallest trypsin inhibitors from sponge gourd seeds. *Agric. Biol. Chem.*, **55**, 2641-2642.

Hatano, K., Kojima, M., Tanokura, M. and Takahashi, K. (1995) Primary structure, sequence-specific 1H-NMR assignments and secondary structure in solution of bromelain inhibitor VI from pineapple stem. *Eur. J. Biochem.*, **232**, 335-343.

Hatano, K., Kojima, M., Tanokura, M. and Takahashi, K. (1996) Solution structure of bromelain inhibitor IV from pineapple stem: structural similarity with Bowman–Birk trypsin/chymotrypsin inhibitor from soybean. *Biochemistry*, **35**, 5379-5384.

Hayashi, K., Takehisa, T., Hamato, N. *et al.* (1994) Inhibition of serine proteases of the blood coagulation system by squash family protease inhibitors. *J. Biochem.*, **116**, 1013-1018.

Heath, R.L., Barton, P.A., Simpson, R.J., Reid, G.E., Lim, G. and Anderson, M.A. (1995) Characterization of the protease processing sites in a multidomain proteinase inhibitor precursor from *Nicotiana alata*. *Eur. J. Biochem.*, **230**, 250-257.

Heinrikson, R.L. and Kezdy, F.J. (1976) Acidic cysteine protease inhibitors from pineapple stem. *Methods Enzymol.*, **45**, 740-751.

Hejgaard, J. (2001) Inhibitory serpins from rye grain with glutamine as P1 and P2 residues in the reactive center. *FEBS Lett.*, **488**, 149-153.

Hejgaard, J., Rasmussen, S.K., Brandt, A. and Svendsen, I. (1985) Sequence homology between barley endosperm protein Z and protease inhibitors of the alpha1-antitrypsin family. *FEBS Lett.*, **180**, 89-94.

Hejgaard, J., Jacobsen, S. and Svendsen, I. (1991) Two antifungal thaumatin-like proteins from barley grain. *FEBS Lett.*, **291**, 127-131.

Helland, R., Berglund, G.I., Otlewski, J. *et al.* (1999) High-resolution structures of three new trypsin-squash-inhibitor complexes: a detailed comparison with other trypsins and their complexes. *Acta Crystallogr. D Biol. Crystallogr.*, **55**, 139-148.

Hengst, U., Albrecht, H., Hess, D. and Monard, D. (2001) The phosphatidylethanolamine-binding protein is the prototype of a novel family of serine protease inhibitors. *J. Biol. Chem.*, **276**, 535-540.

Hernandez, J.F., Gagnon, J., Chiche, L. *et al.* (2000) Squash trypsin inhibitors from *Momordica cochinchinensis* exhibit an atypical macrocyclic structure. *Biochemistry*, **39**, 5722-5730.

Heussen, C., Joubert, F. and Dowdle, E.B. (1984) Purification of human tissue plasminogen activator with *Erythrina trypsin* inhibitor. *J. Biol. Chem.*, **259**, 11,635-11,638.

Hilder, V.A., Gatehouse, A.M.R., Sheerman, S.E., Barker, R.F. and Boulter, D. (1987) A novel mechanism of insect resistance engineered into tobacco. *Nature*, **330**, 160-163.

Hilder, V.A., Barker, R.F., Samour, R.A., Gatehouse, A.M., Gatehouse, J.A. and Boulter, D. (1989) Protein and cDNA sequences of Bowman–Birk protease inhibitors from the cowpea (*Vigna unguiculata* Walp.). *Plant Mol. Biol.*, **13**, 701-710.

Holak, T.A., Gondol, D., Otlewski, J. and Wilusz, T. (1989) Determination of the complete three-dimensional structure of the trypsin inhibitor from squash seeds in aqueous solution by nuclear magnetic resonance and a combination of distance geometry and dynamical simulated annealing. *J. Mol. Biol.*, **210**, 635.

Huang, Q., Liu, S. and Tang, Y. (1993) Refined 1.6-Å resolution crystal structure of the complex formed between porcine beta-trypsin and MCTI-A, a trypsin inhibitor of the squash family. Detailed comparison with bovine beta-trypsin and its complex. *J. Mol. Biol.*, **229**, 1022-1030.

Huffaker, R.C. (1990) Tansley Review No. 25. Proteolytic activity during senescence of plants. *New Phytol. Cambridge*, **116**, 199-231.

Huynh, Q.K., Borgmeyer, J.R. and Zobel, J.F. (1992) Isolation and characterization of a 22 kDa protein with antifungal properties from maize seeds. *Biochem. Biophys. Res. Commun.*, **182**, 1-5.

Initiative (2000) Analysis of the genome sequence of the flowering plant *Arabidopsis thaliana*. The *Arabidopsis* Genome Initiative. *Nature*, **408**, 796-815.

Irving, J.A., Pike, R.N., Lesk, A.M. and Whisstock, J.C. (2000) Phylogeny of the serpin superfamily: implications of patterns of amino acid conservation for structure and function. *Genome Res.*, **10**, 1845-1864.

Ishikawa, A., Ohta, S., Matsuoka, K., Hattori, T. and Nakamura, K. (1994) A family of potato genes that encode Kunitz-type proteinase inhibitors: structural comparisons and differential expression. *Plant Cell Physiol.*, **35**, 303-312.

Iwanaga, S., Nagata, R., Miyamoto, A., Kouzuma, Y., Yamasaki, N. and Kimura, M. (1999) Conformation of the primary binding loop folded through an intramolecular interaction contributes to the strong chymotrypsin inhibitory activity of the chymotrypsin inhibitor from *Erythrina variegata* seeds. *J. Biochem.*, **26**, 162-167.

Jensen, B., Unger, K.K., Uebe, J., Gey, M., Kim, Y.M. and Flecker, P. (1996) Proteolytic cleavage of soybean Bowman–Birk inhibitor monitored by means of high-performance capillary electrophoresis. Implications for the mechanism of proteinase inhibitors. *J. Biochem. Biophys. Methods*, **33**, 171-185.

Johnson, R., Narvaez, J., An, G. and Ryan, C. (1990) Expression of proteinase inhibitor genes from potato and tomato in transgenic plants enhances defence against an insect predator, in *The molecular and cellular biology of the potato* (eds M. E. Vayda and W. D. Park), Biotechnology in Agriculture No. 3, CAB International, Wallingford, UK, pp. 97-102.

Jongsma, M.A. and Bolter, C. (1997) The adaptation of insects to plant protease inhibitors. *J. Insect Physiol.*, **43**, 885-895.

Jongsma, M.A., Bakker, P.L., Peters, J., Bosch, D. and Stiekema, W.J. (1995) Adaptation of *Spodoptera exigua* larvae to plant proteinase inhibitors by induction of gut proteinase activity insensitive to inhibition. Combatting inhibitor-insensitive proteases of insect pests. *Proc. Natl Acad. Sci. USA*, **92**, 8041-8045.

Jongsma, M.A., Stiekema, W.J. and Bosch, D. (1996) Combatting inhibitor-insensitive proteases of insect pests. *Trends Biotech.*, **14**, 331-333.

Joshi, B.N., Sainani, M.N., Bastawade, K.B., Gupta, V.S. and Ranjekar, P.K. (1998) Cysteine protease inhibitor from pearl millet: a new class of antifungal protein. *Biochem. Biophys. Res. Commun.*, **246**, 382-387.

Joshi, B.N., Sainani, M.N., Bastawade, K.B., Deshpande, V.V., Gupta, V.S. and Ranjekar, P.K. (1999) Pearl millet cysteine protease inhibitor. Evidence for the presence of two distinct sites responsible for anti-fungal and anti-feedent activities. *Eur. J. Biochem.*, **265**, 556-563.

Joubert, F.J. (1984) Trypsin isoinhibitors from *Mormodica repens* seeds. *Phytochemistry*, **23**, 1401-1410.

Karrer, E.E., Beachy, R.N. and Holt, C.A. (1998) Cloning of tobacco genes that elicit the hypersensitive response. *Plant Mol. Biol.*, **36**, 681-690.

Katayama, H., Soezima, Y., Fujimura, S., Terada, S. and Kimoto, E. (1994) Property and amino acid sequence of a subtilisin inhibitor from seeds of beach canavalia (*Canavalia lineata*). *Biosci. Biotechnol. Biochem.*, **58**, 2004-2008.

Keegstra, K. and Froehlich, J.E. (1999) Protein import into chloroplasts. *Curr. Opin. Plant Biol.*, **2**, 471-476.

Keilova, H. and Tomasek, V. (1976) Isolation and some properties of cathepsin D inhibitor from potatoes. *Collect. Czech. Chem. Commun.*, **41**, 487-497.

Kervinen, J., Tormakangas, K, Runeberg-Roos, P., Guruprasad, K., Blundell, T. and Teeri, T.H. (1995) Structure and possible function of aspartic proteinases in barley and other plants. *Adv. Exp. Med. Biol.*, **362**, 241-254.

Kim, S.H., Hara, S., Hase, S. *et al.* (1985) Comparative study on amino acid sequences of Kunitz-type soybean trypsin inhibitors, Tia, Tib, and Tic. *J. Biochem.*, **98**, 435-448.

Kimura, M., Kouzuma, Y. and Yamasaki, N. (1993) Amino acid sequence of chymotrypsin inhibitor ECI from the seeds of *Erythrina variegata* (Linn.) var. *Orientalis. Biosci. Biotechnol. Biochem.*, **57**, 102-106.

Koepke, J., Ermler, U., Warkentin, E., Wenzl, G. and Flecker, P. (2000) Crystal structure of cancer chemopreventive Bowman–Birk inhibitor in ternary complex with bovine trypsin at 2.3Å resolution. Structural basis of Janus-faced serine protease inhibitor specificity. *J. Mol. Biol.*, **298**, 477-491.

Kondo, H., Abe, K., Nishimura, I., Watanabe, H., Emori, Y. and Arai, S. (1990) Two distinct cystatin species in rice seeds with different specificities against cysteine proteinases. Molecular cloning, expression, and biochemical studies on oryzacystatin-II. *J. Biol. Chem.*, **265**, 15,832-15,837.

Kreft, S., Ravnikar, M., Mesko, P. *et al.* (1997) Jasmonic acid inducible aspartic proteinase inhibitors from potato. *Phytochemistry*, **44**, 1001-1006.

Krishnamoorthi, R., Gong, Y.X. and Richardson, M. (1990) A new protein inhibitor of trypsin and activated Hageman factor from pumpkin (*Cucurbita maxima*) seeds. *FEBS Lett.*, **273**, 163-167.

Krishnamoorthi, R., Gong, Y.X., Lin, C.L. and VanderVelde, D. (1992) Two-dimensional NMR studies of squash family inhibitors. Sequence-specific proton assignments and secondary structure of reactive-site hydrolyzed *Cucurbita maxima* trypsin inhibitor III. *Biochemistry*, **31**, 898-904.

Krizaj, I., Drobnic-Kosorok, M., Brzin, J., Jerala, R. and Turk, V. (1993) The primary structure of inhibitor of cysteine proteinases from potato. *FEBS Lett.*, **333**, 15-20.

Kuroda, M., Kiyosaki, T., Matsumoto, I., Misaka, T., Arai, S. and Abe, K. (2001) Molecular cloning, characterization, and expression of wheat cystatins. *Biosci. Biotechnol. Biochem.*, **65**, 22-28.

Laskowski, M. and Kato, I. (1980) Protein inhibitors of proteinases. *Annu. Rev. Biochem.*, **49**, 593-626.

Lazaro, A., Rodriguez-Palenzuela, P., Marana, C., Carbonero, P. and Garcia-Olmedo, F. (1988) Signal peptide homology between the sweet protein thaumatin II and unrelated cereal alpha-amylase/trypsin inhibitors. *FEBS Lett.*, **239**, 147-150.

Leah, R. and Mundy, J. (1989) The bifunctional alpha-amylase/subtilisin inhibitor of barley: nucleotide sequence and patterns of seed-specific expression. *Plant Mol. Biol.*, **12**, 673-682.

Lee, M.C., Scanlon, M.J., Craik, D.J. and Anderson, M.A. (1999) A novel two-chain proteinase inhibitor generated by circularization of a multidomain precursor protein. *Nat. Struct. Biol.*, **6**, 526-530.

Lenarcic, B. and Bevec, T. (1998) Thyropins—new structurally related proteinase inhibitors. *Biol. Chem.*, **379**, 105-111.

Lenarcic, B. and Turk, V. (1999) Thyroglobulin type-1 domains in equistatin inhibit both papain-like cysteine proteinases and cathepsin D. *J. Biol. Chem.*, **274**, 563-566.

Lenarcic, B., Ritonja, A., Turk, B., Dolenc, I. and Turk, V. (1992) Characterization and structure of pineapple stem inhibitor of cysteine proteinases. *Biol. Chem. Hoppe-Seyler*, **373**, 459-464.

Leon, J., Rojo, E. and Sanchez-Serrano, J.J. (2001) Wound signalling in plants. *J. Exp. Bot.*, **52**, 1-9.

Ling, M.H., Qi, H.Y. and Chi, C.W. (1993) Protein, cDNA, and genomic DNA sequences of the towel gourd trypsin inhibitor. A squash family inhibitor. *J. Biol. Chem.*, **268**, 810-814.

Luckett, S., Garcia, R.S., Barker, J.J. *et al.* (1999) High-resolution structure of a potent cyclic proteinase inhibitor from sunflower seeds. *J. Mol. Biol.*, **290**, 525-533.

Machleidt, W., Assfalg-Machleidt, I. and Auerswald, A. (1993) Kinetics and molecular mechanisms of inhibition of cysteine proteinases by their protein inhibitors, in *Innovations in proteases and their inhibitors* (ed. F.X. Aviles) Walter de Gruyter & Co, Berlin, pp. 179-196.

Maganja, D.B., Strukelj, B., Pungercar, J., Gubensek, F., Turk, V. and Kregar, I. (1992) Isolation and sequence analysis of the genomic DNA fragment encoding an aspartic proteinase inhibitor homologue from potato (*Solanum tuberosum* L.). *Plant Mol. Biol.*, **20**, 311-313.

Manen, J.F., Simon, P., Van Slooten, J.C., Osteras, M., Frutiger, S. and Hughes, G.J. (1991) A nodulin specifically expressed in senescent nodules of winged bean is a protease inhibitor. *Plant Cell*, **3**, 259-270.

Mares, M., Meloun, B., Pavlik, M., Kostka, V. and Baudys, M. (1989) Primary structure of the cathepsin D inhibitor from potatoes and its structural relationship to soybean trypsin inhibitor family. *FEBS Lett.*, **251**, 94-98.

Margis, R., Reis, E.M. and Villeret, V. (1998) Structural and phylogenetic relationships among plant and animal cystatins. *Arch. Biochem. Biophys.*, **359**, 24-30.

Marino-Buslje, C., Venhudova, G., Molina, M.A. *et al.* (2000) Contribution of C-tail residues of potato carboxypeptidase inhibitor to the binding to carboxypeptidase A. A mutagenesis analysis. *Eur. J. Biochem.*, **267**, 1502-1509.

Markwick, N.P., Laing, W.A., Christeller, J.T., McHenry, J.Z. and Newton, M.R. (1998) Overproduction of digestive enzymes compensates for inhibitory effects of protease and alpha-amylase inhibitors fed to three species of leafrollers (Lepidoptera: Tortricidae). *J. Econ. Entomol.*, **91**, 1265-1276.

Martineau, B., McBride, K.E. and Houck, C.M. (1991) Regulation of metallocarboxypeptidase inhibitor gene expression in tomato. *Mol. Gen. Genet.*, **228**, 281-286.

Maskos, K., Huber-Wunderlich, M. and Glockshuber, R. (1996) RBI, a one-domain alpha-amylase/trypsin inhibitor with completely independent binding sites. *FEBS Lett.*, **397**, 11-16.

McManus, M.T., Laing, W.A. and Christeller, J.T. (1994a) Wounding induces a series of closely related trypsin/chymotrypsin inhibitory peptides in leaves of tobacco. *Phytochemistry*, **37**, 921-926.

McManus, M.T., White, D.W.R and McGregor, P.G. (1994b) Chymotrypsin inhibitors are effective insect resistance factors in transgenic plants. *Transgenic Res.*, **3**, 50-54.

McManus, M.T., Laing, W.A., Christeller, J.T. and White, D.W.R. (1994c) Post-translational modification of an iso-inhibitor from the potato proteinase inhibitor II gene family in transgenic tobacco yields a peptide with homology to potato chymotrypsin I (PCI-1). *Plant Physiol.*, **106**, 771-777.

McManus, M.T., Burgess, E.P.J., Philip, B. *et al.* (1999) Expression of the soybean (Kunitz) trypsin inhibitor in transgenic tobacco: Effects on feeding larvae of *Spodoptera litura. Transgenic Res.*, **8**, 383-395.

McManus, M.T., Ryan, S.N. and Laing, W.A. (2000) Proteinase inhibitors as storage proteins in seeds, in *Seed Research in New Zealand, Special Publication, No. 12* (eds M.T. McManus, H.A. Outred and K.M. Pollock), Agronomy Society of New Zealand, pp. 3-14.

McPhalen, C.A. and James, M.N. (1987) Crystal and molecular structure of the serine proteinase inhibitor CI-2 from barley seeds. *Biochemistry*, **26**, 261-269.

McPhalen, C.A. and James, M.N. (1988) Structural comparison of two serine proteinase-protein inhibitor complexes: eglin-c-subtilisin Carlsberg and CI-2-subtilisin Novo. *Biochemistry*, **27**, 6582-6598.

Melville, J.C. and Ryan, C.A. (1970) Chymotrypsin inhibitor 1 from potatoes: a multisite inhibitor composed of subunits. *Arch. Biochem. Biophys.*, **138**, 700-702.

Melville, J.C. and Ryan, C.A. (1972) Chymotrypsin inhibitor I from potatoes. Large scale preparation and characterization of its subunit components. *J. Biol. Chem.*, **247**, 3445-3453.

Menegatti, E., Tedeschi, G., Ronchi, S. *et al.* (1992) Purification, inhibitory properties and amino acid sequence of a new serine proteinase inhibitor from white mustard (*Sinapis alba* L.) seed. *FEBS Lett.*, **301**, 10-14.

Misaka, T., Kuroda, M., Iwabuchi, K., Abe, K. and Arai, S. (1996) Soyacystatin, a novel cysteine proteinase inhibitor in soybean, is distinct in protein structure and gene organization from other cystatins of animal and plant origin. *Eur. J. Biochem.*, **240**, 609-614.

Mitsumori, C., Yamagishi, K., Fujino, K. and Kikuta, Y. (1994) Detection of immunologically related Kunitz and Bowman–Birk proteinase inhibitors expressed during potato tuber development. *Plant Mol. Biol.*, **26**, 961-969.

Mosolov, V.V. (1998) New studies on natural inhibitors of proteolytic enzymes. *Bioorg. Khim.*, **24**, 332-340.

Murray, C. and Christeller, J.T. (1995) Purification of a trypsin inhibitor (PFTI) from pumpkin fruit phloem exudate and isolation of putative trypsin and chymotrypsin inhibitor cDNA clones. *Biol. Chem. Hoppe-Seyler*, **376**, 281-287.

Murzin, A.G. (1993) Sweet-tasting protein monellin is related to the cystatin family of thiol proteinase inhibitors. *J. Mol. Biol.*, **230**, 689-694.

Mutlul, A. and Gal, S. (1999) Plant aspartic proteinases: enzymes on the way to a function. *Physiologia Plantarum*, **105**, 569-576.

Nagasue, A., Fukamachi, H., Ikenaga, H. and Funatsu, G. (1988) The amino acid sequence of barley rootlet trypsin inhibitor. *Agric. Biol. Chem.*, **52**, 1505-1514.

Nagata, K., Kudo, N., Abe, K., Arai, S. and Tanokura, M. (2000) Three-dimensional solution structure of oryzacystatin-I: a cysteine proteinase inhibitor of the rice, *Oryza sativa* L. *japonica. Biochemistry*, **39**, 14,753-14,760.

Nair, S.A. and Cooperman, B.S. (1998) Antichymotrypsin interaction with chymotrypsin. Reactions following encounter complex formation. *J. Biol. Chem.*, **273**, 17,459-17,462.

Nielsen, S.S. (1988) Degradation of bean proteins by endogenous and exogenous proteases—a review. *Cereal Chem.*, **65**, 435-442.

Nielsen, K.J., Heath, R.L., Anderson, M.A. and Craik, D.J. (1994) The three-dimensional solution structure by 1H NMR of a 6-kDa proteinase inhibitor isolated from the stigma of *Nicotiana alata. J. Mol. Biol.*, **242**, 231-243.

Nielsen, K.J., Heath, R.L., Anderson, M.A. and Craik, D.J. (1995) Structures of a series of 6-kDa trypsin inhibitors isolated from the stigma of *Nicotiana alata. Biochemistry*, **34**, 14,304-14,311.

Oda, Y., Matsunaga, T., Fukuyama, K., Miyazaki, T. and Morimoto, T. (1997) Tertiary and quaternary structures of 0.19 alpha-amylase inhibitor from wheat kernel determined by X-ray analysis at 2.06 Å resolution. *Biochemistry*, **36**, 13,503-13,511.

Odani, S. and Ikenaka, T. (1973) Scission of soybean Bowman–Birk proteinase inhibitor into two small fragments having either trypsin or chymotrypsin inhibitory activity. *J. Biochem. (Tokyo)*, **74**, 857-860.

Odani, S., Koide, T. and Ono, T. (1983) The complete amino acid sequence of barley trypsin inhibitor. *J. Biol. Chem.*, **258**, 7998-8003.

Odani, S., Koide, T. and Ono, T. (1986) Wheat germ trypsin inhibitors. Isolation and structural characterization of single-headed and double-headed inhibitors of the Bowman–Birk type. *J. Biochem. (Tokyo)*, **100**, 975-983.

Odani, S., Yokokawa, Y., Takeda, H. and Abe, S. (1996) The primary structure and characterization of carbohydrate chains of the extracellular glycoprotein proteinase inhibitor from latex of *Carica papaya*. *Eur. J. Biochem.*, **241**, 77-82.

Ohtsubo, K. and Richardson, M. (1992) The amino acid sequence of a 20 kDa bifunctional subtilisin/alpha-amylase inhibitor from bran of rice (*Oryza sativa* L.) seeds. *FEBS Lett.*, **309**, 68-72.

Ojima, A., Shiota, H., Higashi, K. *et al.* (1997) An extracellular insoluble inhibitor of cysteine proteinases in cell cultures and seeds of carrot. *Plant Mol. Biol.*, **34**, 99-109.

Ostergaard, H., Rasmussen, S.K., Roberts, T.H. and Hejgaard, J. (2000) Inhibitory serpins from wheat grain with reactive centers resembling glutamine-rich repeats of prolamin storage proteins. Cloning and characterization of five major molecular forms. *J. Biol. Chem.*, **275**, 33,272-33,279.

Otlewski, J. (1990) The squash inhibitor family of serine proteinases. *Biol. Chem. Hoppe-Seyler*, **371 Suppl.**, 23-28.

Otlewski, J. and Krowarsch, D. (1996) Squash inhibitor family of serine proteinases. *Acta Biochim. Pol.*, **43**, 431-444.

Otlewski, J. and Zbyryt, T. (1994) Single peptide bond hydrolysis/resynthesis in squash inhibitors of serine proteinases. 1. Kinetics and thermodynamics of the interaction between squash inhibitors and bovine beta-trypsin. *Biochemistry*, **33**, 200-207.

Otlewski, J., Whatley, H., Polanowski, A. and Wilusz, T. (1987) Amino-acid sequences of trypsin inhibitors from watermelon (*Citrullus vulgaris*) and red bryony (*Bryonia dioica*) seeds. *Biol. Chem. Hoppe-Seyler*, **368**, 1505-1507.

Pak, J.H., Liu, C.Y., Huangpu, J. and Graham, J.S. (1997) Construction and characterization of the soybean leaf metalloproteinase cDNA. *FEBS Lett.*, **404**, 283-288.

Pearce, G., Johnson, S. and Ryan, C.A. (1993) Purification and characterization from tobacco (*Nicotiana tabacum*) leaves of six small, wound-inducible, proteinase isoinhibitors of the potato inhibitor II family. *Plant Physiol.*, **102**, 639-644.

Peterson, D.M., Forde, J., Williamson, M.S., Rohde, W. and Kreis, M. (1991) Nucleotide sequence of a chymotrypsin inhibitor-2 gene of barley (*Hordeum vulgare* L.). *Plant Physiol.*, **96**, 1389-1390.

Plunkett, G., Senear, D.F., Zuroske, G. and Ryan, C.A. (1982) Proteinase inhibitors I and II from leaves of wounded tomato plants: purification and properties. *Arch. Biochem. Biophys.*, **213**, 463-472.

Rackis, J.J. and Anderson, R.L. (1964) Isolation of four soybean trypsin inhibitors by DEAE-cellulose chromatography. *Biochem. Biophys. Res. Commun.*, **15**, 230-235.

Rasmussen, S.K. (1993) A gene coding for a new plant serpin. *Biochim. Biophys. Acta*, **1172**, 151-154.

Rasmussen, S.K., Dahl, S.W., Norgard, A. and Hejgaard, J. (1996a) A recombinant wheat serpin with inhibitory activity. *Plant Mol. Biol.*, **30**, 673-677.

Rasmussen, S.K., Klausen, J., Hejgaard, J., Svensson, B. and Svendsen, I. (1996b) Primary structure of the plant serpin BSZ7 having the capacity of chymotrypsin inhibition. *Biochim. Biophys. Acta*, **1297**, 127-130.

Rebmann, G., Mauch, F. and Dudler, R. (1991) Sequence of a wheat cDNA encoding a pathogen-induced thaumatin-like protein. *Plant Mol. Biol.*, **17**, 283-285.

Reddy, M.N., Keim, P.S., Heinrikson, R.L. and Kezdy, F.J. (1975) Primary structural analysis of sulfhydryl protease inhibitors from pineapple stem. *J. Biol. Chem.*, **250**, 1741-1750.

Rees, D.C., Lewis, M. and Lipscomb, W.N. (1983) Refined crystal structure of carboxypeptidase A at 1.54 Å resolution. *J. Mol. Biol.*, **168**, 367-387.

Richardson, M. (1974) Chymotryptic inhibitor I from potatoes. The amino acid sequence of subunit A. *Biochem. J.*, **137**, 101-112.

Richardson, M. (1979) The complete amino acid sequence and the trypsin reactive (inhibitory) site of the major proteinase inhibitor from the fruits of aubergine (*Solanum melongena* L.). *FEBS Lett.*, **104**, 322-326.

Richardson, M. and Cossins, L. (1974) Chymotryptic inhibitor I from potatoes: the amino acid sequences of subunits B, C, and D. *FEBS Lett.*, **45**, 11-13.

Richardson, M., Valdes-Rodriguez, S. and Blanco-Labra, A. (1987) A possible function for thaumatin and a TMV-induced protein suggested by homology to a maize inhibitor. *Nature*, **327**, 432-434.

Ritonja, A., Krizaj, I., Mesko, P. *et al.* (1990) The amino acid sequence of a novel inhibitor of cathepsin D from potato. *FEBS Lett.*, **267**, 13-15.

Rohrmeier, T. and Lehle, L. (1993) WIP1, a wound-inducible gene from maize with homology to Bowman–Birk proteinase inhibitors. *Plant Mol. Biol.*, **22**, 783-792.

Rosenkrands, I., Hejgaard, J., Rasmussen, S.K. and Bjorn, S.E. (1994) Serpins from wheat grain. *FEBS Lett.*, **343**, 75-80.

Ryan, C.A. (1984) Defense responses of plants, in *Genes involved in microbe plant interactions* (ed. D.P.S. Verma and Th. Hohn), Springer Verlag, Wein, pp. 375-386.

Ryan, C.A. (1989) Proteinase inhibitor gene families: strategies for transformation to improve plant defenses against herbivores. *BioEssays*, **10**, 1020-1024.

Ryan, C.A. (1990) Protease inhibitors in plants: genes for improving defenses against insects and pathogens. *Annu. Rev. Phytopathol.*, **28**, 425-449.

Ryan, C.A. (2000) The systemin signaling pathway: differential activation of plant defensive genes. *Biochim. Biophys. Acta*, **1477**, 112-121.

Ryan, C.A., Hass, G.M. and Kuhn, R.W. (1974) Purification and properties of a carboxypeptidase inhibitor from potatoes. *J. Biol. Chem.*, **249**, 5495-5499.

Ryan, S.N., Laing, W.A. and McManus, M.T. (1998) A cysteine proteinase inhibitor purified from apple fruit. *Phytochemistry*, **49**, 957-963.

Saitoh, E., Isemura, S., Sanada, K. and Ohnishi, K. (1991) The human cystatin gene family: cloning of three members and evolutionary relationship between cystatins and Bowman–Birk type proteinase inhibitors. *Biomed. Biochim. Acta*, **50**, 599-605.

Sanchez de la Hoz, P., Castagnaro, A. and Carbonero, P. (1994) Sharp divergence between wheat and barley at loci encoding novel members of the trypsin/alpha-amylase inhibitors family. *Plant Mol. Biol.*, **26**, 1231-1236.

Scanlon, M.J., Lee, M.C., Anderson, M.A. and Craik, D.J. (1999) Structure of a putative ancestral protein encoded by a single sequence repeat from a multidomain proteinase inhibitor gene from *Nicotiana alata*. *Structure Fold Des.*, **7**, 793-802.

Schick, C., Pemberton, P.A., Shi, G.P. *et al.* (1998) Cross-class inhibition of the cysteine proteinases cathepsins K, L, and S by the serpin squamous cell carcinoma antigen 1: a kinetic analysis. *Biochemistry*, **37**, 5258-5266.

Schimoler-O'Rourke, R., Richardson, M. and Selitrennikoff, C.P. (2001) Zeamatin inhibits trypsin and alpha amylase activities. *Appl. Environ. Microbiol.*, **67**, 2365-2366.

Schirra, H.J., Scanlon, M.J., Lee, M.C., Anderson, M.A. and Craik, D.J. (2001) The solution structure of C1-T1, a two-domain proteinase inhibitor derived from a circular precursor protein from Nicotiana alata. *J. Mol. Biol.*, **306**, 69-79.

Schu, P., Suarez Rendueles, P. and Wolf, D.H. (1991) The proteinase yscB inhibitor (PB12 gene of yeast and studies on the function of its protein product. *Eur. J. Biochem.*, **197**, 1-7.

Seemuller, U., Eulitz, M., Fritz, H. and Strobl, A. (1980) Structure of the elastase-cathepsin G inhibitor of the leech Hirudo medicinalis. *Hoppe-Seylers Z. Physiol. Chem.*, **361**, 1841-1846.

Shimazaki, A., Makino, Y., Omichi, K., Odani, S. and Hase, S. (1999) A new sugar chain of the proteinase inhibitor from latex of *Carica papaya*. *J. Biochem. (Tokyo)*, **125**, 560-565.

Skern, T., Zorn, M., Blaas, D., Kuechler, E. and Sommergruber, W. (1990) Protease or protease inhibitor? *Nature*, **344**, 26.

Song, H.K. and Suh, S.W. (1998) Kunitz-type soybean trypsin inhibitor revisited: refined structure of its complex with porcine trypsin reveals an insight into the interaction between a homologous inhibitor from *Erythrina caffra* and tissue-type plasminogen activator. *J. Mol. Biol.*, **275**, 347-363.

Song, H.K., Kim, Y.S., Yang, J.K., Moon, J., Lee, J.Y. and Suh, S.W. (1999) Crystal structure of a 16 kDa double-headed Bowman–Birk trypsin inhibitor from barley seeds at 1.9 Å resolution. *J. Mol. Biol.*, **293**, 1133-1144.

Sottrup-Jensen, L. (1989) Alpha-macroglobulins: structure, shape, and mechanism of proteinase complex formation. *J. Biol. Chem.*, **264**, 11,539-11,542.

Srinivasan, A., Raman, A. and Singh, T.P. (1991) Preliminary X-ray investigation of a bifunctional inhibitor from Indian finger millet (ragi). *J. Mol. Biol.*, **222**, 1-2.

Strobl, S., Muhlhahn, P., Bernstein, R. *et al.* (1995) Determination of the three-dimensional structure of the bifunctional alpha-amylase/trypsin inhibitor from ragi seeds by NMR spectroscopy. *Biochemistry*, **34**, 8281-8293.

Strukelj, B., Pungercar, J., Ritonja, A. *et al.* (1990) Nucleotide and deduced amino acid sequence of an aspartic proteinase inhibitor homologue from potato tubers (*Solanum tuberosum* L.). *Nucleic Acids Res.*, **18**, 4605.

Strukelj, B., Pungercar, J., Mesko, P. *et al.* (1992) Characterization of aspartic proteinase inhibitors from potato at the gene, cDNA and protein levels. *Biol. Chem. Hoppe-Seyler*, **373**, 477-482.

Strukelj, B., Ravnikar, M., Mesko, P. *et al.* (1995) Molecular cloning and immunocytochemical localization of jasmonic acid inducible cathepsin D inhibitors from potato. *Adv. Exp. Med. Biol.*, **362**, 293-298.

Stubbs, M.T., Laber, B., Bode, W. *et al.* (1990) The refined 2.4 Å X-ray crystal structure of recombinant human stefin B in complex with the cysteine proteinase papain: a novel type of proteinase inhibitor interaction. *EMBO J.*, **9**, 1939-1947.

Svendsen, I., Jonassen, I., Hejgaard, J. and Boisen, S. (1980) Amino acid sequence homology between a serine protease inhibitor from barley and potato inhibitor I. *Carlsberg Research Communications*, **45**, 389-395.

Svendsen, I., Jonassen, I., Hejgaard, J. and Boisen, S. (1981) Amino acid sequence homology between a serine protease inhibitor from barley and potato inhibitor I. *Biochem. Soc. Trans.*, **9**, 265p.

Svendsen, I., Boisen, S. and Hejgaard, J. (1982) Amino acid sequence of serine protease inhibitor CI-1 from barley. Homology with barley inhibitor CI-2, potato inhibitor I, and leech eglin. *Carlsberg Research Communications*, **47**, 45-53.

Svendsen, I., Hejgaard, J. and Mundy, J. (1986) Complete amino acid sequence of the alpha-amylase/subtilisin inhibitor from barley. *Carlsberg Research Communications*, **51**, 43-50.

Tai, H., McHenry, L., Fritz, P.J. and Furtek, D.B. (1991) Nucleic acid sequence of a 21 kDa cocoa seed protein with homology to the soybean trypsin inhibitor (Kunitz) family of protease inhibitors. *Plant Mol. Biol.*, **16**, 913-915.

Tashiro, M., Hashino, K., Shiozaki, M., Ibuki, F. and Maki, Z. (1987) The complete amino acid sequence of rice bran trypsin inhibitor. *J. Biochem. (Tokyo)*, **102**, 297-306.

Tasneem, M., Cornford, C.A. and McManus, M.T. (1994) Characterisation of serine proteinase inhibitors in dry seeds of cultivated pasture grass seeds. *Seed Sci. Res.* **4**, 231-242.

Tasneem, M., Cornford, C.A., Laing, W.A. and McManus, M.T. (1996) Two dual trypsin/chymotrypsin iso-inhibitors purified from *Festuca arundinacea* seed. *Phytochemistry*, **43**, 983-988.

Taylor, M.A., Baker, K.C., Briggs, G.S. *et al.* (1995) Recombinant pro-regions from papain and papaya proteinase IV are selective high affinity inhibitors of the mature papaya enzymes. *Protein Eng.*, **8**, 59-62.

Terada, S., Fujimura, S., Katayama, H., Nagasawa, M. and Kimoto, E. (1994) Purification and characterization of two Kunitz family subtilisin inhibitors from seeds of *Canavalia lineata*. *J. Biochem.*, **115**, 392-396.

Thompson, J.D., Gibson, T.J., Plewniak, F., Jeanmougin, F. and Higgins, D.G. (1997) The ClustalX windows interface: flexible strategies for multiple sequence alignment aided by quality analysis tools. *Nucleic Acids Res.*, **24**, 4876-4882.

Trexler, M., Bányai, L. and Patthy, L. (2001) A human protein containing multiple types of protease-inhibitory modules. *Proc. Natl Acad. Sci. USA*, **98**, 3705-3709.

Trumper, S., Follmann, H. and Haberlein, I. (1994) A novel-dehydroascorbate reductase from spinach chloroplasts homologous to plant trypsin inhibitor. *FEBS Lett.*, **352**, 159-162.

Turk, V., Krizaj, I., Drobnic-Kosorok, M., Brzin, J. and Jerala, R. (1993) The primary structure of inhibitor of cysteine proteinases from potato. *FEBS Lett.*, **333**, 15-20.

Turk, B., Turk, V. and Turk, D. (1997) Structural and functional aspects of papain-like cysteine proteinases and their protein inhibitors. *Biol. Chem.*, **378**, 141-150.

Valueva, T.A., Revina, T.A., Kladnitskaya, G.V. and Mosolov, V.V. (1998) Kunitz-type proteinase inhibitors from intact and *Phytophthora*-infected potato tubers. *FEBS Lett.*, **426**, 131-134.

Vendrell, J., Querol, E. and Aviles, F.X. (2000) Metallocarboxypeptidases and their protein inhibitors. Structure, function and biomedical properties. *Biochim. Biophys. Acta*, **1477**, 284-298.

Verissimo, P., Ramalho-Santos, M., Faro, C. and Pires, E. (1998) A comparative study on the aspartic proteinases from different species of *Cynara*. *Adv. Exp. Med. Biol.*, **436**, 459-463.

Vierstra, R.D. (1993) Protein degradation in plants. *Annu. Rev. Plant Physiol. Plant. Mol. Biol.*, **44**, 385-410.

Vierstra, R.D. (1996) Proteolysis in plants: mechanisms and functions. *Plant Mol. Biol.*, **32**, 275-302.

Vigers, A.J., Roberts, W.K. and Selitrennikoff, C.P. (1991) A new family of plant antifungal proteins. *Mol. Plant Microbe Interact*, **4**, 315-323.

Villanueva, J., Canals, F., Prat, S., Ludevid, D., Querol, E. and Aviles, F.X. (1998) Characterization of the wound-induced metallocarboxypeptidase inhibitor from potato. cDNA sequence, induction of gene expression, subcellular immunolocalization and potential roles of the C-terminal propeptide. *FEBS Lett.*, **440**, 175-182.

Visal, S., Taylor, M.A. and Michaud, D. (1998) The proregion of papaya proteinase IV inhibits Colorado potato beetle digestive cysteine proteinases. *FEBS Lett.*, **434**, 401-405.

Vodkin, L.O. and Scandalios, J.G. (1981) Genetic control, developmental expression, and biochemical properties of plant peptidases. Isozymes. *Curr. Topics Biol. Med. Res.*, **5**, 1-25.

Volpicella, M., Schipper, A., Jongsma, M.A., Spoto, N., Gallerani, R. and Ceci, L.R. (2000) Characterization of recombinant mustard trypsin inhibitor 2 (MTI2) expressed in *Pichia pastoris*. *FEBS Lett.*, **468**, 137-141.

Waldron, C., Wegrich, L.M., Merlo, P.A. and Walsh, T.A. (1993) Characterization of a genomic sequence coding for potato multicystatin, an eight-domain cysteine proteinase inhibitor. *Plant Mol. Biol.*, **23**, 801-812.

Walsh, T.A. and Strickland, J.A. (1993) Proteolysis of the 85-kilodalton crystalline cysteine proteinase inhibitor from potato releases functional cystatin domains. *Plant Physiol.*, **103**, 1227-1234.

Wen, L., Huang, J.K., Zen, K.C. *et al.* (1992) Nucleotide sequence of a cDNA clone that encodes the maize inhibitor of trypsin and activated Hageman factor. *Plant Mol. Biol.*, **18**, 813-814.

Werner, M.H. and Wemmer, D.E. (1992) Three-dimensional structure of soybean trypsin/chymotrypsin Bowman–Birk inhibitor in solution. *Biochemistry*, **31**, 999-1010.

Wieczorek, M., Otlewski, J., Cook, J. *et al.* (1985) The squash family of serine proteinase inhibitors. Amino acid sequences and association equilibrium constants of inhibitors from squash, summer squash, zucchini, and cucumber seeds. *Biochem. Biophys. Res. Commun.*, **126**, 646-652.

Wiederanders, B. (2000) The function of propeptide domains of cysteine proteinases. *Adv. Exp. Med. Biol.*, **477**, 261-270.

Williamson, M.S., Forde, J., Buxton, B. and Kreis, M. (1987) Nucleotide sequence of barley chymotrypsin inhibitor-2 (CI-2) and its expression in normal and high-lysine barley. *Eur. J. Biochem.*, **165**, 99-106.

Wilusz, T., Wieczorek, M., Polanowski, A., Denton, A., Cook, J. and Laskowski, M.J. (1983) Amino-acid sequence of two trypsin isoinhibitors, ITD I and ITD III from squash seeds (*Cucurbita maxima*). *Hoppe-Seyler's Z. Physiol. Chem.*, **364**, 93-95.

Wright, H.T. (1996) The structural puzzle of how serpin serine proteinase inhibitors work. *BioEssays*, **18**, 453-464.

Yamada, T., Ohta, H., Shinohara, A. *et al.* (2000) A cysteine protease from maize isolated in a complex with cystatin. *Plant Cell Physiol.*, **41**, 185-191.

Yamamoto, Y., Watabe, S., Kageyama, T. and Takahashi, S.Y. (1999) A novel inhibitor protein for *Bombyx* cysteine proteinase is homologous to propeptide regions of cysteine proteinases. *FEBS Lett.*, **448**, 257-260.

Yoo, B.C., Aoki, K., Xiang, Y. *et al.* (2000) Characterization of cucurbita maxima phloem serpin-1 (CmPS-1). A developmentally regulated elastase inhibitor. *J. Biol. Chem.*, **275**, 35,122-35,128.

Zhu, Y., Huang, Q., Qian, M., Jia, Y. and Tang, Y. (1999) Crystal structure of the complex formed between bovine beta-trypsin and MCTI-A, a trypsin inhibitor of squash family, at 1.8 Å resolution. *J. Protein. Chem.*, **18**, 505-509.

5 Multienzyme complexes involved in the Benson–Calvin cycle and in fatty acid metabolism

Brigitte Gontero, Sandrine Lebreton
and Emmanuelle Graciet

5.1 Introduction

In the past, enzymological studies have been devoted largely to the characterization of individual enzymes. This was a period of accumulating data on the main properties of enzymes (oligomerization state, kinetic properties, etc.). However, this very useful exercise still needs to be completed to allow a better understanding of living cells—in particular, the concept of spatial organization. It is now clear that enzymes are not separate and independent molecules in the cell, but interact with many components, including membranes or other proteins, to form more complex structures. Because of the extremely high concentration of proteins, the cell appears very crowded, similar to a concentrated 'soup'. Enzymes are therefore packed together and the mean distance between them is lower than the mean diameter of a protein (Goodsell 1991). Thus specific interactions between enzymes, involving both spatial and electrostatic elements, are likely to occur *in vivo*. These enzyme–enzyme interactions lead to supramolecular structures, often referred to as multienzyme complexes. However, they are also called 'protein machines', 'clusters', 'supramolecular complexes', 'aggregates' and 'metabolons'. Srere (1985) proposed the use of metabolons for supramolecular complexes of sequential metabolic enzymes and cellular structural elements. This term has now been extensively used for the tricarboxylic acid cycle or Krebs cycle.

A distinction should also be made between multienzyme complexes and multifunctional proteins. Multienzyme complexes are associations of two or more polypeptide chains bearing different catalytic centers, while multifunctional proteins consist of a single polypeptide chain but with two or more different catalytic centers (Hawkins and Lamb 1995).

These multienzymes and multifunctional proteins have been isolated from prokaryotic organisms and in many organelles from eukaryotes. Their molecular masses range from a few hundred thousands to several million Daltons. They are, therefore, quite ubiquitous. However, multifunctional proteins are more representative of the eukaryotic world.

Evidence of multienzyme complexes in sequential metabolic pathways is more compelling for some pathways (e.g. glycolysis) than for others (Srere 1987). In particular, only recently has the concept that metabolic pathways, consisting of multienzymes, been applied to plants. Therefore, further studies of such enzyme complexes are required in plants.

If enzymes that catalyze consecutive reactions are embedded into a complex, and if their active sites are oriented in a favorable manner, a process called channeling can occur. Channeling is the direct transfer of reaction intermediates from one active site to the next active site thus avoiding diffusion in the cell milieu. One may therefore expect the efficiency of the overall process to be considerably increased owing to the supramolecular organization of these enzymes. However, there is a vast, and often conflicting, literature about the existence and the role of these channeling effects (Srivastava and Bernhard, 1986; Easterby, 1989; Srere and Ovadi, 1990; Wu *et al.*, 1991; Sainis and Jawali, 1994; Mendes *et al.*, 1996; Purcarea *et al.*, 1999; Ovadi and Srere, 2000). Channeling has only been proved clearly for tryptophan synthase (Pan *et al.*, 1997). Indeed this enzyme, involved in the two last steps of tryptophan biosynthesis, is made up of two pairs of two subunits (α and β). The X-ray structure at 2.5 Å of *Salmonella typhimurium* shows a 25–30 Å tunnel between each α and β pair (Hyde *et al.* 1988). In this tunnel, the diffusion of indole from its production site (α-subunit) to its utilization site (β-subunit) occurs (Dunn *et al.* 1990; Schlichting *et al.* 1994).

Although there is no doubt that quite a significant number of multienzyme complexes do catalyze consecutive reactions, others do not. The most significant example of this are the aminoacyl-tRNA synthetases. Whereas in prokaryotic cells these enzymes often appear as physically distinct entities, they occur as multienzyme complexes in eukaryotes (Deutscher, 1984; Mirande *et al.*, 1992; Agou and Mirande, 1997). As will be documented in this chapter, many purified complexes from plants are also made up of enzymes that do not catalyze consecutive reactions. It is evident that in the cases where enzymes belonging to multienzyme complexes do not catalyze consecutive reactions, channeling cannot occur. Therefore, if there is a functional advantage in the physical association of different enzymes, this advantage should be an alteration of the intrinsic properties of the enzymes within the complex. This means that information, or rather an instruction, is transferred from protein to protein within the multienzyme complex. Statistical mechanics shows that the interaction of an enzyme with other enzymes would lead to an increase in the free energy stored by this enzyme and that this energy may be used to alter the intrinsic catalytic properties of the protein (Ricard *et al.*, 1998).

Some enzymes may exist either in a 'free' isolated state or embedded in a multienzyme complex within the same compartment; a regulatory function on pathway flux by association–dissociation may exist, this concept being the so-called 'ambiquity'. Another consequence of association is that one of the enzymes upon release from the complex structure may transitorily retain an imprinting from the other protein and this imprinting may in turn alter its properties (Lebreton *et al.*, 1997b; Lebreton and Gontero, 1999). Also, the dynamic interactions between two proteins may orientate the metabolic flux into a preferential pathway if one of these two proteins belongs to more than one pathway. For instance, the interaction between aldolase and either glyceraldehyde phosphate dehydrogenase or glycerol phosphate dehydrogenase will direct the metabolic flux toward glycolysis or to lipid biosynthesis as glyceraldehyde phosphate or dihydroxyacetone phosphate, will be produced, respectively (Vertessy and Ovadi, 1987).

The purpose of this chapter is to illustrate the importance of protein–protein interactions within the Calvin cycle and in fatty acid metabolism. We realize how difficult it is to compile an exhaustive list of all multienzyme complexes involved in these two metabolic pathways. Nonetheless, we have tried to discuss the best characterized examples that will lead to a better understanding of the physical, enzymatic and regulatory aspects introduced through heterologous interactions between different enzyme molecules. We apologize in advance to those authors whose work has not been described or not sufficiently emphasized here.

5.2 Supramolecular complexes involved in the Benson–Calvin cycle

The reductive pentose–phosphate pathway or Benson–Calvin cycle (figure 5.1) occurs in the stroma of chloroplasts and is responsible for CO_2 assimilation to produce carbohydrates, starch in the chloroplast and sucrose in the cytosol. The cycle is regulated by light and does not operate in the dark. The notion that the enzymes involved in this metabolic pathway are not randomly distributed but interact to give multienzyme complexes was proposed because some of these proteins could not be isolated by conventional purification protocols. Indeed, several multienzyme complexes have been isolated from chloroplasts (Müller, 1972; Sainis and Harris, 1986; Nicholson *et al.*, 1987; Gontero *et al.*, 1988; Sainis *et al.*, 1989; Clasper *et al.*, 1991; Giudici-Orticoni *et al.*, 1992; Gontero *et al.*, 1993; Süss *et al.*, 1993; Clasper *et al.*, 1994; Gontero *et al.*, 1994; Sainis and Jawali, 1994; Süss *et al.*, 1995; Avilan *et al.*, 1997a,b; Lebreton *et al.*, 1997a,b; Wedel *et al.*, 1997; Wedel and Soll, 1998; Lebreton and Gontero, 1999; Avilan *et al.*, 2000) and different authors have reported different compositions for these supramolecular associations.

5.2.1 *Phosphoribulokinase–glyceraldehyde-3-phosphate dehydrogenase*

A bi-enzyme complex has been purified from the green alga, *Chlamydomonas reinhardtii* (Avilan *et al.*, 1997b). This complex is made up of phosphoribulokinase (PRK, EC 2.7.1.19) and glyceraldehyde-3-phosphate dehydrogenase

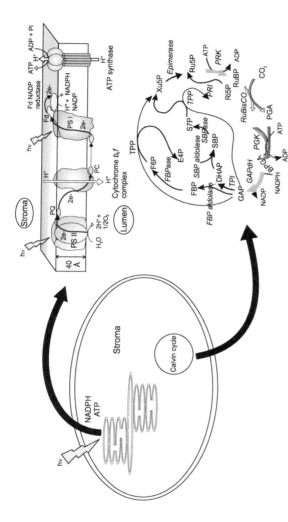

Figure 5.1 Schematic illustration of the photosynthetic electron transport chain and of the Benson–Calvin cycle. The electron transport chain that occurs in the thylakoids comprises three main complexes, photosystem II or PSII, photosystem I or PS I and the cytochrome b_6f complex. The water oxidizing complex (WOC) or oxygen evolving complex (OEC) is responsible for the photolysis of water to produce oxygen and the ATP synthase for the ATP synthesis. Plastoquinones (PQs) connect PS II to the cytochrome b_6f, and plastocyanin (PC) connect the cytochrome b_6f complex to PS I. The ferredoxin (Fd) NADP reductase is responsible for the reduction of NADP into NADPH. ATP and NADPH are then in part used in the reductive pentose phosphate pathway or Benson–Calvin cycle that occurs in the stroma. The assimilation of CO_2 can be divided into three main phases: carboxylation, reduction and regeneration. In the first phase, CO_2 fixation is accomplished by ribulose 1,5-bisphosphate carboxylase (Rubisco) that catalyzes the conversion of ribulose-1,5-bisphosphate (RuBP) into two molecules of 3-phosphoglycerate (PGA). In the second phase, PGA is converted into BPGA (1,3-bisphosphoglycerate) via PGK (phosphoglycerate kinase) and then via GAPDH (glyceraldehyde-3-phosphate dehydrogenase) into glyceraldehyde-3-phosphate (GAP). In the third phase, RuBP is regenerated and the cycle is completed when PRK (phosphoribulokinase) reforms RuBP from ribulose-5-phosphate (Ru5P). FBPase and SBPase are the two phosphatases of the cycle, namely fructose-1,6-bisphosphatase and sedoheptulose-1,7-bisphosphatase that convert fructose-1,6-bisphosphate (FBP) and sedoheptulose-1,7-bisphosphate (SBP) into fructose-6-phosphate (F6P) and sedoheptulose-7-phosphate (S7P). TPP, thiamine pyrophosphate; TPI, triose phosphate isomerase; PRI, phosphoribose isomerase; Ri5P and Xu5P are ribose-5-phosphate and xylulose-5-phosphate respectively.

(GAPDH, EC 1.2.1.13). The enzyme PRK catalyzes the ATP-dependent phosphorylation of ribulose-5-phosphate to form ribulose-1,5-bisphosphate, the CO_2 acceptor in photosynthetic organisms. The enzyme GAPDH catalyzes the reversible reduction and dephosphorylation of 1,3-bisphosphoglycerate into glyceraldehyde-3-phosphate using β-nicotinamide adenine dinucleotide (NADPH) generated by the photosystem I in the light (Buchanan 1980) (figure 5.1). This bi-enzyme complex has a molecular mass of 460 kDa. As the chloroplast GAPDH from algae is homotetrameric (A_4) and the PRK dimeric (R_2), using densitometric analysis it has been shown that the complex is made up of two molecules of GAPDH $(2 \times A_4)$ and two of PRK $(2 \times R_2)$. Although different in its subunit composition, this bi-enzyme complex is quite reminiscent of complexes isolated from either spinach leaves (Gontero et al., 1988) or from another green alga, Scenedesmus obliquus (Nicholson et al., 1987). A bi-enzyme complex with both latent PRK and GAPDH activity has also been characterized from Scenedesmus obliquus. It has a molecular mass of 560 kDa and its apparent subunit composition is A_8R_6. For higher plants, it is known that GAPDH is made up of two different subunits of molecular masses close to 39.5 kDa (subunit A) and 41.5 kDa (subunit B). This is contrary to the enzyme from green alga where this enzyme only exists as a homotetramer (A_4). In higher plants, the GAPDH also exists as a heterotetramer (A_2B_2) with an even higher molecular mass of 600 kDa (Yonuschot et al. 1970). The B subunit presents a C-terminal extension of about 30 amino acids when compared with the A subunit, and this extension seems to be responsible for the aggregated state observed with the heteromeric form (AB) of GAPDH (Baalmann et al., 1996; Scheibe et al., 1996). When GAPDH is associated with PRK in spinach, structures such as $(A_2B_2)2A_4R_2$ or $(A_2B_2)(A_4)2R_2$ have been proposed (Clasper et al., 1991).

PRK and GAPDH are light-regulated due to several mechanisms and play a key role in the Calvin cycle (Leegood, 1990). These two enzymes do not catalyze consecutive reactions but use substrates that are products of the photochemical reactions of chloroplasts: ATP for the kinase and NADPH for the dehydrogenase. Their association could, therefore, provide a control unit for regulation of energy consumption.

It has been shown that all the bi-enzyme complexes (GAPDH–PRK) described above spontaneously dissociate in the presence of either physiological or non-physiological reducing agents, such as thioredoxin or dithiothreitol. In parallel with the dissociation of the complex, the PRK becomes more active and GAPDH, which shows a dual specificity towards NAD(H) or NADP(H) in the complex, shows specificity towards NADP(H) when reduced.

These two enzymes, PRK and GAPDH, each may be obtained in a free independent state. Not associated with each other, they form dimeric and tetrameric forms, respectively. Some authors have also reported the existence of aggregated forms for these two enzymes. In Sinapsis alba, it has been proposed that the polymerization of higher molecular mass of GAPDH only occurs in the presence of a binding fraction that has been lost during purification (Cerff, 1978; Cerff and Chambers, 1978). Subsequently, Easterby's group has proposed that this binding fraction could be PRK (Nicholson *et al.*, 1987). As highly purified PRK does not form aggregates by itself, it has also been suggested that other components from chloroplasts are also required (Porter, 1990).

In *Chlamydomonas reinhardtii* cells, enzyme dissociation induced by reduction is reversible under oxidizing conditions. Indeed, the oxidized partners are able to spontaneously reconstitute the complex *in vitro*, which is quite similar to the native state (Lebreton *et al.*, 1997a). In only a few cases has it been possible to assemble particles from their separate parts *in vitro* that resemble the native complexes (Reed *et al.*, 1975). Therefore this complex is a very suitable model with which to study protein–protein interactions.

The gene coding PRK from *Chlamydomonas reinhardtii* has been isolated, cloned and expressed in *Escherichia coli*. The recombinant protein forms a complex with GAPDH, which is apparently indistinguishable from that extracted from *Chlamydomonas* cells. A PRK has been isolated from the 12-2B mutant of the same organism. In this mutant, the Arg_{64} residue has been replaced by a cysteine and the bi-enzyme complex does not form. These results suggest that this arginine residue plays a key role in the formation of the complex and this conclusion has been demonstrated experimentally by site-directed mutagenesis (Avilan *et al.*, 1997a). In these experiments, Arg_{64} has been replaced by Ala_{64}, Lys_{64}, or Glu_{64}. Whereas the Ala_{64} and Glu_{64} containing PRKs are unable to form a complex with GAPDH, the Lys_{64} mutant does, although to a lesser extent. The recombinant PRK associated with GAPDH displays the same type of kinetic behavior as does the wild-type enzyme in the complex. While oxidized PRK in a free state is devoid of any activity, the one inserted in the bi-enzyme complex is active (Lebreton *et al.*, 1997b).

In the reaction catalyzed by the PRK embedded in the complex, examination of the progress curves of the reaction reveals an initial lag. It has been shown that this lag corresponds with the dissociation of the complex. If this complex is incubated for 15 min in the same reaction mixture but lacking the substrates of PRK, then no lag is observed with initiation of the reaction (Lebreton *et al.*, 1997b).

A simple model that represents the dissociation of the complex together with PRK activity has been proposed. The equation that derived from the model fits the experimental results best only if one assumes (as postulated in the model) that both the free PRK released upon dissociation of the complex and the bound PRK of this complex are active. If only the free PRK is assumed to be active, then the fit is biased. One may therefore measure the reaction rate catalyzed by the bound PRK of the complex by monitoring the reaction rate immediately after mixing this complex with its substrates in a suitable reaction medium. The active oxidized form of PRK, which is released upon the dissociation of the complex, is not stable and it loses activity slowly until it becomes identical to the stable and almost inactive form. The metastable conformation of the free oxidized

PRK may be characterized by fluorescence spectroscopy. Thus there are three different forms that have very different conformations and activities: the stable, essentially inactive, enzyme form; the enzyme form bound to GAPDH; and the metastable free active enzyme form. These three forms differ in their K_m and k_{cat} values (Lebreton et al., 1997b; Lebreton and Gontero, 1999). These three forms are also present in the reduced state and differ in their K_m and k_{cat} values (Lebreton and Gontero, 1999) (table 5.1).

From these data it appears that GAPDH may provide an instruction to PRK and, as a consequence of their interaction, an increase of activity of PRK is observed. Mixing the stable, oxidized and almost inactive PRK with GAPDH indeed results in the formation of the bi-enzyme complex and the catalytic activity of PRK increases. Moreover, the free PRK, which is released after dissociation of the complex, is in a metastable state that slowly returns to the stable and almost inactive state. This metastable state is very active and therefore retains the imprinting exerted on PRK by GAPDH. Why then is this isolated form of PRK in a metastable state more active than the same enzyme bound to GAPDH? The reason is probably that the catalytic activity requires mobility of the enzyme molecule which is favored if this molecule is in a free state.

Table 5.1 Kinetic parameters for the different forms of phosphoribulokinase (PRK) in oxidized and reduced states. The parameters were obtained by fitting the experimental data to the Michaelis–Menten equation. Experiments were repeated at least ten times and the standard errors on all parameters were less than 10%. Ru5P, ribulose-5-phosphate

Kinetic parameters	Oxidation–reduction states	Enzyme forms	Ru5P	ATP
K_m (µM)	oxidized	stable PRK	115	89
		PRK inserted in the complex	30	46
		metastable PRK	59	48
	reduced	stable PRK	55	55
		PRK inserted in the complex	61	60
		metastable PRK	94	51
k_{cat} (s^{-1} site^{-1})	oxidized	stable PRK	0.062	0.065
		PRK inserted in the complex	3.25	3.25
		metastable PRK	56.3	56.5
	reduced	stable PRK	23.3	23
		PRK inserted in the complex	32.4	32.7
		metastable PRK	300	303.8
k_{cat}/K_m (mM^{-1}s^{-1} site^{-1})	oxidized	stable PRK	0.54	0.73
		PRK inserted in the complex	108	70
		metastable PRK	954	1177
	reduced	stable PRK	424	418
		PRK inserted in the complex	531	545
		metastable PRK	3191	5957

Quite recently, it has been shown that a new protein, CP12 may be oligomerized together with PRK and GAPDH into a stable 600 kDa hetero-oligomeric complex. This three-protein complex CP12–PRK–GAPDH is found in higher plants (*Spinacia oleracea*), in cyanobacterium *Synechocystis* (PCC6803) and in green algae (*Chlamydomonas reinhardtii*) (Wedel and Soll, 1998). CP12 is a small nuclear-encoded protein of 75 amino acids. It has a molecular mass of 8.2 kDa (Pohlmeyer *et al.*, 1996). However after dissociation from the complex, it has a molecular mass of 70 kDa, indicating a probable aggregation state. The stoichiometry of this complex in spinach was proposed to be two *N*-terminally dimerized CP12 peptides, each carrying one dimeric PRK and one heterotetramer of GAPDH (A_2B_2) on the *C*-terminal peptide loop. To understand the physiological role of this complex, Wedel and colleagues have incubated this entity plus 2.5 mM of different cofactors such as NAD(P) and NAD(P)H for 1 h at 4°C. The incubated samples were then chromatographed through a Sephacryl S400 size exclusion column and the water-incubated sample showed a partial dissociation, although this dissociation is more pronounced using NADP and NADPH. It has also been shown that NAD(P) inhibits the PRK activity, NADPH increases it and no significant effect is detected with NAD(H) (Wedel *et al.*, 1997). Since nothing was known of CP12 at the time of this study, it is possible that Ricard and co-workers failed to notice the protein when purifying the bi-enzyme complex. We have re-investigated these results to confirm the existence of CP12 in the PRK–GAPDH complex. However, as it is possible to reconstitute the complex from even purified recombinant PRK and purified GAPDH, it seems very probable that CP12 is not an absolute requirement for the assembly of this bi-enzyme complex.

The use of immunoelectron cryomicroscopy reveals that the PRK and the GAPDH interact *in vivo* and are associated with the thylakoids of *Chlamydomonas reinhardtii*. In addition to these enzymes, phosphoribose isomerase, Rubisco, fructose-1,6-bisphosphate aldolase, sedoheptulose bisphosphatase, nitrite reductase, ferredoxin–NADP oxidoreductase and ATP synthase were also found (Süss *et al.*, 1995). Probably the interactions of the PRK and the GAPDH enzymes is stronger than with the other enzymes, suggesting that these enzymes form the core of a bigger complex *in vivo*.

5.2.2 *Phosphoglycerate kinase–glyceraldehyde-3-phosphate dehydrogenase*

Hybridization experiments (immobilization of one protein (X) onto a membrane, incubation with another protein (Y) and probing with antibodies raised against protein (Y)) have shown that GAPDH and phosphoglycerate kinase (EC 2.7.2.3) from pea (*Pisum sativum*) can interact *in vitro*. This result has been confirmed using ultrafiltration, isoelectric focusing and fluorescence anisotropy. The K_d or dissociation constants values range from 1 to 5 μM, depending on pH (Wang *et al.*, 1996). Association between these two enzymes has also been suggested by kinetic experiments. At this point, it is interesting to notice that these enzymes, which are not unique to the reductive pentose–phosphate pathway, have also been described as interacting partners in the mammalian, microbial and plant cytosol (Maciozek *et al.*, 1990).

5.2.3 Phosphoribose isomerase–phosphoribulokinase–ribulose-1,5-bisphosphate carboxylase-oxygenase

Sainis and Harris reported the existence of a complex comprising phosphoribose isomerase (EC 5.3.1.6), PRK and Rubisco (ribulose-1,5-bisphosphate carboxylase–oxygenase, EC 4.1.1.39) in spinach (Sainis and Harris, 1986). Later, they reported a two-enzyme complex (phosphoribose isomerase was not present in this complex) in pea chloroplasts and suggested that ammonium sulfate, used in purification, may be responsible for the rupture of the complex into two components. In 1994, they obtained some evidence for channeling in the three-enzyme complex from spinach, and reported a better crystallization of this structure at pH 7 than at pH 8. The overall molecular mass of this complex was close to 800–850 kDa. The crystals, $0.2 \times 0.2 \times 0.3$ mm in size, were subjected to X-ray analysis at 3.5 Å. The crystalline structure within this complex in the presence of ribulose-1,5-bisphosphate revealed the hexadecameric structure (eight large and eight small subunits) characteristic of the purified enzyme, but with a quite different packing mode (Hosur *et al.*, 1993). All four axes of symmetry, observed in the crystals from the highly purified enzyme in the presence of 2-carboxyarabinitol phosphate, were parallel, while in the crystals from the complex, the axes were inclined at an angle of 70°. Probably this difference is due to the presence of the other proteins in this complex. Recently, a complex with a molecular mass of 600 kDa has been described from tobacco (*Nicotiana tabacum*) (Jebanathirajah and Coleman, 1998). Beside the isomerase, the kinase and the carboxylase, a fourth protein, carbonic anhydrase was reported as a component of the multienzyme complex.

Other results clearly demonstrate that phosphoribose isomerase from yeast and PRK from pea may physically interact. Using the countercurrent distribution (CCD) technique, it has been shown that the partitioning of these enzymes in an aqueous two-phase system (dextran, polyethylene glycol and aqueous buffer) is affected by the presence of 0.4 mM ribose-5-phosphate (Skrukrud and Anderson, 1991). The presence of this metabolite is required to allow association between these two enzymes. This result may explain why in many preparations the isomerase is lost, probably because its interaction with the kinase is not tight enough. It seems likely that the interaction between these two enzymes might be stabilized in the presence of ribose-5-phosphate or when other proteins are included in the complex. The use of the same technique (CCD) has also been used to show interactions between six enzymes of the

Calvin cycle (Rubisco, phosphoglycerate kinase, triose phosphate isomerase, GAPDH, aldolase and fructose bisphosphatase). In these experiments, the distribution coefficient G of the purified enzymes was compared with the distribution coefficients of a crude extract from spinach leaves (Persson and Johansson, 1989). While purified enzymes (phosphoglycerate kinase, aldolase) showed a single homogeneous peak, with a unique G value, the same enzymes within a crude extract showed at least two peaks. This heterogeneity was linked to the existence of two populations of enzymes and probably to the existence of protein complexes. As the other enzymes previously mentioned also showed heterogeneity in their distribution coefficients, the same conclusion was drawn and suggests protein–protein interaction between these enzymes (Persson and Johansson, 1989).

5.2.4 *Phosphoribose isomerase–phosphoribulokinase–ribulose-1,5-bisphosphate carboxylase/oxygenase–phosphoglycerate kinase–glyceraldehyde-3-phosphate dehydrogenase*

A five-enzyme complex purified from spinach leaves (Gontero *et al.* 1988) has been studied in detail. This complex has an overall mass of about 520 kDa. It has been shown by immunochemical (ELISA) and densitometric analysis that this complex is made up of two subunits of PRK, two A and two B subunits of GAPDH, and two large and four small subunits of Rubisco. The conclusion that Rubisco has an L_2S_4 structure was confirmed by kinetic titration of active sites by a competitive inhibitor (6-phosphogluconate) (Rault *et al.* 1993). Based on the overall mass of this complex, phosphoribose isomerase was suggested to be dimeric and phosphoglycerate kinase monomeric. Evidence that PRK may be activated by thioredoxins was found and, interestingly, this activation depended on whether the kinase was free in solution or embedded in the five-enzyme complex (Rault *et al.*, 1991). This activation process has been studied and a kinetic model has been proposed (Gontero *et al.*, 1993). Pseudo-dissociation constants (K_d) for reduced thioredoxins for free PRK or PRK inserted in the complex were $0.8\,\mu M \pm 0.1$ and $0.11\,\mu M \pm 0.01$, respectively, suggesting that the affinity of thioredoxin for phosphoribulokinase is higher if the enzyme is embedded in the complex. Moreover, the time constant of the activation process of the free enzyme decreases with thioredoxin concentrations, whereas the time constant of the activation process of the complex increases. The time required to get half-maximal activation with the complex and the free enzyme was about 2 min and 20 min, respectively. The time taken to form the complex is therefore compatible with the induction time of the Calvin cycle upon dark–light transitions.

The kinetic constants of free Rubisco were compared with those measured for Rubisco inserted in the five-enzyme complex. The values of the K_m were $0.140\,mM \pm 0.03$ and $70\,\mu M \pm 6$, respectively whereas V (s^{-1} site^{-1}) were

1.57 ± 0.1 and 7.13 ± 0.2. Therefore the ratio $V{:}K_m$ is about tenfold higher for the complex than for the free enzyme. Statistical thermodynamics of information transfer has been developed and applied to this supramolecular structure (Gontero *et al.*, 1994; Ricard *et al.*, 1994). This complex seems to be associated with the thylakoids; $MgCl_2$ and KCl enhance the strength of this association while EDTA stimulates its release from the membranes. It was suggested that this spatial organization could permit a link between the Calvin cycle and the light reactions of photosynthesis. The light-dependent association-dissociation process of the complex with the membrane would provide another possible mechanism of regulation of carbon metabolism in plants (Giudici-Orticoni *et al.*, 1992).

Quite recently, multienzyme complexes with molecular mass of 520 kDa from cotton (*Gossypium hirsutum*), which comprise phosphoribose isomerase, PRK, Rubisco, phosphoglycerate kinase, GAPDH and fructose bisphosphatase have been identified (Babadzhanova *et al.*, 2000).

5.3 Supramolecular complexes involved in fatty acid metabolism

The biosynthesis of saturated fatty acids from acetyl CoA occurs in all organisms and in mammals it is prominent in liver, adipocyte tissue and mammary glands. While biosynthesis occurs in the cytosol, the reverse process (β-oxidation of fatty acids) occurs in the mitochondria by a series of quite distinct reactions. In plants, overall fatty acid synthesis, and consequently its regulation, is more complicated than in any other organism. In plants, fatty acid synthesis is localized in the plastid, where a fraction of the newly synthesized acyl chains is used for lipid synthesis (mainly galactolipids). However, the major portion is exported into the cytosol. In turn, some of the extraplastidial glycerolipids may come back to the plastid. Because the site of synthesis (plastid) and the site of esterification (mainly outside the chloroplast) are different, fine regulation of the biosynthetic pathway should occur (Ohlrogge and Jaworski, 1997).

Plant lipids are constituents of cellular membrane and of surface layers such as wax, cutin and suberin. They or their metabolite derivatives (phosphoinositides) have biological activity and are energy store (acyl lipids). The acyl lipids are the largest parts of the lipids produced by plants. The major fatty acids in plants are palmitate (16 carbon), oleate (18 carbon), linoleate (18:2 carbon) and α-linolenate (18:3 carbon). Their formation starts by the synthesis *de novo* of long chain fatty acids catalyzed by the concerted action of acetyl CoA carboxylase and fatty acid synthase.

It has been suggested that plastids may be autonomous in their ability to generate acetyl CoA from pyruvate (Johnston *et al.*, 1997). Indeed, acetyl CoA could be supplied directly by a plastidic pyruvate decarboxylase/dehydrogenase or indirectly from mitochondrial pyruvate dehydrogenase via release of free acetate, transport into the chloroplast and conversion in acetyl CoA via a chloroplast acetyl CoA synthase. This last enzyme exists within the chloroplast and has been purified from spinach and *Arabidopsis* (Kuhn *et al.*, 1981; Harwood, 1996; Ke *et al.*, 2000b). It has a molecular mass of approximately 73 kDa. In *Arabidopsis*, its localization is mainly chloroplastic (97.7±5.5% of the activity resides within the plastid).

Recently, Ke *et al.* (2000b) suggested that pyruvate was the main precursor of acetyl CoA in plastids of developing *Arabidopsis* seeds. The involvement of either acetyl CoA synthase or pyruvate dehydrogenase may vary between cell types, development stages, or with respect to environmental cues. Multiplicity of these acetyl CoA generating mechanisms would enable a fine regulation of the supply of this precursor into the fatty acid biosynthesis.

5.3.1 Acetyl CoA carboxylases

5.3.1.1 Reaction mechanism of acetyl CoA carboxylases

The acetyl CoA:bicarbonate ligase (EC 6.4.1.2) carries out the first committed step in fatty acid biosynthesis. It is a soluble class 1 biotin-containing enzyme that catalyzes the ATP-dependent formation of malonyl CoA from acetyl CoA and bicarbonate. Plant acetyl CoA carboxylase, like its bacterial or animal counterparts and other biotin-containing carboxylases, catalyzes its reaction in two steps (figure 5.2):

> *Step 1*: ATP is used in the carboxylation of the biotin prosthetic group of biotin carboxyl carrier protein (BCCP). Bicarbonate is the primary source of carbon and carboxylation occurs in two steps with a carboxyphosphate intermediate. This reaction is catalyzed by biotin carboxylase.
>
> *Step 2*: Via a carboxyltransferase, the carboxyl group is transferred to acetyl CoA to yield malonyl CoA.

The reaction mechanism is a ping-pong type, i.e. the binding of the first substrate (ATP) to the biotin enzyme is followed by the release of the first product (ADP). Then, the second substrate (acetyl CoA) binds to the carboxybiotin enzyme and the last product (malonyl CoA) is released.

$$\text{BCCP–biotin} + \text{HCO}_3^- + \text{ATP} \overset{①}{\rightleftharpoons} \text{BCCP–biotin-CO}_2 + \text{ADP} + \text{P}_i$$

$$\text{BCCP–biotin-CO}_2 + \text{acetyl CoA} \overset{②}{\rightleftharpoons} \text{BCCP–biotin} + \text{malonyl CoA}$$

① biotin carboxylase

② carboxyl transferase

Figure 5.2 The first committed step in fatty acid synthesis: the reaction catalyzed by acetyl CoA carboxylase. BCCP, biotin carboxyl carrier protein.

5.3.1.2 Enzyme structure

There are two main structures: acetyl CoA carboxylases (ACCase) are either multienzyme complexes or multifunctional proteins.

In *Escherichia coli*, the enzyme is made up of at least three proteins: biotin carboxylase, BCCP and carboxyltransferase (Guchhait *et al.*, 1974). Unlike multifunctional eukaryotic enzymes, the bacterial ACCase is a multienzyme complex (prokaryotic type) that readily dissociates with retention of biochemical activities (Kondo *et al.*, 1991) and the biotin carboxylase structure has been solved at 2.4 Å (Waldrop *et al.*, 1994). The chloroplastic ACCase from pea is also a multienzyme complex. The biotin carboxyl carrier protein has a molecular mass of 38 kDa and another protein coded by the accD gene has a molecular mass of 87 kDa and is identified as a component of the pea acetyl-CoA carboxylase complex. The four subunits are assembled into a complex with a molecular size of 600–700 kDa. The carboxyltransferase component is made up of non identical α and β subunits. The α subunit was shown to be equivalent to the IEP96 of pea chloroplast ACCase, a membrane associated polypeptide (Shorrosh *et al.*, 1996).

In animal and yeast (Witters and Watts, 1990) ACCases, all three functional domains are present on a single polypeptide chain. They are multifunctional proteins with a molecular mass of 200–240 kDa. In these multifunctional proteins (also named eukaryotic-type ACCase), the arm of biotin carboxyl carrier protein has to be flexible to react both with the biotin carboxylase and the carboxyl transferase. The flexibility is given by a high number of Ala and Pro in this region. In liver or adipocyte tissues, the ACCase can exist in two forms: an active polymer (10^6 Da) or an inactive protomer (2 to 5×10^5 Da). Dissociation to the protomeric form is favored by low protein concentration, Cl^-, pH greater than 7.5, palmityl CoA and carboxylation of the enzyme to produce carboxybiotin. The aggregated form is favored by citrate, acetyl CoA, high protein concentration and pH 6.5 to 7. The protomeric form appears as particles of 100–300 Å using electron microscopy, while the polymeric form appears as a network of filaments 70–100 Å in width and 4000 Å in length (Volpe and Vagelos, 1976).

The chloroplast acetyl CoA carboxylases from maize, wheat and rice have been shown to be multifunctional proteins. Many dicotyledonous species contain two distinct types of ACCases. One is a multienzyme form in the plastids in both epidermal and mesophyll cells and the second is a multifunctional form in the cytosol, mainly in epidermal cells. The kinetic properties of these two forms differ and it seems that the multienzyme form has a lower K_m for bicarbonate (0.85 mM) and a higher K_m for acetyl CoA (0.25 mM) when compared with the multifunctional form (2.5 mM, 0.015 mM) (Alban *et al.*, 1994). In the Gramineae, such as rice and wheat, the multisubunit form seems to be absent and isozymes of the multifunctional type are found both in plastids and the cytosol (Konishi and Sasaki, 1994; Schulte *et al.*, 1997). It has been shown that selective grass herbicides of the diphenoxypropionic acid type and the cyclohex-anedione type inhibited, *in vitro*, the plastidic eukaryotic form from wheat but did not inhibit that of the prokaryotic form from pea (Konishi and Sasaki, 1994).

5.3.1.3   Regulation

Plant fatty acids are mainly synthesized in plastids, although ACCase as mentioned previously also exists in the cytosol. These two types of ACCases both generate pools of malonyl CoA. In the chloroplast, the fate of this metabolite is to produce 16 and 18 carbon fatty acids. In the cytosol, it will generate fatty acids of 20 carbons or even longer. Cytosolic malonyl CoA is also used for the synthesis of flavonoids, stilbenoids, malonic acid and malonyl derivatives. A regulatory system different from that of other eukaryotes and characteristic of plastids has been shown to operate (Sasaki *et al.*, 1997). Furthermore, there is coordinated regulation in the expression of genes coding cytosolic and chloro-plastic ACCases between the cytosol and the chloroplast (Ke *et al.*, 2000a).

    In yeast and animals, enzyme activity is controlled by metabolites, enzyme protomer polymerization and by phosphorylation/dephosphorylation so that the excess energy is stored in the form of fatty acids in response to environmental conditions (Iverson *et al.*, 1990). However, in plants this type of regulation has not yet been found. Nonetheless, it has been shown that fatty acid synthesis *de novo* increases in chloroplasts upon illumination and decreases under dark conditions. Indeed, the synthesis of palmitate requires 14 molecules of NADPH and seven of ATP, which in chloroplasts are produced by the light reactions of photosynthesis. Therefore, a link between photosynthesis and fatty acid synthe-sis is clearly needed. It has been shown that the activity of ACCase is modulated by dark–light transitions. This enzyme is activated by thioredoxins, has an optimum pH of 8 and requires 2.5 mM Mg^{2+} for optimal activity, conditions that prevail under light conditions (Sasaki *et al.*, 1997). The first demonstration that ACCase is activated by light, via disulfide bridge reduction, was shown in pea by Kozaki and Sasaki (1999). Moreover, it was shown that the cysteine residues responsible for redox regulation reside on the carboxyltransferase component and not on the biotin carboxyl carrier protein (Kozaki *et al.*, 2000). To understand the activation process of carboxyltransferase at a molecular level, heterologous expression of this enzyme was achieved in *E. coli* (Kozaki *et al.*, 2000). The resulting recombinant enzyme was shown to be redox-regulated and activated by thioredoxin or dithiothreitol (Kozaki *et al.*, 2000).

5.3.2   Fatty acid synthases

Fatty acid synthase (EC 2.3.1.85) is the enzyme system that catalyzes the synthesis of long chain fatty acids from acetyl CoA, malonyl CoA and NADPH. The principal reactions in fatty acid synthesis are shown in figure 5.3. Two types of fatty acid synthase have been reported. Some (type I) are multifunctional, and

ACAT	acetyl CoA + ACP $\leftrightarrows$ acetyl-ACP + CoA
MCAT	malonyl CoA + ACP $\leftrightarrows$ malonyl-ACP + CoA
KAS I	acyl-ACP + malonyl-ACP $\leftrightarrows$ β-ketoacyl-ACP + CO_2 + ACP
KAS II	palmitoyl–ACP + malonyl-ACP $\leftrightarrows$ β-keto-octadecanoyl-ACP + CO_2 + ACP
KAS III	acetyl CoA + malonyl-ACP $\leftrightarrows$ acetoacyl-ACP + CO_2 + ACP
β-ketoacylACP reductase	β-ketoacyl-ACP + NAD(P)H $\leftrightarrows$ β-hydroxyacyl-ACP + NAD(P)
β-hydroxyacylACP dehydratase	β-hydroxyacyl-ACP $\leftrightarrows$ enoyl-ACP + H_2O
Enoyl-ACP reductase	enoyl-ACP + NAD(P)H $\leftrightarrows$ acyl-ACP + NAD(P)
Acyl-ACP thioesterase	acyl-ACP + H_2O $\leftrightarrows$ non-esterified fatty acid + ACP

palmitoyl-ACP

or stearoyl-ACP

Figure 5.3 Reactions catalyzed by fatty acid synthase in plants. The synthesis of fatty acid starts with transacylase reactions catalyzed by ACAT (acetyl CoA:ACP transacylase) and MCAT (malonyl CoA:ACP transacylase). The β-keto-ACP synthases or KAS enzymes catalyzed condensation reactions. The addition of two carbon units is then accomplished by the two reductases and one dehydratase. ACP, acyl carrier protein.

have been described in mammals, fungi and certain Mycobacteria. In plants and bacteria, a series of seven enzymes independently catalyze fatty acid synthesis. These sets of proteins are often referred as type II or dissociated fatty acid synthases.

5.3.2.1 Type II fatty acid synthases

Plant fatty acid synthase is characteristic of type II and shows similarity with the *E. coli* fatty acid synthase in many respects. There is a strong conservation of protein structure and enzyme mechanisms between the fatty acid synthesis pathways of *E. coli* and plants. Fatty acid metabolism in *E. coli* thus provides a useful model for plants where synthesis *de novo* occurs in the chloroplast and results obtained with *E. coli* will therefore be discussed here.

5.3.2.2 Acyl carrier protein

Acyl carrier protein (ACP) was the first protein of lipid metabolism to be purified to homogeneity (Simoni *et al.*, 1967), and detailed studies on the three-dimensional structure, NMR analysis and functional aspects of ACP have been reported (Holak *et al.*, 1988; Kim and Prestegard, 1989). ACP has a molecular mass of 8.9 kDa and is a rod-shaped protein rich in acidic residues. These acidic residues are mainly grouped in three α-helices, but a short fourth helix has also been reported.

A unique feature of type II fatty acid synthases is that all the intermediates are covalently bound to ACP. The acyl intermediates of fatty acid biosynthesis are bound to the protein through a thioester linkage attached to the terminal sulfhydryl of the 4′-phosphopantetheine prosthetic group (figure 5.4). Nonetheless, functionality of recombinant rat ACP has been shown in distantly related and very different enzyme systems (such as *E. coli*) and therefore suggests that type I and type II ACPs have similar conformations (Tropf *et al.*, 1998).

Isoforms of ACP have been identified in plants. In barley, there are three ACPs in which ACP I and ACP II are chloroplastic. In *Arabidopsis*, seeds, leaves and roots express one or more tissue-specific forms (Hlousek-Radojcic *et al.*, 1992). ACPs from cyanobacteria (*Anabaena variabilis*, *Synechocystis* 6803) have been purified and their structure determined to be more similar to *E. coli* than to plant ACP, although the phosphopantetheine prosthetic group region is highly conserved in all ACPs (Froehlich *et al.*, 1990).

5.3.2.3 Transacylase reactions

Condensing enzymes, such as β-keto-ACP synthases III from *E. coli* (Tsay *et al.*, 1992) and spinach (Clough *et al.*, 1992), have been shown to possess intrinsic acetyl CoA:ACP transacylase activity (ACAT). Nonetheless, it has been shown that a separate ACAT exists in barley, avocado and pea (Harwood 1996), but probably has less importance for overall fatty acid synthesis.

The crystal structure of malonyl CoA:acyl carrier protein transacylase (MCAT) from *E. coli* at 1.5 Å resolution has been reported (Serre *et al.*, 1995). In a fashion similar to serine hydrolases, Ser_{92} is hydrogen bonded to His_{101}. However, instead of a carboxyl acid, Glu_{250} serves as a hydrogen bond acceptor. Quite recently, identification of a cDNA for this gene from *Brassica napus* was reported. The cDNA clone encodes an open reading frame (ORF) corresponding to a protein of 351 amino acids, having 47% homology to the *E. coli* MCAT. (Simon and Slabas, 1998). While the bacterial MCAT is highly specific for malonyl CoA and does not participate in the priming transfer of the acetyl moiety onto ACP, the MCAT from multifunctional fatty acid synthase (FAS) possesses a dual specificity. This dual specificity has been investigated by mutagenesis (Rangan and Smith, 1997). Malonyl transacylase activity of the Arg_{606}-Ala and Arg_{606} Lys mutant enzymes was reduced by 100-fold and tenfold, respectively. In contrast, acetyl transacylase activity was increased 6.6-fold in the Arg_{606} Ala mutant and 1.7-fold in the Arg_{606} Lys mutant. Kinetic studies revealed that selectivity of the enzyme for acetyl CoA was increased by greater than 16,000-fold by the Ala mutation and 16-fold by the Lys mutation. These results indicate that Arg_{606} plays an important role in the binding of malonyl moieties to the transacylase domain, but is not required for binding of acetyl moieties.

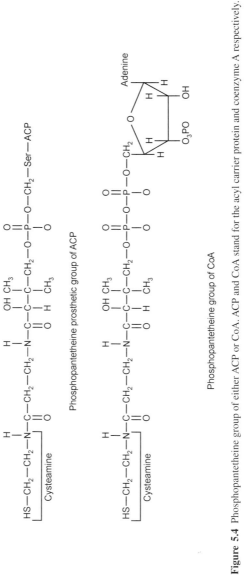

Figure 5.4 Phosphopantetheine group of either ACP or CoA. ACP and CoA stand for the acyl carrier protein and coenzyme A respectively.

5.3.2.4   Condensation reactions

Condensing enzymes comprise a structurally and functionally related family that are found in various metabolic pathways. In fatty acid synthesis, the chain elongation step consists of the condensation of acyl groups, derived from acyl ACP or acyl CoA with malonyl ACP. The two substrates are produced by the acetyl and malonyl CoA–ACP transacylases. These condensation reactions are catalyzed by a group of enzymes, the β-keto-ACP synthases or KAS (EC 2.3.1.41) that differ in their specificity. The reaction can be divided in three steps:

> *Step I*: The acyl group of acyl ACP is transferred to a Cys residue at the active site of the enzyme leading to the formation of a thioester.
>
> *Step II*: There is a decarboxylation of malonyl ACP generating a carbanion.
>
> *Step III*: There is a formation of a carbon–carbon bond via a nucleophilic attack of the carbanion to the carbon atom of the thioester.

The β-keto-ACP synthases (KAS) can be part of multienzyme complexes, or domains of large multifunctional polypeptide chains such as the mammalian fatty acid synthase, or single enzymes, such as in plants and most bacteria. Most of these enzymes are active as dimers made up of two apparently identical subunits of molecular mass around 42–46 kDa. The type II fatty acid synthases of higher plants and *E. coli* contain three condensing enzymes called KAS I, II and III. They differ in amino acid sequence, in chain-length specificity and in sensitivity to an inhibitor of condensing enzymes, namely cerulenin.

KAS I (β-keto-ACP synthase I) is responsible for the bulk of the condensation reactions during successive rounds of the fatty acid synthase chain-lengthening process. It may be responsible for the elongation of C10:1 that cannot be catalyzed by KAS II. The KAS I subunit has a molecular mass of 42.6 kDa, is a dimer in its native state and this dimer possesses both malonyl and acyl ACP binding sites. It is inhibited by cerulenin via covalent modification of the active site sulfhydryl (Rock and Cronan, 1996). The crystal structure of FabB, the *E. coli* homolog, was obtained at 2.3Å resolution (Olsen *et al.*, 1999).

KAS II (β-keto-ACP synthase II) is responsible exclusively for the conversion of palmitate to stearate. In *E. coli*, KAS II is responsible for the elongation of palmitoleic acid C16:1 to *cis*-vaccenic acid C18:1, and is also sensitive to cerulenin. It was the first β-keto-ACP synthase (FabF) from *E. coli* to be crystallized (Huang *et al.*, 1998). The crystal structure was resolved at 2.4Å, and the subunit has the shape of a disk with overall dimensions $30 \times 58 \times 60$ Å. The quaternary structure of this enzyme reveals a homodimer state.

Siggaard-Andersen identified one conserved cysteine residue in a comparison of 42 condensing enzymes (Siggaard-Andersen, 1993). This residue corresponds to Cys_{163} in KAS II and this residue was assigned as the active site cysteine. Comparison of the structure of this KAS with thiolase, an enzyme that catalyzes the degradation of β-ketoacyl CoA structures in the β-oxidation pathway, shows that these enzymes contain the same fold, suggesting that these enzymes might be evolutionary related.

The reaction catalyzed by KAS III (condensing enzyme, β-keto-ACP synthases III; FabH in *E. coli*) is of great importance since the substrate specificity and particularly its high degree of specificity towards acetyl CoA (Tsay *et al.*, 1992), was shown to be the determining factor in the biosynthesis of branched-chain fatty acids by type II fatty acid synthases (Choi *et al.*, 2000). Overexpression of the enzyme leads to an increase in shorter-chain fatty acids in the membrane phospholipids (Tsay *et al.*, 1992). This enzyme is also capable of catalyzing acetyl CoA transacetylation (Tsay *et al.*, 1992). In common with its bacterial counterpart, KAS III from spinach possesses acetyl CoA transacetylase activity but this activity is 90-fold slower than the condensation rate with malonyl ACP (Clough *et al.*, 1992). The spinach enzyme has been purified and is thought to be a dimer (2 subunits of 40.5 kDa) (Clough *et al.*, 1992).

FabH has also been crystallized (Qiu *et al.*, 1999; Janson *et al.*, 2000). As with other components of the dissociated enzyme of type II fatty acid synthase, it exists as a 'free' enzyme in the dimeric form with a molecular mass of around 70 kDa. In eukaryotes, the enzymatic activity is encoded in the *N*-terminus of a large multifunctional type I fatty acid synthase, which displays no overall sequence homology to the bacterial enzyme. The crystal structure of FabH in the presence or in the absence of ligand (acetyl CoA) (Qiu *et al.*, 1999) reveals a quasi twofold symmetry and a small binding pocket, which is in good agreement with the fact that FabH accepts acetyl, propionyl and butyryl CoA, but no long-chain acyl CoA (Ohlrogge and Benning, 2000). Three amino acids are conserved (Cys_{112}, His_{244}, Asn_{274}) and are associated with the active site. This pocket is formed by the convergence of two α-helices and is accessed via a narrow hydrophobic tunnel (Davies *et al.*, 2000). It is a dimeric slab-shaped protein with overall dimensions $45 \times 55 \times 80 Å$. However, despite the lack of overall sequence similarities between FabF and FabH, the FabH core is similar to that of FabF.

A notable difference between KAS III and the other KASs is the absence in KAS III (Jaworski *et al.*, 1994) of a conserved Gly-Pro region, seven residues upstream from the cysteine in other KASs (Revill and Leadlay, 1991). The enzyme from *E. coli*, nonetheless, possesses the Ala-Cys-Ala tripeptide characteristic of condensing enzymes (Tsay *et al.*, 1992). KAS III proteins either from *E. coli* and from spinach, both have a clear specificity for acetyl CoA, suggesting that the physiological function of KAS III is to catalyze the initial reaction in fatty acid biosynthesis.

In Gram-positive bacteria, such as *Bacillus subtilis*, FabH enzymes are less active with acetyl CoA and instead use a branched-chain acyl CoA. In this organism, two enzymes FabH1 and FabH2 have been found; the physiological significance of these two forms has not yet been elucidated (Choi *et al.*, 2000). Streptomycetes are known to produce branched-chain fatty acids and a minor proportion of straight-chain fatty acids. Preliminary studies suggest that fatty acids are synthesized by type I fatty acid synthases (Rossi and Corcoran, 1973), but recently it was shown that a streptomycetes has a type II fatty acid synthase. The FabH enzyme from *Streptomyces glaucescens* was purified and it was shown that this enzyme is both responsible for initiating branched and straight-chain fatty acid biosynthesis. It is a homodimer of 72 kDa (Han *et al.*, 1998). In *Streptomyces*, a FabH-independent pathway for straight-chain fatty acid biosynthesis also operates. This pathway is not yet been fully characterized but the possibility that a type I fatty acid synthase operates cannot yet be ruled out.

5.3.2.5 Further reactions of fatty acid synthesis

The successive addition of two carbon units to the growing fatty acyl chain requires the participation of two reductases and a dehydratase, in addition to the condensation reactions.

β-ketoacyl-ACP reductase catalyzes the first reductive step and has been purified to homogeneity from spinach, avocado and rape (Slabas and Fawcett, 1992). Two forms have been found, differing in their reduced nucleotide preference with the isoform utilizing NADPH identified as a 130 kDa protein. A second enzyme, β-hydroxyacyl ACP dehydrase has been purified from spinach (Shimakata and Stumpf, 1982). The enoyl-ACP reductase (ENR) catalyzes the fourth reaction involved in chain lengthening, the last reductive step of fatty acid synthesis. Two forms are present which utilize either NADH or NADPH, and the crystal structure of the *E. coli* ENR has been determined at 2.1 Å. It is a tetramer and has a domain reminiscent of the Rossman fold (Baldock *et al.*, 1998). A close functional relationship between the plant and bacterial ENR has been inferred from genetic complementation of an *envM* (enoyl-ACP reductase gene) mutant of *E. coli* by the plant ENR (Kater *et al.*, 1994). The 3D structure (Rafferty *et al.*, 1994) of the ENR from *Brassica napus* has also been determined and the kinetic mechanism of this enzyme elucidated. It follows a compulsory-order mechanism resulting in a ternary complex with NADH binding before the acyl substrate (Fawcett *et al.*, 2000).

5.3.2.6 Termination of fatty acid synthesis

It has long been known that the chloroplast stroma contains a long chain acyl-ACP thioesterase which shows high activity with oleyl-ACP, lower activity with palmitoyl-ACP and little activity with stearoyl-ACP. This substrate specificity determines the normal pattern of products synthesized by the plastid fatty acid synthase. In higher plants, these enzymes can be classified into two families: FatA represents the commonly found 18:1 ACP thioesterase and FatB includes thioesterases preferring acyl-ACP with saturated acyl groups (Jones *et al.*, 1995). Plant thioesterases (TEs) have no similarity with animal or bacterial TEs (Voelker *et al.*, 1992). Moreover while these latter enzymes have a serine residue in the active site, the plant TE has cysteine and histidine in their active site (Yuan *et al.*, 1995; 1996a; 1996b). In addition to TEs, acyltransferases have also been identified in plants. They are responsible for the transfer of the fatty acid from ACP to glycerol-3-phosphate (Ohlrogge and Browse, 1995). Whether the fatty acid is released from ACP by TE or an acyltransferase determines whether the acid is exported from the plastid.

5.3.2.7   Type I fatty acid synthases

Fatty acid synthase (FAS) from the budding yeast, *Saccharomyces cerevisiae* (Lynen, 1980) was purified a long time ago and its molecular mass is about 2.3×10^6 Da. It is a multifunctional enzyme consisting of six copies each of a multifunctional α polypeptide (207,863 Da) and a multifunctional β polypeptide (220,077 Da) with the subunits arranged as $\alpha_6\beta_6$. Dissociation into its inactive subunits has been performed by repeated freeze-thawed cycles in 1 M NaCl and LiCl. Reassociation into an active complex can be achieved by decreasing the ionic strength (Volpe and Vagelos, 1976). This enzyme has been extensively studied; it has been crystallized, and studied by electron microscopy and small-angle X-ray analysis. Single particle images from stain and cryoelectron microscopy indicate that the molecule has a shape similar to a prolate ellipsoid (Stoops *et al.*, 1992). The 25 Å resolution structure of the yeast enzyme reveals an unusual feature in that, apparently, all the catalytic domain is inside this barrel-like structure. In contrast, most other enzymes of this type have their catalytic sites disposed in crevices or pits on the outside surface of the structure (Kolodziej *et al.*, 1996). FAS from *Schizosaccharomyces pombe* also forms a heterododecameric structure and electron micrographs of negatively stained molecule suggest that the complex adopts a unique barrel-shaped structure (Niwa *et al.*, 1998).

   If the yeast enzyme differs with regard to spatial organization (multifunctional enzyme) from the prokaryotic type (series of discrete enzymes), it also appears to differ in the reactions that it catalyzes, in particular the transacetylation reactions (Stoops *et al.*, 1990). In yeast, the transacetylation occurs from CoA to a Ser-OH (acetyl transacylase site) to a Cys-SH (β-ketoacyl synthase site), which differs from transfer from CoA to Ser-OH to 4′-phosphopantetheine-SH (ACP site) to Cys-SH, the transfers catalyzed by the prokaryotic synthases.

   In common with the yeast enzyme, animal FASs are multifunctional enzymes that consist of two multifunctional polypeptides (Wakil, 1989). Although all of the functional domains are present on a single polypeptide, coupling of the individual activities necessary for catalysis of the overall reaction of fatty acid synthesis is performed only by the dimer. Native FAS in animal and human cells is a homodimer arranged in a head-to-tail manner (Jayakumar *et al.*, 1996; Witkowski *et al.*, 1999). In this long-established structural model, the two subunits are arranged in this head-to-tail orientation, so that the β-ketoacyl synthase of one subunit is juxtaposed with the ACP domain of the opposite subunit and two sites for palmitate synthesis are formed per dimer. However recently, it has been shown that the β-ketoacyl synthase of one subunit can interact with either subunit. Thus the old model should be revisited to accommodate the findings that head-to-tail and functional contacts are possible both between and within subunits (Witkowski *et al.*, 1999).

The different domains of FAS have also been well studied. The component activities of human FAS were initially grouped into three domains: domain I (ketosynthase and transacylases); domain II (dehydratase, enoyl reductase, ketoreductase and ACP); and domain III (thioesterase), although later the dehydratase activity was assigned to domain I (Chirala *et al.*, 1997). Cloning and expression of the different domains of the multifunctional FAS will clearly elucidate the organization of the so-called 'proficient' enzyme (Wakil, 1989).

It has been shown that a high degree of homology exists between animal FASs (particularly in the active sites of chicken and rat enzymes), but the chicken and yeast FASs show a lower degree of homology (Chang and Hammes, 1989). Differences between FASs from animal origin and of yeast center around the enzyme responsible for the second reduction step, i.e. the reduction of the 2,3-enoyl intermediate to the saturated acyl residue. In many FASs, hydrogen is transferred directly from NADPH to the double bond, whereas in the enzyme from yeast, FMN acts as an intermediate hydrogen carrier (Lynen, 1980).

Another interesting model for the study of FAS is the phytoflagellate *Euglena gracilis*, as it occupies an intermediate position between higher plants and non-plant eukaryotes in having both cytoplasmic type I and chloroplast type II FAS. Surprisingly, there is no evidence for a β-keto-ACP synthase III for priming, as has been reported in type II FAS of higher plants and bacteria (Worsham *et al.*, 1993). The *E. gracilis* type I FAS, induced in streptomycin-bleached cells, has been characterized. It has a molecular mass of 270 kDa. The enzyme, *in vitro*, mainly synthesizes palmitic acid or its CoA esters from acetyl- and malonyl-CoA as substrates, with K_m values of 20 and 31 µM, respectively. In common with FASs of other lower eukaryotes, the *Euglena* FAS is a flavoprotein but, in contrast to yeast, the flavin cofactor is covalently attached to the enzyme protein (Siebenlist *et al.*, 1991).

5.4 Kinetic aspects of multienzyme complexes

It has been mentioned that the association between enzymes may result in the change of their properties. The aim of this section is to explain the origin of the energy required to generate these modifications. The information content (I) based on the sequence of a protein has been well defined. It is:

$$I = \log_2 \Omega \tag{5.1}$$

where Ω represents the number of messages that may be conveyed through sequences of n amino acids of different nature. Recently, a new concept (Ricard *et al.*, 1998) has been introduced in which the information content may be defined by the following equation:

$$I_\varepsilon = \log_2 \Omega_\varepsilon \tag{5.2}$$

where Ω_ε represents the number of complexions or number of messages that may be conveyed by a population of the same protein molecules distributed, according to Boltzmann's law, over all the different energy states associated with different conformations. Whether an enzyme is bound (to a surface or to another protein) or is free, two numbers of complexions Ω_ε^b and Ω_ε^f may thus be defined. Association of an enzyme with a rigid body, results in the molecules being partitioned over a smaller number of energy states, and the value Ω decreases. It can be deduced that:

$$\Omega_\varepsilon^f > \Omega_\varepsilon^b \tag{5.3}$$

The difference in the number of complexions explains the loss of entropy (S) and of the information content (i.e. a decrease of the number of messages) of the bound enzyme. This loss is equivalent to:

$$S^b - S^f = k_B \ln \frac{\Omega_\varepsilon^b}{\Omega_\varepsilon^f} \tag{5.4}$$

where k_B is the Boltzmann constant.

Moreover the free energy content of the bound enzyme, during the binding process, can be shown to increase relative to that of the free state. The change of free energy of the enzyme during the binding process is:

$$\Delta G = G^b - G^f = \Delta H - k_B T \ln \frac{\Omega_\varepsilon^b}{\Omega_\varepsilon^f} \tag{5.5}$$

where G, H and T represent Gibbs' free energy, enthalpy and the absolute temperature, respectively. As ΔH is very small, ΔG can be expected to be positive. This change does not correspond with the free energy of association of the enzyme with the rigid body but represents the free energy stored in the enzyme upon its association with the rigid body. This situation could arise if some catalytic conformation were imposed on one enzyme by association with another. The stored energy may then be used and results in an alteration of the properties of the enzymes. An extreme situation may be illustrated by complexes in which the activity of one component depends absolutely on the presence of another. The activity of oxidized PRK within the bi-enzyme complex, previously described in *Chlamydomonas reinhardtii* cells, and the observation that, in the yeast FAS, the dissociated components have no enzymatic activity, support this contention.

Statistical mechanics offers the possibility to express how this energy is used and the extent of the instruction and energy transfer (Gontero *et al.*, 1994; Lebreton *et al.*, 1997a,b; Ricard *et al.*, 1998; Lebreton and Gontero, 1999).

Channeling, described as a way to increase catalytic efficiency in metabolic fluxes, has been extensively discussed previously. Moreover, it has been observed that even if an effect of channeling on reaction intermediate concentrations exists in the steady state, this is too small to fulfill a useful regulatory role (Cornish-Bowden and Cardenas, 1993). Therefore, it will not be further developed here.

5.5   Conclusion

Multienzyme complexes have obvious importance in carbon metabolism. The differences in the composition of the complexes obtained in different laboratories probably results from the diversity of purification protocols used (ammonium sulfate precipitation, use of polyethylene glycol, etc). Also, the non-covalent protein–protein interactions holding the diverse enzymes together are relatively weak and probably do not survive the conditions of cell disruption. Some associations to the thylakoid membranes have been reported by different groups and may reflect a link between carbon metabolism and the energy-transducing machinery of photosynthesis.

In plants, as in bacteria, the fatty acid synthesis seems to be catalyzed by a series of separate enzymes and an ACP, while in fungi and animals the enzymes and the acyl carrier protein are part of a multifunctional proficient enzyme. Also, the three components of the ACCase in animals and fungi are on a single polypeptide while in bacteria the enzyme is a multienzyme dissociable into three components. In plants, the situation is more complex since both types, a multienzyme form in the plastid and a multifunctional form in the cytosol, are present. These data support the view that in metabolic pathways in which the intermediates are not required for other metabolic purposes, evolution seems to have proceeded toward the formation of covalent linkage of sequential metabolic reactions. In plants, some intermediates (acyl chains) are metabolized within the chloroplast, but a major portion is exported outside this organelle. Thus it might be suggested that if a supramolecular structure exists, the association between its components are likely to be loose. It might therefore explain why no multienzyme plant FASs has been purified so far. Nonetheless, if biochemical evidence of multienzyme complexes in FAS have not yet been reported, some functional evidence suggests that they might exist. Indeed, the *in vivo* stromal concentrations (31–51 µM) of some metabolites (e.g acetyl CoA and its esters) fall far short of those required to account for observed rates of fatty acid synthesis from acetate in intact chloroplasts (Roughan, 1997). Moreover, Roughan and Ohlrogge (1996) have reported that all of the components (enzymes and metabolites) required to produce high rates of fatty acid synthesis from acetate remain functional and within the chloroplast, even under conditions where the bulk of the soluble protein content was released. This result strongly supports the existence of a multienzyme complex that, furthermore, seems to be associated with the membranes (thylakoids) and therefore could be a metabolon. In the near future, scientists will probably be able to isolate such structures.

Supramolecular structures, even if they are substructures of existing complexes *in vivo*, are good candidates with which to study the regulation of cell metabolism. Their study, although far from trivial, will clearly help in the understanding of how metabolic pathways indeed are regulated. We believe that the study of multienzyme complexes or metabolons will be developed and extended in the future on the basis of data that have been accumulated to the present time.

References

Agou, F. and Mirande, M. (1997) Aspartyl-tRNA synthetase from rat: *in vitro* functional analysis of its assembly into the multisynthetase complex. *Eur. J. Biochem.*, **243**, 259-267.

Alban, C., Baldet, P. and Douce, R. (1994) Localization and characterization of two structually different forms of acetyl-CoA carboxylase in young pea leaves, of which one is sensitive to aryloxyphenoxypropionate herbicides. *Biochem. J.*, **300**, 557-565.

Avilan, L., Gontero, B., Lebreton, S. and Ricard, J. (1997a) Information transfer in multienzyme complexes—2. The role of Arg64 of *Chlamydomonas reinhardtii* phosphoribulokinase in the information transfer between glyceraldehyde-3-phosphate dehydrogenase and phosphoribulokinase. *Eur. J. Biochem.*, **250**, 296-302.

Avilan, L., Gontero, B., Lebreton, S. and Ricard, J. (1997b) Memory and imprinting effects in multienzyme complexes—I. Isolation, dissociation, and reassociation of a phosphoribulokinase-glyceraldehyde-3-phosphate dehydrogenase complex from *Chlamydomonas reinhardtii* chloroplasts. *Eur. J. Biochem.*, **246**, 78-84.

Avilan, L., Lebreton, S. and Gontero, B. (2000) Thioredoxin activation of phosphoribulokinase in a bi-enzyme complex from *Chlamydomonas reinhardtii* chloroplasts. *J. Biol. Chem.*, **275**, 9447-9451.

Baalmann, E., Scheibe, R., Cerff, R. and Martin, W. (1996) Functional studies of chloroplast glyceraldehyde-3-phosphate dehydrogenase subunits A and B expressed in *Escherichia coli*: formation of highly active A4 and B4 homotetramers and evidence that aggregation of the B4 complex is mediated by the B subunit carboxy terminus. *Plant Mol. Biol.*, **32**, 505-513.

Babadzhanova, M.A., Bakaeva, N.P. and Babadzhanova, M.P. (2000) Functional properties of the multienzyme complex of Calvin cycle key enzymes. *Russian J. of Plant Physiol.*, **47**, 23-31.

Baldock, C., Rafferty, J.B., Stuitje, A.R., Slabas, A.R. and Rice, D.W. (1998) The X-ray structure of *Escherichia coli* enoyl reductase with bound NAD+ at 2.1Å resolution. *J. Mol. Biol.*, **284**, 1529-1546.

Buchanan, B.B. (1980) Role of light in the regulation of chloroplast enzymes. *Annu. Rev. Plant Physiol.*, **31**, 341-374.

Cerff, R. (1978) Glyceraldehyde-3-phosphate dehydrogenase (NADP) from *Sinapis alba* L. NAD(P)-induced conformation changes of the enzyme. *Eur. J. Biochem.*, **82**, 45-53.

Cerff, R. and Chambers, S.E. (1978) Glyceraldehyde-3-phosphate dehydrogenase (NADP) from *Sinapis alba* L. Isolation and electrophoretic characterization of isoenzymes. *Hoppe-Seylers Z. Physiol. Chem.*, **359**, 769-772.

Chang, S.I. and Hammes, G.G. (1989) Homology analysis of the protein sequences of fatty acid synthases from chicken liver, rat mammary gland, and yeast. *Proc. Natl Acad. Sci. USA*, **86**, 8373-8376.

Chirala, S.S., Huang, W.Y., Jayakumar, A., Sakai, K. and Wakil, S.J. (1997) Animal fatty acid synthase: functional mapping and cloning and expression of the domain I constituent activities. *Proc. Natl Acad. Sci. USA*, **94**, 5588-5593.

Choi, K.H., Heath, R.J. and Rock, C.O. (2000) beta-ketoacyl-acyl carrier protein synthase III (FabH) is a determining factor in branched-chain fatty acid biosynthesis. *J. Bacteriol.*, **182**, 365-370.

Clasper, S., Easterby, J.S. and Powls, R. (1991) Properties of two high-molecular-mass forms of glyceraldehyde-3-phosphate dehydrogenase from spinach leaf, one of which also possesses latent phosphoribulokinase activity. *Eur. J. Biochem.*, **202**, 1239-1246.

Clasper, S., Chelvarajan, R.E., Easterby, J.S. and Powls, R. (1994) Isolation of multiple dimeric forms of phosphoribulokinase from an alga and a higher plant. *Biochim. Biophys. Acta*, **1209**, 101-106.

Clough, R.C., Matthis, A.L., Barnum, S.R. and Jaworski, J.G. (1992) Purification and characterization of 3-ketoacyl-acyl carrier protein synthase III from spinach. A condensing enzyme utilizing acetyl-coenzyme A to initiate fatty acid synthesis. *J. Biol. Chem.*, **267**, 20,992-20,998.

Cornish-Bowden, A. and Cardenas, M.L. (1993) Channelling can affect concentrations of metabolic intermediates at constant net flux: artefact or reality? [published erratum appears in *Eur. J. Biochem.*, 1993, **216**, 879]. *Eur. J. Biochem.*, **213**, 87-92.

Davies, C., Heath, R.J., White, S.W. and Rock, C.O. (2000) The 1.8 Å crystal structure and active-site architecture of beta-ketoacyl-acyl carrier protein synthase III (FabH) from *Escherichia coli*. *Structure Fold. Des.*, **8**, 185-195.

Deutscher, M.P. (1984) The eucaryotic aminoacyl-tRNA synthetase complex: suggestions for its structure and function. *J. Cell Biol.*, **99**, 373-377.

Dunn, M.F., Aguilar, V., Brzovic, P. *et al.* (1990) The tryptophan synthase bienzyme complex transfers indole between the alpha- and beta-sites via a 25-30 Å long tunnel. *Biochemistry*, **29**, 8598-8607.

Easterby, J.S. (1989) The analysis of metabolite channelling in multienzyme complexes and multifunctional proteins [published erratum appears in *Biochem. J.*, 1990, **267**, 843]. *Biochem. J.*, **264**, 605-607.

Fawcett, T., Copse, C.L., Simon, J.W. and Slabas, A.R. (2000) Kinetic mechanism of NADH-enoyl-ACP reductase from *Brassica napus*. *FEBS Lett.*, **484**, 65-68.

Froehlich, J.E., Poorman, R., Reardon, E., Barnum, S.R. and Jaworski, J.G. (1990) Purification and characterization of acyl carrier protein from two cyanobacteria species. *Eur. J. Biochem.*, **193**, 817-825.

Giudici-Orticoni, M.T., Gontero, B., Rault, M. and Ricard, J. (1992) Organisation structurale et fonctionnelle d'enzymes du cycle de Benson-Calvin à la surface des thylakoïdes des chloroplastes d'Epinard. *C. R. Acad. Sci. III*, **314**, 477-483.

Gontero, B., Cardenas, M.L. and Ricard, J. (1988) A functional five-enzyme complex of chloroplasts involved in the Calvin cycle. *Eur. J. Biochem.*, **173**, 437-443.

Gontero, B., Mulliert, G., Rault, M., Giudici-Orticoni, M.T. and Ricard, J. (1993) Structural and functional properties of a multi-enzyme complex from spinach chloroplasts. 2. Modulation of the kinetic properties of enzymes in the aggregated state. *Eur. J. Biochem.*, **217**, 1075-1082.

Gontero, B., Giudici-Orticoni, M.T. and Ricard, J. (1994) The modulation of enzyme reaction rates within multi-enzyme complexes. 2. Information transfer within a chloroplast multi-enzyme complex containing ribulose bisphosphate carboxylase-oxygenase. *Eur. J. Biochem.*, **226**, 999-1006.

Goodsell, D.S. (1991) Inside a living cell. *Trends Biochem. Sci.*, **16**, 203-206.

Guchhait, R.B., Polakis, S.E., Dimroth, P., Stoll, E., Moss, J. and Lane, M.D. (1974) Acetyl coenzyme A carboxylase system of *Escherichia coli*. Purification and properties of the biotin carboxylase, carboxyltransferase, and carboxyl carrier protein components. *J. Biol. Chem.*, **249**, 6633-6645.

Han, L., Lobo, S. and Reynolds, K.A. (1998) Characterization of beta-ketoacyl-acyl carrier protein synthase III from *Streptomyces glaucescens* and its role in initiation of fatty acid biosynthesis. *J. Bacteriol.*, **180**, 4481-4486.

Harwood, J.L. (1996) Recent advances in the biosynthesis of plant fatty acids. *Biochim. Biophys. Acta*, **1301**, 7-56.

Hawkins, A.R. and Lamb, H.K. (1995) The molecular biology of multidomain proteins. Selected examples. *Eur. J. Biochem.*, **232**, 7-18.

Hlousek-Radojcic, A., Post-Beittenmiller, D. and Ohlrogge, J. (1992) Expression of constitutive and tissue specific acyl carrier protein isoforms in *Arabidopsis*. *Plant Physiol.*, **98**, 206-14.

Holak, T.A., Kearsley, S.K., Kim, Y. and Prestegard, J.H. (1988) Three-dimensional structure of acyl carrier protein determined by NMR pseudoenergy and distance geometry calculations. *Biochemistry*, **27**, 6135-6142.

Hosur, M.V., Sainis, J.K. and Kannan, K.K. (1993) Crystallization and X-ray analysis of a multienzyme complex containing RUBISCO and RuBP. *J. Mol. Biol.*, **234**, 1274-1278.

Huang, W., Jia, J., Edwards, P., Dehesh, K., Schneider, G. and Lindqvist, Y. (1998) Crystal structure of beta-ketoacyl-acyl carrier protein synthase II from *E. coli* reveals the molecular architecture of condensing enzymes. *EMBO J.*, **17**, 1183-1191.

Hyde, C.C., Ahmed, S.A., Padlan, E.A., Miles, E.W. and Davies, D.R. (1988) Three-dimensional structure of the tryptophan synthase alpha 2 beta 2 multienzyme complex from *Salmonella typhimurium*. *J. Biol. Chem.*, **263**, 17,857-17,871.

Iverson, A.J., Bianchi, A., Nordlund, A.C. and Witters, L.A. (1990) Immunological analysis of acetyl-CoA carboxylase mass, tissue distribution and subunit composition. *Biochem. J.*, **269**, 365-371.

Janson, C.A., Konstantinidis, A.K., Lonsdale, J.T. and Qiu, X. (2000) Crystallization of *Escherichia coli* beta-ketoacyl-ACP synthase III and the use of a dry flash-cooling technique for data collection. *Acta Crystallogr. D Biol. Crystallogr.*, **56**, 747-748.

Jayakumar, A., Huang, W.Y., Raetz, B., Chirala, S.S. and Wakil, S.J. (1996) Cloning and expression of the multifunctional human fatty acid synthase and its subdomains in *Escherichia coli*. *Proc. Natl Acad. Sci. USA*, **93**, 14,509-14,514.

Jaworski, J.G., Tai, H., Ohlrogge, J.B. and Post-Beittenmiller, D. (1994) The initial reactions of fatty acid biosynthesis in plants. *Prog. Lipid Res.*, **33**, 47-54.

Jebanathirajah, J.A. and Coleman, J.R. (1998) Association of carbonic anhydrase with a Calvin cycle enzyme complex in *Nicotiana tabacum*. *Planta*, **204**, 177-182.

Johnston, M.L., Luethy, M.H., Miernyk, J.A. and Randall, D.D. (1997) Cloning and molecular analyses of the *Arabidopsis thaliana* plastid pyruvate dehydrogenase subunits. *Biochim. Biophys. Acta*, **1321**, 200-206.

Jones, A., Davies, H.M. and Voelker, T.A. (1995) Palmitoyl-acyl carrier protein (ACP) thioesterase and the evolutionary origin of plant acyl-ACP thioesterases. *Plant Cell*, **7**, 359-371.

Kater, M.M., Koningstein, G.M., Nijkamp, H.J. and Stuitje, A.R. (1994) The use of a hybrid genetic system to study the functional relationship between prokaryotic and plant multi-enzyme fatty acid synthetase complexes. *Plant Mol. Biol.*, **25**, 771-790.

Ke, J., Behal, R.H., Back, S.L., Nikolau, B.J., Wurtele, E.S. and Oliver, D.J. (2000a) The role of pyruvate dehydrogenase and acetyl-coenzyme A synthetase in fatty acid synthesis in developing *Arabidopsis* seeds. *Plant Physiol.*, **123**, 497-508.

Ke, J., Wen, T.N., Nikolau, B.J. and Wurtele, E.S. (2000b) Coordinate regulation of the nuclear and plastidic genes coding for the subunits of the heteromeric acetyl-coenzyme A carboxylase. *Plant Physiol.*, **122**, 1057-1071.

Kim, Y. and Prestegard, J.H. (1989) A dynamic model for the structure of acyl carrier protein in solution. *Biochemistry*, **28**, 8792-8797.

Kolodziej, S.J., Penczek, P.A., Schroeter, J.P. and Stoops, J.K. (1996) Structure-function relationships of the *Saccharomyces cerevisiae* fatty acid synthase. Three-dimensional structure. *J. Biol. Chem.*, **271**, 28,422-28,429.

Kondo, H., Shiratsuchi, K., Yoshimoto, T. *et al.* (1991) Acetyl-CoA carboxylase from *Escherichia coli*: gene organization and nucleotide sequence of the biotin carboxylase subunit. *Proc. Natl Acad. Sci. USA*, **88**, 9730-9733.

Konishi, T. and Sasaki, Y. (1994) Compartmentalization of two forms of acetyl-CoA carboxylase in plants and the origin of their tolerance toward herbicides. *Proc. Natl Acad. Sci. USA*, **91**, 3598-3601.

Kozaki, A. and Sasaki, Y. (1999) Light-dependent changes in redox status of the plastidic acetyl-CoA carboxylase and its regulatory component. *Biochem. J.*, **339**, 541-546.

Kozaki, A., Kamada, K., Nagano, Y., Iguchi, H. and Sasaki, Y. (2000) Recombinant carboxyltransferase responsive to redox of pea plastidic acetyl-CoA carboxylase. *J. Biol. Chem.*, **275**, 10,702-10,708.

Kuhn, D.N., Knauf, M. and Stumpf, P.K. (1981) Subcellular localization of acetyl-CoA synthetase in leaf protoplasts of *Spinacia oleracea* leaf cells. *Arch. Biochem. Biophys.*, **209**, 441-450.

Lebreton, S. and Gontero, B. (1999) Memory and imprinting in multienzyme complexes. Evidence for information transfer from glyceraldehyde-3-phosphate dehydrogenase to phosphoribulokinase under reduced state in *Chlamydomonas reinhardtii*. *J. Biol. Chem.*, **274**, 20,879-20,884.

Lebreton, S., Gontero, B., Avilan, L. and Ricard, J. (1997a) Information transfer in multienzyme complexes—1. Thermodynamics of conformational constraints and memory effects in the bienzyme glyceraldehyde-3-phosphate-dehydrogenase-phosphoribulokinase complex of *Chlamydomonas reinhardtii* chloroplasts. *Eur. J. Biochem.*, **250**, 286-295.

Lebreton, S., Gontero, B., Avilan, L. and Ricard, J. (1997b) Memory and imprinting effects in multienzyme complexes—II. Kinetics of the bienzyme complex from *Chlamydomonas reinhardtii* and hysteretic activation of chloroplast oxidized phosphoribulokinase. *Eur. J. Biochem.*, **246**, 85-91.

Leegood, R.C. (1990) Enzymes in the Calvin Cycle, in *Methods in Plant Biochemistry, Vol. 3, Enzymes of Primary Metabolism*, (ed. P.J. Lea), Academic Press, New York, pp. 15-37.

Lynen, F. (1980) On the structure of fatty acid synthetase of yeast. *Eur. J. Biochem.*, **112**, 431-442.

Maciozek, J., Anderson, J.B. and Anderson, L.E. (1990) Isolation of chloroplastic phosphoglycerate kinase. Kinetics of the two enzyme phosphoglycerate kinase/glyceraldehyde-3-phosphate dehydrogenase couple. *Plant Physiol.*, **94**, 291-296.

Mendes, P., Kell, D.B. and Westerhoff, H.V. (1996) Why and when channelling can decrease pool size at constant net flux in a simple dynamic channel. *Biochim. Biophys. Acta*, **1289**, 175-186.

Mirande, M., Lazard, M., Martinez, R. and Latreille, M.T. (1992) Engineering mammalian aspartyl-tRNA synthetase to probe structural features mediating its association with the multisynthetase complex. *Eur. J. Biochem.*, **203**, 459-466.

Müller, B. (1972) A labile CO_2-fixing enzyme complex in spinach chloroplasts. *Z. Naturforsch.*, **27b**, 925-932.

Nicholson, S., Easterby, J.S. and Powls, R. (1987) Properties of a multimeric protein complex from chloroplasts possessing potential activities of NADPH-dependent glyceraldehyde-3-phosphate dehydrogenase and phosphoribulokinase. *Eur. J. Biochem.*, **162**, 423-431.

Niwa, H., Katayama, E., Yanagida, M. and Morikawa, K. (1998) Cloning of the fatty acid synthetase beta subunit from fission yeast, coexpression with the alpha subunit, and purification of the intact multifunctional enzyme complex. *Protein Expr. Purif.*, **13**, 403-413.

Ohlrogge, J. and Benning, C. (2000) Unraveling plant metabolism by EST analysis. *Curr. Opin. Plant Biol.*, **3**, 224-228.

Ohlrogge, J. and Browse, J. (1995) Lipid biosynthesis. *Plant Cell*, **7**, 957-970.

Ohlrogge, J. and Jaworski, J.G. (1997) Regulation of fatty acid synthesis. *Annu. Rev. Plant Physiol. Plant Mol. Biol.*, **48**, 109-136.

Olsen, J.G., Kadziola, A., von Wettstein-Knowles, P., Siggaard-Andersen, M., Lindquist, Y. and Larsen, S. (1999) The X-ray crystal structure of beta-ketoacyl [acyl carrier protein] synthase I. *FEBS Lett.*, **460**, 46-52.

Ovadi, J. and Srere, P.A. (2000) Macromolecular compartmentation and channeling. *Int. Rev. Cytol.*, **192**, 255-280.

Pan, P., Woehl, E. and Dunn, M.F. (1997) Protein architecture, dynamics and allostery in tryptophan synthase channeling. *Trends Biochem. Sci.*, **22**, 22-27.

Persson, L.O. and Johansson, G. (1989) Studies of protein–protein interaction using countercurrent distribution in aqueous two-phase systems. Partition behaviour of six Calvin-cycle enzymes from a crude spinach (*Spinacia oleracea*) chloroplast extract. *Biochem J.*, **259**, 863-870.

Pohlmeyer, K., Paap, B.K., Soll, J. and Wedel, N. (1996) CP12: a small nuclear-encoded chloroplast protein provides novel insights into higher-plant GAPDH evolution. *Plant Mol. Biol.*, **32**, 969-978.

Porter, M.A. (1990) The aggregation states of spinach phosphoribulokinase. *Planta*, **181**, 349-357.

Purcarea, C., Evans, D.R. and Herve, G. (1999) Channeling of carbamoyl phosphate to the pyrimidine and arginine biosynthetic pathways in the deep sea hyperthermophilic archaeon Pyrococcus abyssi. *J. Biol. Chem.*, **274**, 6122-6129.

Qiu, X., Janson, C.A., Konstantinidis, A.K. *et al.* (1999) Crystal structure of beta-ketoacyl-acyl carrier protein synthase III. A key condensing enzyme in bacterial fatty acid biosynthesis. *J. Biol. Chem.*, **274**, 36,465-36,471.

Rafferty, J.B., Simon, J.W., Stuitje, A.R., Slabas, A.R., Fawcett, T. and Rice, D.W. (1994) Crystallization of the NADH-specific enoyl acyl carrier protein reductase from *Brassica napus*. *J. Mol. Biol.*, **237**, 240-242.

Rangan, V.S. and Smith, S. (1997) Alteration of the substrate specificity of the malonyl-CoA/acetyl-CoA:acyl carrier protein *S*-acyltransferase domain of the multifunctional fatty acid synthase by mutation of a single arginine residue. *J. Biol. Chem.*, **272**, 11,975-11,978.

Rault, M., Gontero, B. and Ricard, J. (1991) Thioredoxin activation of phosphoribulokinase in a chloroplast multi-enzyme complex. *Eur. J. Biochem.*, **197**, 791-797.

Rault, M., Giudici-Orticoni, M.T., Gontero, B. and Ricard, J. (1993) Structural and functional properties of a multi-enzyme complex from spinach chloroplasts. 1. Stoichiometry of the polypeptide chains. *Eur. J. Biochem.*, **217**, 1065-1073.

Reed, L.J., Pettit, F.H., Eley, M.H., Hamilton, L., Collins, J.H. and Oliver, R.M. (1975) Reconstitution of the *Escherichia coli* pyruvate dehydrogenase complex. *Proc. Natl Acad. Sci. USA*, **72**, 3068-3072.

Revill, W.P. and Leadlay, P.F. (1991) Cloning, characterization, and high-level expression in *Escherichia coli* of the *Saccharopolyspora erythraea* gene encoding an acyl carrier protein potentially involved in fatty acid biosynthesis. *J. Bacteriol.*, **173**, 4379-4385.

Ricard, J., Giudici-Orticoni, M.T. and Gontero, B. (1994) The modulation of enzyme reaction rates within multi-enzyme complexes. 1. Statistical thermodynamics of information transfer through multi-enzyme complexes. *Eur. J. Biochem.*, **226**, 993-998.

Ricard, J., Gontero, B., Avilan, L. and Lebreton, S. (1998) Enzymes and the supramolecular organization of the living cell. Information transfer within supramolecular edifices and imprinting effects. *Cell. Mol. Life Sci.*, **54**, 1231-1248.

Rock, C.O. and Cronan, J.E. (1996) Escherichia coli as a model for the regulation of dissociable (type II) fatty acid biosynthesis. *Biochim. Biophys. Acta*, **1302**, 1-16.

Rossi, A. and Corcoran, J.W. (1973) Identification of a multienzyme complex synthesizing fatty acids in the actinomycete *Streptomyces erythraeus*. *Biochem. Biophys. Res. Commun.*, **50**, 597-602.

Roughan, P.G. (1997) Stromal concentrations of coenzyme A and its esters are insufficient to account for rates of chloroplast fatty acid synthesis: evidence for substrate channelling within the chloroplast fatty acid synthase. *Biochem. J.*, **327**, 267-273.

Roughan, P.G. and Ohlrogge, J.B. (1996) Evidence that isolated chloroplasts contain an integrated lipid-synthesizing assembly that channels acetate into long-chain fatty acids. *Plant Physiol.*, **110**, 1239-1247.

Sainis, J.K. and Harris, G.C. (1986) The association of ribulose-1,5-bisphosphate carboxylase with phosphoriboisomerase and phosphoribulokinase. *Biochem. Biophys. Res. Commun.*, **139**, 947-954.

Sainis, J.K. and Jawali, N. (1994) Channeling of the intermediates and catalytic facilitation to Rubisco in a multienzyme complex of Calvin cycle enzymes. *Indian J. Biochem. Biophys.*, **31**, 215-220.

Sainis, J.K., Merriam, K. and Harris, G.C. (1989) The association of D-ribulose bisphosphate carboxylase-oxygenase with phosphoribulokinase. *Plant Physiol.*, **89**, 368-374.

Sasaki, Y., Kozaki, A. and Hatano, M. (1997) Link between light and fatty acid synthesis: thioredoxin-linked reductive activation of plastidic acetyl-CoA carboxylase. *Proc. Natl Acad. Sci. USA*, **94**, 11,096-11,101.

Scheibe, R., Baalmann, E., Backhausen, J.E., Rak, C. and Vetter, S. (1996) C-terminal truncation of spinach chloroplast NAD(P)-dependent glyceraldehyde-3-phosphate dehydrogenase prevents inactivation and reaggregation. *Biochim. Biophys. Acta*, **1296**, 228-234.

Schlichting, I., Yang, X.J., Miles, E.W., Kim, A.Y. and Anderson, K.S. (1994) Structural and kinetic analysis of a channel-impaired mutant of tryptophan synthase. *J. Biol. Chem.*, **269**, 26,591-26,593.

Schulte, W., Topfer, R., Stracke, R., Schell, J. and Martini, N. (1997) Multi-functional acetyl-CoA carboxylase from *Brassica napus* is encoded by a multi-gene family: indication for plastidic localization of at least one isoform. *Proc. Natl Acad. Sci. USA*, **94**, 3465-3470.

Serre, L., Verbree, E.C., Dauter, Z., Stuitje, A.R. and Derewenda, Z.S. (1995) The *Escherichia coli* malonyl-CoA:acyl carrier protein transacylase at 1.5-Å resolution. Crystal structure of a fatty acid synthase component. *J. Biol. Chem.*, **270**, 12,961-12,964.

Shimakata, T. and Stumpf, P.K. (1982) Fatty acid synthetase of *Spinacia oleracea* leaves. *Plant Physiol.*, **69**, 1257-1262.

Shorrosh, B.S., Savage, L.J., Soll, J. and Ohlrogge, J.B. (1996) The pea chloroplast membrane-associated protein, IEP96, is a subunit of acetyl-CoA carboxylase. *Plant J.*, **10**, 261-268.

Siebenlist, U., Wohlgemuth, S., Finger, K. and Schweizer, E. (1991) Isolation of a novel type-I fatty-acid synthetase from *Euglena gracilis*. Specific derepression in streptomycin-bleached cells. *Eur. J. Biochem.*, **202**, 515-519.

Siggaard-Andersen, M. (1993) Conserved residues in condensing enzyme domains of fatty acid synthases and related sequences. *Prot. Seq. Data Anal.*, **5**, 325-335.

Simon, J.W. and Slabas, A.R. (1998) cDNA cloning of *Brassica napus* malonyl-CoA:ACP transacylase (MCAT) (fab D) and complementation of an *Escherichia coli* MCAT mutant. *FEBS Lett.*, **435**, 204-206.

Simoni, R.D., Criddle, R.S. and Stumpf, P.K. (1967) Fat metabolism in higher plants: XXXI. Purification and properties of plant and bacterial acyl carrier proteins. *J. Biol. Chem.*, **242**, 573-578.

Skrukrud, C.L. and Anderson, L.E. (1991) Chloroplast phosphoribulokinase associates with yeast phosphoriboisomerase in the presence of substrate. *FEBS Lett.*, **280**, 259-261.

Slabas, A.R. and Fawcett, T. (1992) The biochemistry and molecular biology of plant lipid biosynthesis. *Plant Mol. Biol.*, **19**, 169-191.

Srere, P.A. (1985) The metabolon. *Trends Biochem. Sci.*, **10**, 109-110.

Srere, P.A. (1987) Complexes of sequential metabolic enzymes. *Annu. Rev. Biochem.*, **56**, 89-124.

Srere, P.A. and Ovadi, J. (1990) Enzyme-enzyme interactions and their metabolic role. *FEBS Lett.*, **268**, 360-364.

Srivastava, D.K. and Bernhard, S.A. (1986) Metabolite transfer via enzyme–enzyme complexes. *Science*, **234**, 1081-1086.

Stoops, J.K., Singh, N. and Wakil, S.J. (1990) The yeast fatty acid synthase. Pathway for transfer of the acetyl group from coenzyme A to the Cys-SH of the condensation site. *J. Biol. Chem.*, **265**, 16,971-16,977.

Stoops, J.K., Kolodziej, S.J., Schroeter, J. P., Bretaudiere, J.P. and Wakil, S.J. (1992) Structure–function relationships of the yeast fatty acid synthase: negative-stain, cryo-electron microscopy, and image analysis studies of the end views of the structure. *Proc. Natl Acad. Sci. USA*, **89**, 6585-6589.

Süss, K.H., Arkona, C., Manteuffel, R. and Adler, K. (1993) Calvin cycle multienzyme complexes are bound to chloroplast thylakoid membranes of higher plants *in situ*. *Proc. Natl Acad. Sci. USA*, **90**, 5514-5518.

Süss, K.H., Prokhorenko, I. and Adler, K. (1995) *In situ* association of Calvin cycle enzymes, ribulose-1,5-bisphosphate carboxylase/oxygenase activase, ferredoxine-NADP reductase, and nitrite reductase with thylakoid and pyrenoid membranes of *Chlamydomonas reinhardtii* chloroplasts as revealed by immunoelectron microscopy. *Plant Physiol.*, **107**, 1387-1397.

Tropf, S., Revill, W.P., Bibb, M.J., Hopwood, D.A. and Schweizer, M. (1998) Heterologously expressed acyl carrier protein domain of rat fatty acid synthase functions in *Escherichia coli* fatty acid synthase and *Streptomyces coelicolor* polyketide synthase systems. *Chem. Biol.*, **5**, 135-146.

Tsay, J.T., Oh, W., Larson, T.J., Jackowski, S. and Rock, C.O. (1992) Isolation and characterization of the beta-ketoacyl-acyl carrier protein synthase III gene (fabH) from *Escherichia coli* K-12. *J. Biol. Chem.*, **267**, 6807-6814.

Vertessy, B. and Ovadi, J. (1987) A simple approach to detect active-site-directed enzyme–enzyme interactions. The aldolase/glycerol-phosphate-dehydrogenase enzyme system. *Eur. J. Biochem.*, **164**, 655-659.

Voelker, T.A., Worrell, A.C., Anderson, L. *et al.* (1992) Fatty acid biosynthesis redirected to medium chains in transgenic oilseed plants. *Science*, **257**, 72-74.

Volpe, J.J. and Vagelos, P.R. (1976) Mechanisms and regulation of biosynthesis of saturated fatty acids. *Physiol. Rev.*, **56**, 339-417.

Wakil, S.J. (1989) Fatty acid synthase, a proficient multifunctional enzyme. *Biochemistry*, **28**, 4523-4530.

Waldrop, G.L., Rayment, I. and Holden, H.M. (1994) Three-dimensional structure of the biotin carboxylase subunit of acetyl-CoA carboxylase. *Biochemistry*, **33**, 10,249-10,256.

Wang, X., Tang, X.Y. and Arderson, L.E. (1996) Enzyme-enzyme interaction in the chloroplast: physical evidence for association between phosphoglycerate kinase and glyceraldehyde-3-phosphate dehydrogenase *in vitro*. *Plant Sci.*, **117**, 45-53.

Wedel, N. and Soll, J. (1998) Evolutionary conserved light regulation of Calvin cycle activity by NADPH-mediated reversible phosphoribulokinase/CP12/glyceraldehyde-3-phosphate dehydrogenase complex dissociation. *Proc. Natl Acad. Sci. USA*, **95**, 9699-9704.

Wedel, N., Soll, J. and Paap, B.K. (1997) CP12 provides a new mode of light regulation of Calvin cycle activity in higher plants. *Proc. Natl Acad. Sci. USA*, **94**, 10,479-10,484.

Witkowski, A., Joshi, A.K., Rangan, V.S., Falick, A.M., Witkowska, H.E. and Smith, S. (1999) Dibromopropanone cross-linking of the phosphopantetheine and active-site cysteine thiols of the animal fatty acid synthase can occur both inter- and intrasubunit. Reevaluation of the side-by-side, antiparallel subunit model. *J. Biol. Chem.*, **274**, 11,557-11,563.

Witters, L.A. and Watts, T.D. (1990) Yeast acetyl-CoA carboxylase: *in vitro* phosphorylation by mammalian and yeast protein kinases. *Biochem. Biophys. Res. Commun.*, **169**, 369-376.

Worsham, L.M., Williams, S.G. and Ernst-Fonberg, M.L. (1993) Early catalytic steps of *Euglena gracilis* chloroplast type II fatty acid synthase. *Biochim. Biophys. Acta*, **1170**, 62-71.

Wu, X.M., Gutfreund, H., Lakatos, S. and Chock, P.B. (1991) Substrate channeling in glycolysis: a phantom phenomenon. *Proc. Natl Acad. Sci. USA*, **88**, 497-501.

Yonuschot, G.R., Ortwerth, B.J. and Koeppe, O.J. (1970) Purification and properties of a nicotinamide adenine dinucleotide phosphate-requiring glyceraldehyde 3-phosphate dehydrogenase from spinach leaves. *J. Biol. Chem.*, **245**, 4193-4198.

Yuan, L., Voelker, T.A. and Hawkins, D.J. (1995) Modification of the substrate specificity of an acyl–acyl carrier protein thioesterase by protein engineering. *Proc. Natl Acad. Sci. USA*, **92**, 10,639-10,643.

Yuan, L., Nelson, B.A. and Caryl, G. (1996a) Additions and corrections to the catalytic cysteine and histidine in the plant acyl–acyl carrier protein thioesterases. *J. Biol. Chem.*, **271**, 11034A.

Yuan, L., Nelson, B.A. and Caryl, G. (1996b) The catalytic cysteine and histidine in the plant acyl–acyl carrier protein thioesterases. *J. Biol. Chem.*, **271**, 3417-3419.

6 Self-compartmentalizing proteolytic complexes

Bjarke Veierskov and Christina Ingvardsen

6.1 Introduction

Self-compartmentalizing proteolytic complexes have been identified in all living cells. The most complex one is the proteasome that is present in all eukaryotic cells. The 20S form of the proteasome is, however, also present in Archaea and some Eubacteria. The Clp and Lon proteases are found in prokaryotic cells and eukaryotic organelles of a prokaryotic origin (the chloroplast and the mitochondria). The similarity in structure between these proteolytic complexes indicates that the basis for this structure has developed early in evolution (Murzin, 1998). Much of the information presented in this chapter is based on studies in animals and yeast systems, with studies on plant systems highlighted were appropriate.

6.2 Physical properties of the proteosome

The biochemical properties and cellular localization of this proteolytic system have been elucidated during the last ten years. Early attempts to classify the proteolytic activity of the proteasome showed that it did not belong to any of the four classic mechanistic groups of proteases—serine, cysteine, aspartate or metalloproteases—but to the threonine proteinases, closely linked to the serine proteases (Seemüller *et al.*, 1995). Proteins cannot be degraded unless they are targeted for breakdown by ubiquitin, which is a reusable signal. Although all elements of this proteolytic system have been identified in plants, only few specifically regulated processes have been identified.

The proteasome is a large cylinder-shaped ATP-dependent multisubunit protease that mostly degrades polyubiquitylated proteins. The proteasome is prefixed as 26S after its sedimentation coefficient, although a more accurate value is 30.3S (Yoshimura *et al.*, 1993). The 26S proteasome comprises a 20S proteolytic core and the 19S regulatory complex that binds to the core (figure 6.1). The 20S proteasome is a closed barrel-shaped particle (11×15 nm of ca. 700 kDa) with narrow axial pores (≈ 1.3 nm) through which only unfolded proteins are able to pass (Löwe *et al.*, 1995; Wang *et al.*, 1997).

The 19S regulatory complex (ca. 900–1000 kDa) has been given different names in the literature such as 'the ball', the 'μ particle', 'PA700', '19S cap' and the '19S regulatory complex'. Early electron micrographs revealed a structure

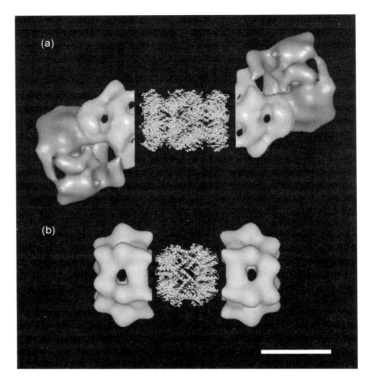

Figure 6.1 A comparison of the structures of the 26S proteasome and the ClpAP protease. (a) Model of the 26S proteasome obtained after combining the 3D reconstruction of the 19S regulator—the lid (distal) and base (proximal) subcomplexes are indicated by different shades of grey—from *Drosophila* with the crystal structure of the 20S core from *Thermoplasma*. (b) A model of the ClpAP protease from *Escherichia coli* derived from the combination of the 3D reconstruction of the ClpAP protease with the crystal structure of the ClpP protease. The scale bar represents 10 nm. (Reprinted from Zwickl, *et al.*, 2000).

resembling a 'Chinese dragon head' but accurate 3D electron microscopy has shown a flexible linkage between the 19S regulatory complex and the 20S core causing the proteasome to exhibit a wagging-type movement of the 19S regulatory complex (Walz *et al.*, 1998).

In vertebrates, the 20S proteasome also interacts with another regulator, the PA28 or 11S regulator. This regulator, which takes part in class I antigen presentation (Rechsteiner *et al.*, 2000), has not been identified in plants. A novel proteasome activator called PA 200 has recently been identified in the rabbit (V. Ustrell, personal communication) and others might be identified in the future.

6.2.1 The 20S core

The 20S core of the 26S proteasome is found in all living domains: Eukarya, Archaea as well as in some Eubacteria (Lupas *et al.*, 1994; Pühler *et al.*, 1994; Tamura *et al.*, 1995; Maupin-Furlow *et al.*, 1998). The 20S core particle contains 14 different subunits divided into two related families, α and β. The subunits are arranged into four seven-membered rings. The two outer rings consist of α-type subunits and the two inner rings of β-type subunits, collectively forming the barrel-shaped complex. Yeast and plants have proteasomes with seven different α- and β-type subunits, all having a defined location (Groll *et al.*, 1997; Tanaka and Tsurumi, 1997; Dahlmann *et al.*, 1999; Fu *et al.*, 1999). Vertebrates have seven different α-type and ten different β-type subunits (Tanaka and Tsurumi, 1997), with the extra β-type subunits involved in antigen presentation (Altuvia and Margalit, 2000). The subunits have been given a great variety of names, making it difficult to compare subunits from different species (table 6.1). According to the new nomenclature by Groll *et al.* (1997), the eukaryotic proteasome subunits are named α1–7 and β1–7, based on their location in the proteasomal rings. All seven α-subunits have been isolated from *Arabidopsis thaliana*. Six out of seven were encoded by families of at least two genes (Fu *et al.*, 1998). α-Subunits have also been reported from spinach and tobacco (Bahrami and Gray, 1999). All β-subunits have also been isolated in *A. thaliana*, most encoded by a least a pair of paralogs (Fu *et al.*, 1999).

The α-subunits have a highly conserved *N*-terminal extension which is close to the entrance to the central core of the proteasome (Groll *et al.*, 2000). The *N*-terminal tail of the α3 subunit regulates access to the core by interaction with the *N*-terminals of the other α-subunits, and deleting the *N*-terminal tail of the α3-subunit opens the gateway to the core. The activation of the proteasome by sodium dodecyl sulfate (SDS) is presumably caused by an opening of the gate in a similar manner rather than an activation of the proteolytic sites (Groll *et al.*, 2000). The mammalian 11S regulatory complex is known to open the gate to the 20S core by inserting *C*-terminal extensions into pockets of the α ring (Pickart and VanDemark, 2000). A similar mechanism has not been ascribed to the 19S regulatory complex. The closed core of the 20S proteasome seems to be able to open in an oscillating manner when substrate is present (Osmulski and Gaczynska, 2000). This oscillation may explain reported proteolytic activity of the 20S core towards non-ubiquitnylated substrates such as oxidized proteins (Reinheckel *et al.*, 1998).

The proteasome has three distinct proteolytic activities: chrymotrypsin-like (ChT-L) activity; peptidylglutamyl-peptid hydrolyzing (PGPH) activity; and trypsin-like (T-L) activity, all located on the inside of the β-subunits. They belong to a class of hydrolases termed *N*-terminal nucleophile hydrolases, involving the *N*-terminal Thr residue (Duggleby *et al.*, 1995). ChT-L cleaves after hydrophobic amino acid residues, preferring branched chain amino acids

Table 6.1 Subunits of the 26S proteasome from eukaryotes

Systematic name	Arabidopsis[a]	Yeast[b]	Human[c]		Activity[c,d]
20S core particle					
α1	PAA1	PRS2	Iota/PRS2		RNase
	PAA2				
α2	PAB1	PRS4/Y7	C3		
	PAB2				
α3	PAC1	PRS5/Y13	C9		
	PAC2				
α4	PAD1	PRE6	C6/XAPC7		
	PAD2				
α5	PAE1	PUP2	ZETA		RNase
	PAE2				
α6	PAF1	PRE5	C2/PROS30		
	PAF2				
α7	PAG1	PRS1	C8		
β1	PBA1	PRE3	Y/delta		PGPH
β2	PBB1	PUP1	Z		T-L
	PBB2				
β3	PBC1	PUP3	C10		
	PBC2				
β4	PBD1	PRE1	C7		
	PBD2				
β5	PBE1	PRE2	X/MB1		ChT-L
β6	PBF1	PRS3	C5		
β7	PBG1	PRE4	N3		
β1-i			LMP2		ChT-L
β2-i			LMP10/MECL-1		T-L
β5-i			LMP7		ChT-L
19S regulatory particle					
Rpt1	RPT1	Cim/Yta3	S7	Mss1	ATPase
Rpt2	RPT2	Yta5	S4	Mts2	ATPase
Rpt3	RPT3	Yta2/Ynt1	S6	Tbp7	ATPase
Rpt4	RPT4	Crl13/Sug2/ Pcs1	S10b		ATPase
Rpt5	RPT5	Yta1	S6a	Tbp1	ATPase
Rpt6	RPT6	Sug1/Cim3/ Crl3	S8	Trip1	ATPase
Rpn1	RPN1	Hrd2/Nas1	S2	Trap2	
Rpn2	RPN2	Sen3	S1		
Rpn3	RPN3	Sun2	S3		
Rpn4	RPN4	Son1/Ufd5			
Rpn5	RPN5	Nas5			
Rpn6	RPN6	Nas6	S9		
Rpn7	RPN7		S10a		
Rpn8	RPN8	Nas3	S12		
Rpn9	RPN9	Nas7	S11		

Table 6.1 (continued)

Systematic name	Arabidopsis[a]	Yeast[b]	Human[c]		Activity[c,d]
Rpn10	RPN10	Mcb1/Sun1	S5a		
Rpn11	RPN11	Mpr1	S13	Poh1	
Rpn12	RPN12	Nin1	S14		
		Nas2	S15		

[a]Fu *et al.* (1999).
[b]Glickman *et al.* (1999).
[c]Orlowski and Wilk (2000).
[d]See text for further information.

and is located on the β5 subunit (Eleuteri *et al.*, 1997). The β1 subunit harbors the PGPH activity that cleaves after acidic residues, whereas the T-L activity on the β2 subunit cleaves after basic residues. All three activities have been identified in plants (Murray *et al.*, 1997). Although proteolytic activity has only been ascribed to three of the β-subunits, their activity depends on their specific location in connection to the other subunits in the β-ring. The *N* terminal Thr of the catalytically active β-subunits are susceptible to inactivation by N^{α}-acetylation, and so the subunit is synthesized with a propeptide. The propeptides also seem to be important for the assembly of the 20S core complex, and are first cleaved when the 20S proteasome is fully assembled (Arendt and Hochstrasser, 1999).

The mechanism of substrate degradation by the *N*-terminal proteases of the proteasome has been described by Orlowski and Wilk (2000). The proteasome degrades its substrates in a slow, progressive manner where the entire protein is degraded before the next is attacked (Akopian *et al.*, 1997). The degradation of a polypeptide may take several seconds (Bogyo *et al.*, 1997). The polypeptides are cut into peptides shorter than 20 residues. These small peptides had earlier been suggested to escape the central core of the proteasome through small openings known to be present in the core complex, but it is now generally accepted that the peptide leaves the proteasome through the opposite end. It has been proposed originally, that the length of the degradation products was due to a molecular ruler (Wenzel *et al.*, 1994) but recent evidence shows that this is not the case (Kisselev *et al.*, 1998).

During assembly of the eukaryotic proteasome, each α-subunit is believed to form a dimer with the corresponding β-subunit before assembly into a 15S precursor/half-proteasome (Gerards *et al.*, 1998). The two half-proteasomes assemble into the 20S proteasome in an autocatalytic process, in which pro-sequences are removed. This process requires a short-lived chaperone, named Ump1p that is destroyed during the assembly process (Ramos *et al.*, 1998). Unlike the eukaryotes, the α-subunits in *Thermoplasma* associate into a ring on which the β-subunits assemble (Zwickl *et al.*, 1994).

It has been suggested that the 20S proteasome also posseses RNase activity (Pamnani *et al.*, 1994; Pouch *et al.*, 1995). It has been shown that the 20S proteasome is involved in the cleavage of viral RNA due to RNase activity of the zeta (α5) subunit (Jarrousse *et al.*, 1999; Jørgensen and Hendil, 1999). As the RNase activity of the 20S proteasome is very low, it is not known whether or not it has any biological significance.

As the 20S proteasome on its own can utmost only degrade unfolded proteins (see section 6.2.1), the degradation of polyubiquitylated proteins requires an association of the 20S proteasome with the 19S regulatory complex.

6.2.2   The 19S regulatory complex

Addition of the 19S regulatory complex gives the proteasome the ability to degrade polyubiquitylated proteins in an ATP- and Mg^{2+}-dependent process (Adams *et al.*, 1997; DeMartino and Slaughter, 1999). It is involved in substrate recognition, substrate unfolding and the feeding of polypeptide chains into the 20S proteasome (Adams *et al.*, 1997). It also exhibits chaperone-like activity (Braun *et al.*, 1999). It is found in all eukaryotes, and 18 subunits have been identified, ranging in size from 30 to 110 kDa (Glickman *et al.*, 1999). The 3D structure is not known in detail.

The subunits of the 19S regulatory complex can be divided into two groups, the ATPase subunits and the non-ATPase subunits. The nomenclature has been very impenetrable (subunit Rpt6 is known under such diverse names as Sug1, Cim3, Crl3, S8, trip1, p45 and m56). However, new nomenclature has been introduced (Glickman *et al.*, 1998a), making it easier to compare subunits from different species. The ATPase subunits are thus named Rpt# for regulatory particle triple-A protein and the non-ATPase subunits Rpn# for regulatory particle non-ATPase (see table 1). Each regulatory complex contains six different ATPase subunits belonging to the AAA family (ATPases associated with a variety of cellular activities) of ATPases (Confalonieri and Duguet, 1995; Glickman *et al.*, 1998a; Rubin *et al.*, 1998). The ATPase subunits are conserved, being 66 to 76% identical between yeast and humans and 65 to 75% identical between yeast and plants (Glickman *et al.*, 1998a; Fu *et al.*, 1999). It has been shown that the ATPase subunits interact in specific pairs, presumably via their coiled regions (Richmond *et al.*, 1997). This interaction is believed to be important for the specific location of the ATPase subunits in the regulatory complex.

The 19S regulatory complex has at least 11 non-ATPase subunits, but not much is known about their function (Glickman *et al.*, 1998a). The best characterized is Rpn10 (S5a) which binds polyubiquitin chains (Deveraux *et al.*, 1994; Haracska and Udvardy, 1995). Surprisingly, this subunit is not essential for the degradation of polyubiquitylated substrates in yeast, indicating that other subunits of the 19S regulatory complex are also involved in the recognition of ubiquitin chains (van Nocker *et al.*, 1996). Isopeptidase activity (de-ubiquitylation) has also been reported, but the subunit(s) responsible for this activity have not been identified (Lam *et al.*, 1997).

The 19S regulatory complex can be divided into the lid and the base. The base contains all six proteasomal ATPases plus Rpn1, Rpn2 and Rpn10 (Glickman *et al.*, 1998b). The base associates with 20S and is sufficient to activate 20S, suggesting a role of the base in opening the channel of the 20S proteasome. The lid consists of eight subunits (Rpn3, Rpn5, Rpn6, Rpn7, Rpn8, Rpn9, Rpn11 and Rpn12) and is, together with the base, required for ubiquitin-dependent degradation (Glickman *et al.*, 1998b).

All six ATPase subunits have been identified in *Arabidopsis*, two of which have paralogs (Fu *et al.*, 1999). ATPase subunits have also been isolated in rice, tomato and *Brassica rapa* (Suzuka *et al.*, 1994; Prombona *et al.*, 1995; Kitashiba and Toriyama, 1997). Of the non-ATPase subunits, Rpn3 has been isolated from carrot and Rpn10 from rice (Smith *et al.*, 1999), whereas six different non-ATPase subunits have been identified in *Arabidopsis* (Rpn1, Rpn2, Rpn6, Rpn8, Rpn10 and Rpn11) (Fu *et al.*, 1999).

6.2.3 Mechanisms to target identification—ubiquitylation

The 26S proteasome is not able to degrade proteins without a targeting signal. This signal is a polyubiquitin chain that is recognized by the 19S regulatory complex with ODC being the exception (see section 6.2.7). Ubiquitin is conjugated to the substrates in an isopeptide bond between the *C*-terminal glycine of ubiquitin and a lysine residue of the target protein. This reaction requires three classes of enzymes (E1, E2 and E3, figure 6.2). In the first step, ubiquitin is activated by a ubiquitin-activating enzyme (E1). The activated ubiquitin is transferred from E1 to E2, the ubiquitin-conjugating enzyme. The substrate is recognized by the E3 enzyme, which also recognizes E2. The ubiquitylation of proteins is presumably controlled either by the level of E2/E3 enzymes or by phosphorylation or other modifications of the substrates.

6.2.3.1 Ubiquitin-activating enzymes (E1)

Ubiquitin-activating enzymes are very conserved and only one to three different genes are found in various organisms such as yeast, humans and plants (Hatfield *et al.*, 1997). In yeast, where only a single ubiquitin-activating enzyme is found, deletion is lethal (McGrath *et al.*, 1991). Plants have more ubiquitin-activating enzymes when compared with yeast—two have been found in *Arabidopsis* and three in wheat (Hatfield *et al.*, 1997).

The E1 enzyme first directs the formation of an acyl phosphoanhydride bond between the AMP moiety of ATP and the *C*-terminal glycine of ubiquitin. Ubiquitin is subsequently bound to a cysteine residue on E1 via a high-energy thiolester linkage with the concomitant release of AMP.

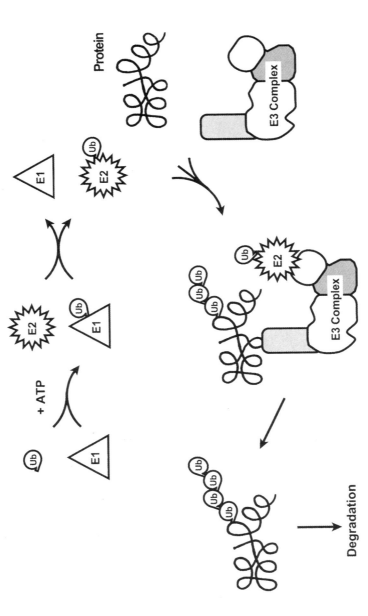

Figure 6.2 Ubiquitylation of substrate proteins. The C-terminal glycine of ubiquitin (Ub) first forms a thioester bond to E1 (ubiquitin-activating enzyme) in an ATP-dependent process. Then ubiquitin is transfered to the cysteine residue of E2, the ubiquitin-conjugating enzyme. Subsequently E3, the ubiquitin ligase, attaches ubiquitin to the ε-amino group of the lysine residue of the target protein or lysine$_{63}$ on ubiquitin itself through an isopeptide bond.

6.2.3.2   Ubiquitin-conjugating enzymes (E2)

Ubiquitin is transferred from the E1 enzyme, to the ubiquitin-conjugating enzyme (E2) by a transthiolesterification to a cysteine on E2. In some cases, E2 is thought to be able to transfer ubiquitin to the substrate on its own, but in other cases the process involves a ubiquitin ligase (E3). E2s are generally designated by UBC or Ubc followed by a number, but numbering of the homologous enzymes in the different species does not always correlate (Haas and Siepmann, 1997).

Phylogenetically, E2 enzymes have been divided into two major groups, Ubc4-like and Ubc2-like (Haas and Siepmann, 1997). The numbering in this system is based on the yeast ubiquitin-conjugating enzymes. Each of the two groups has at least six families, and several of these families are large and seem to harbor enzymes whose functions are related to specific roles of the pathway. The largest family characterized so far is the Ubc4/5 family, which has members from a range of organisms, including yeast, human, rice, tomato, pea and *Arabidopsis*. The enzymes in this family are involved in the stress response and regulation of transcription factors. Also ubiquitin-conjugating enzymes from the Ubc1 family, like Ubc1 from tomato (Ubc4-like) seem to be involved in stress responses (Feussner *et al.*, 1997). Of the Ubc2-like families, four (Ubc2, Ubc3, Ubc9 and Ubc) have reported plant members. The Ubc2 family has been found in alfalfa and *Arabidopsis*. Enzymes from two families, Ubc3 and Ubc9, are involved in regulation of the cell cycle. Ubiquitin-conjugating enzymes belonging to the Ubc3 family have been reported from wheat and *Arabidopsis*, while several Ubcs from *Arabidopsis* belong to the Ubc9 family.

The E2 enzymes have also been divided into four classes according to their structure. Class I enzymes are small (16–18 kDa), consisting almost entirely of the conserved UBC domain. Examples of class I E2 enzymes are UBC4 and UBC5 from yeast. Class II Ubcs have *C*-terminal extensions of various length (as CDC34 and UBC6), class III Ubcs have *N*-terminal extensions, and class IV ubiquitin-conjugating enzymes, of which only a few have been found, have both *C*- and *N*-terminal extensions. The *N*- and/or *C*-terminal extensions are believed to be important for substrate specificity and cellular localization.

6.2.3.3   Ubiquitin ligases (E3)

The E3 enzymes perform substrate recognition and subsequent assembly of the multi-ubiquitin chain on the substrate. The E3s uses different protein–protein interaction domains to couple with the ubiquitin-conjugating enzyme and the substrate. Two major groups of E3 ligases have been identified: the HECT domain and the really interesting new gene (RING) finger E3s.

The HECT (homologous to E6-AP *C*-terminus) domain proteins are often large (90–200 kDa). This group of E3 ligases forms a thioester between ubiquitin and a conserved cysteine in the 350 amino acid HECT domain. The domain is located at the *C*-terminal whereas the *N*-terminal extension takes part in the substrate binding. The N-terminal of Rsp5p from budding yeast contains several domains called 'ww' (Wang *et al.*, 1999), which are crucial for interaction with its target protein, the large subunit of RNA polymerase II. The N-terminal of the human homolog NEDD4 allows interaction with a sodium channel (Rotin, 1998).

The other group is named the RING finger E3s. They all have a RING finger protein necessary for E2 interaction. The RING finger domain has eight residues of ordered cysteines and histidines and binds two zinc atoms. Two subtypes of RING finger domains have been identified: the RING-HC containing only one histidine at position four, and RING-H2s that contains histidine at position four and five (Jackson *et al.*, 2000). To the RING finger group of E3s belong the SCF (Skp1–cullin–F-box protein) complex, the VBC complex and the APC complex (anaphase promoting), as well as Ubr1p, which is important for the N-end rule (Callis and Vierstra, 2000), and for the interaction with SKP1.

The SCF family (Skp1–cullin–F-box protein), consists of at least four proteins: cullin (Cdc53), Skp1, Rbx1/Roc1/Hrt1 and one of several F-box proteins (Patton *et al.*, 1998; Deshaies, 1999). The F box motif is ca. 45 residues long and is important for binding to the substrate (Gray *et al.*, 1999). The number of F-box proteins is very large; 33 have been reported in mammals (Winston *et al.*, 1999) and at least 20 can be identified in *Arabidopsis* (del Pozo and Estelle, 2000). The cullin proteins of the SCF complex are regulated by ubiquitylation or by conjugation of Rub1 (ubiquitin-like protein) (Lammer *et al.*, 1998; Zhou and Howley, 1998). It has been shown that Cdc34 and Ubc5 are the E2s interacting with the SCF complex (Banerjee *et al.*, 1995). One of the SCF complexes degrades Sic1p, an inhibitor of a set of cyclin/cyclin dependent kinase complexes as well as G1 cyclins, and is thus involved in the G1/S transition in the cell cycle (Skowyra *et al.*, 1997). In plants, the SCF complexes are implicated in response to auxin and jasmonic acid as well as in floral development and regulation of the circadian clock (del Pozo and Estelle, 2000).

The VBC ubiquitin ligase contains the van Hippel–Lindau (VHL) tumor suppressor (Kaelin *et al.*, 1998). The structure of VBC is similar to SCF, containing elongin B, elongin C, a Skp1-like protein and an adaptor protein VHL that binds to elongin C through a special motif, the BC box. The elongins were originally identified in a complex controlling transcriptional elongation (Aso *et al.*, 1995).

The APC belongs to its own family of E3 enzymes. This large complex is built of at least 11 subunits (Grossberger *et al.*, 1999; Kurasawa and Todokoro, 1999). The APC triggers anaphase, by degradation of the anaphase inhibitor Pds1p, and the exit from mitosis by degradation of B-type cyclins (Glotzer *et al.*, 1991; CohenFix *et al.*, 1996; King *et al.*, 1996). The activity of the APC is regulated by phosphorylation of several of the APC subunits. Two destruction boxes, the D-box and the KEN-box have been identified in substrates targeted for recognition by APC (Glotzer *et al.*, 1991; Pfleger and Kirschner, 2000). The D-box is a nine-residue motif, whereas the KEN-box has seven residues.

Often ancillary factors, such as Hsp70, are mentioned in connection with ubiquitin ligases, as they are needed for the E3 to perform its function (Huibregtse *et al.*, 1991). As some ubiquitin ligases are large complexes consisting of many subunits, the division between ancillary factors and E3 complexes is rather diffuse.

6.2.3.4   The ubiquitin-chain assembly factor (E4)

The E1, E2 enzymes are essential for initiating the ubiquitylation process. However, to make polyubiquitin chains, many of the E2s need an E3. It seems that at least in some cases an additional factor, named E4, is needed to make polyubiquitin chains with more than three ubiquitin moieties (Koegl *et al.*, 1999). This factor, isolated in yeast as UFD2, has homologes in other organisms. The *C*-terminus (U-box) of these proteins is very conserved, suggesting that this part of the protein is important for its function, and this conserved U-box is also found in plant genes. Deletion of this protein is not lethal in yeast, but it has importance for stress tolerance (Koegl *et al.*, 1999). This indicates that although it is not needed for all types of poly-ubiquitylation, it has specific functions. It is possible that the different types of ubiquitylation are part of a less well understood signal process, similar to ubiquitin-like proteins. Mono-ubiquitylation has long been known not to initiate degradation by the proteasome, and recently it has been shown that mono-ubiquitylated membrane proteins are endocytosed and transported to the vacuole for degradation (Terrell *et al.*, 1998).

6.2.4   Recognition sites

For proteins to be degraded by the proteasome, the substrate proteins must have an exposed recognition site and a lysine residue outside the recognition site. Furthermore, the corresponding E2s and E3s have to be present and active. The recognition sites may be exposed when the proteins denature, when subunits are disintegrated, or they might be exposed in the native form which may be the case with short-lived proteins. The E2 and/or E3 enzymes can be regulated at the transcriptional level or their activity can be regulated by phosphorylation as in APC (Peters *et al.*, 1996). The E2/E3 complexes are believed to recognize specific degradation signals on the substrates, but only a few of these signals are known.

The first reported degradation site was the *N*-degron. Proteins degraded by the *N*-end rule require specific destabilizing residues at the *N*-terminus (the *N*-degron) as well as a proximal lysine residue (Varshavsky *et al.*, 2000). Although only a few natural substrates of this system are known, components of it are found in yeast and mammals (Byrd *et al.*, 1998; Kwon *et al.*, 1999) as well as in plants (Potuschak *et al.*, 1998; Worley *et al.*, 1998). It has been shown in yeast that the absence of the *N*-rule pathway causes significant up- or downregulation of a diverse number of gene products (Kwon *et al.*, 1999).

The B-type cyclins have a degradation site called the 'destruction box' (D-box) (Glotzer *et al.*, 1991; King *et al.*, 1996). If fused to otherwise stable proteins, the D-box sequence (RXXLXXIXN) signals degradation independent of the tertiary structure of the proteins (Glotzer *et al.*, 1991). The D-box containing proteins are poly-ubiquitylated by the APC (Gmachl *et al.*, 2000). Another signal is found in the transcription factor inhibitor IκBα where a short phosphorylation site (DSGLDS) acts as recognition site after its phosphorylation (Yaron *et al.*, 1997; Laney and Hochstrasser, 1999). A similar sequence has also been identified in the transcription factor β-catenin, which is also degraded by the ubiquitin-dependent pathway (Aberle *et al.*, 1997). Furthermore, different domains with importance for ubiquitin-dependent degradation have been found in c-myc, Gcn4 and c-Jun (Musti *et al.*, 1996; Flinn *et al.*, 1998).

Gilon *et al.* (1998) found several destabilizing domains in yeast proteins. The most common feature was strong hydrophobicity, suggesting that hydrophobicity can be more important than the actual sequence (Gilon *et al.*, 1998; Laney and Hochstrasser, 1999). In agreement with this, a hydrophobic sequence is involved in the ubiquitin-dependent degradation of MATα2 (Johnson *et al.*, 1998). A nine amino acid recognition signal (SINNDAKSS) is required for endocytosis of Ste2p, indicating that recognition sites are also of importance for ubiquitylation of proteins not destined for 26S degradation (Hicke and Riezman, 1996).

Other degradation signals probably exist to cover the broad range of different substrates degraded by the ubiquitin-dependent pathway. Unfolded proteins might be recognized by chaperones as the Hsp70 or Ydj1, which, if not able to refold the protein, deliver the protein to the degradation machinery (Lee *et al.*, 1996; Bercovich *et al.*, 1997). The specific recognition signals involved in chaperone mediated degradation by the ubiquitin-dependent pathway are not known.

6.2.5 *Regulation of substrates by phosphorylation*

In a number of cases, phosphorylation is needed before the substrate can be ubiquitylated (Musti *et al.*, 1996). One phosphorylation site is the PEST-sequence (enriched in proline (P), glutamic acid (E), serine (S) and threonine (T) residues). Such PEST-sequences, which may vary considerably in sequence and length, are found in Gnc4, Fos, G1-cyclins, phytochrome A, Matα2, p53 and fructose-1,6-bisphosphatase, which are all known substrates of the ubiquitin-dependent pathway (Rechsteiner and Rogers, 1996; Deshaies, 1997). Additional recognition sites on the substrate might also be needed, as the PEST-sequence is not always sufficient (Salama *et al.*, 1994). The PEST-sequence is also involved in ubiquitin-dependent endocytosis (Roth *et al.*, 1998). It is known that phosphorylation often causes drastic conformational changes in a protein as has been shown for the P-type ATPases (Yaffe and Elia, 2001). This conformational change may expose otherwise hidden recognition sites. PEST-sequences are not the only phosphorylation sites causing ubiquitylation of proteins. In IκBα, the sequence DSGLDS is a phosphorylation site. In contrast to most PEST-regions, this phosphorylated site also acts as the recognition site for a ubiquitin ligase (Yaron *et al.*, 1997).

6.2.6 De-ubiquitylating enzymes

The de-ubiquitylating enzymes (DUBs) are believed to be involved in the processing of ubiquitin precursors, in the proofreading and recycling of poly-ubiquitin chains and in the removal of ubiquitin from adducts such as amides or glutathiones (Wilkinson and Hochstrasser, 1998). All these functions are very important to ensure that ubiquitin is reused so that the cells' ubiquitin pool is not drained. As many as 17 different DUBs have been found in yeast, emphasizing the importance of these proteins (Amerik *et al.*, 2000). In the literature, de-ubiquitylating enzymes are also called isopeptidases, ubiquitin *C*-terminal hydrolases, ubiquitin thioesterases or ubiquitin-specific processing proteases. The DUBs can be divided into two groups: the UCH family (ubiquitin *C*-terminal hydrolase) and the UBP family (ubiquitin-specific processing protease) (Wilkinson and Hochstrasser, 1998).

As early as in 1983, the first member of the UCH family was found in rabbit reticulocytes but so far no members of this family have been found in plants (Johnston *et al.*, 1999). UBP family members are present in *Arabidopsis*— two small ubiquitin-specific proteases able to cleave ubiquitin fusion proteins have been found (Yan *et al.*, 2000). The *Arabidopsis* sequencing project has revealed another putative de-ubiquitylating enzyme, indicating that plants may have several different de-ubiquitylating enzymes.

6.2.7 Ornithine decarboxylase

Protein degradation by the 26S proteasome normally requires poly-ubiquitylation, but in a few cases the 26S proteasome degrades proteins without the involvement of a poly-ubiquitin chain. Ornithine decarboxylase (ODC), a key enzyme in the synthesis of polyamines, is one such enzyme that has been shown to be degraded by the 26S, but not the 20S, proteosome, in a process that is ATP- and antizyme-dependent but ubiquitin-independent (Murakami *et al.*, 2000). Like ubiquitin, antizyme is recycled. Labile proteins, normally degraded in an ubiquitin-dependent manner by the 26S proteasome, are degraded ubiquitin-independently when the *N*-terminus of antizyme is fused to them (Li *et al.*, 1996). Since the *N*-terminus of antizyme is not required for the association of the ODC/antizyme with the 26S proteasome, the mechanism still remains unresolved.

Ornithine decarboxylase is not the only protein known to be degraded by the proteasome without being ubiquitylated since p21^{Cip1} has also been shown to be degraded by the proteasome without being ubiquitylated (Sheaff *et al.*, 2000). This ubiquitin independent degradation of proteins by the 26S proteasome might happen in other cases as well.

6.2.8 Oxidized proteins

In mammals, the ubiquitin–proteasome dependent pathway is involved in the response to mild forms of oxidative stress (Shang *et al.*, 1997; Figueiredo-Pereira and Cohen, 1999). However, proteins may form aggregates if the oxidizing conditions are too strong, which makes them poor substrates for degradation (Grune *et al.*, 1996). There is evidence that degradation of oxidized proteins is performed by the 20S core of the proteasome (Grune *et al.*, 1996). This core is normally regarded as being unable to perform proteolysis by itself due to the inability of substrates to enter the terminal pores. The 26S proteasome is more susceptible to oxidative stress when compared with the 20S proteasome, which seems to be protected from oxidative damage by Hsp90 (Conconi and Friguet, 1997; Grune *et al.*, 1997; Reinheckel *et al.*, 1998). It is possible that it is oxidative modifications of the 20S core that enables oxidized proteins to enter the proteolytic core without being ubiquitylated. The conjugating enzymes of the ubiquitin pathway are susceptible to oxidation like most other proteins (Grune *et al.*, 1995).

Although the oxidative state of proteins in plants has been given only little attention, there are reports showing that protein turnover is regulated or facilitated by the oxidative state of proteins (Mehta *et al.*, 1996). In addition, cytokinins, which retard senescence, also prevent formation of free radicals.

6.3 Biological processes regulated by the proteasome

6.3.1 Biotic stress

Plants infected with a virus react by upregulating the expression of ubiquitin genes, indicating a function of the ubiquitin-dependent pathway during biotic stress (Genschik *et al.*, 1992a; Aranda *et al.*, 1996). The virus itself may have ubiquitin-like genes that are believed to interfere with ubiquitin-dependent reactions in the plant. The increased level of plant ubiquitin could be the consequence of trying to overcome the virus infection by dilution of the virus ubiquitin. Plants with perturbations in the ubiquitin-dependent pathway spontaneously form necrotic lesions and accumulate defence-related compounds (Conrath *et al.*, 1998). It is not known if this is due to lack of degradation of a single important regulatory protein or due to a general accumulation of substrates for ubiquitin-dependent proteolysis. It is of importance for the plant to degrade viral proteins since virus-encoded movement protein facilitates the cell-to-cell spread of virus (Reichel and Beachy, 2000). This movement protein is believed to be poly-ubiquitylated and subsequently degraded by the 26S proteasome, allowing the plant to repress virus spread.

Recent investigations have shown that ubiquitin as well as the proteasome are important for the resistance reactions occurring after incompatible fungus–plant interactions (Mazeyrat *et al.*, 1999; Becker *et al.*, 2000; Etienne *et al.*, 2000). Increased expression of ubiquitin genes has been found in infected plants and in the infecting fungi, as has been shown during plant infection by *Phytophthora infestans* and *Magnaporthe grisea* (Pieterse *et al.*, 1991; McCafferty and Talbot, 1998). The function of the ubiquitin-dependent pathway might be to remove a repressor, thereby turning on the signaling pathway for plant defense responses. Although plants infected with fungi have changes in the ubiquitin conjugation pattern (Handke *et al.*, 1993), the origin of the ubiquitin, the conjugating enzymes and the substrate proteins might be from either the plant or the fungus, which complicates interpretation of the results.

6.3.2   Abiotic stress

Abiotic stresses, such as high and low temperature, oxidative stress, exposure to heavy metals and ozone, mechanical injury, starvation and drought, interact with the ubiquitin-dependent proteolysis. As an enhanced level of malfunctioning proteins is the consequence of most abiotic stresses, a fine-tuned removal of these proteins by upregulation of some ubiquitin genes and downregulation of others should be expected.

6.3.2.1   Temperature stress

To observe a temperature dependent change in gene expression, the temperature must rise drastically as a slow increase in temperature normally does not give rise to a heat shock response (Rickey and Belknap, 1991). Heat has been shown to influence the expression of ubiquitin genes in many plant species, such as sunflower, potato, maize, apple, rice, *Arabidopsis thaliana* and *Nicotiana sylvestris*. Mostly, gene expression increases, but downregulation or unchanged levels have also been found (Binet *et al.*, 1991; Garbarino *et al.*, 1992; Sun and Callis, 1997). In yeast, several different elements in the promotor region contribute to the stress control of the poly-ubiquitin gene *UBI4*. This may also occur in plants, since they have several poly-ubiquitin genes.

Whether or not the ubiquitin-conjugating enzymes from plants are also induced by heat shock is not clear. Five ubiquitin-conjugating enzymes from *A. thaliana*, AtUBC8–12, belonging to the same group of conjugating enzymes, are not induced by heat shock (Girod and Vierstra, 1993). Likewise, the expression of the *Ubc4* gene in *N. sylvestris* is not induced by heat (Genschik *et al.*, 1994a), whereas a ubiquitin-conjugating enzyme from tomato was induced by different kinds of stress, including heat shock (Feussner *et al.*, 1997). In yeast, the ubiquitin-conjugating enzymes UBC4 and UBC5 are heat inducible (Seufert and Jentsch, 1990).

The level of ubiquitin–protein conjugates in mammals as well as in plants has been found to increase after heat shock (Ferguson *et al.*, 1990; Fujimuro *et al.*, 1997). This higher level of conjugates may be due to an increase in the expression of ubiquitin-conjugating enzymes and/or an increased activity of these enzymes. A higher level of ubiquitin does not necessarily result in a higher level of ubiquitin conjugates or a faster turnover of ubiquitylated proteins, as this is controlled by alterations in the level and activity of conjugating enzymes and the proteasome. Furthermore, it has been found that an increase in the level of ubiquitin-protein conjugates can lead to a decrease in the level of free ubiquitin (Ferguson *et al.*, 1990).

The expression level of the proteasome is also influenced by heat shock. Our investigations of *Pharbitis nil* showed that the level of mRNA expression of an α-subunit of the proteasome increased after heat shock (Ingvardsen and Veierskov, 2001). This increase was most pronounced after 4 h and thus is not as fast as the normal heat shock response that is observed in less than 30 min. It is not known whether this increase in expression results in an increase in the proteasome activity. The level of proteasome protein did not increase during the first hours, but might have increased at a later stage.

6.3.2.2 Other types of abiotic stress

The ubiquitin–proteasome dependent pathway also participates in the reaction to abiotic stress such as wounding. Injury of potato tubers increased ubiquitin gene expression (Rickey and Belknap, 1991) and slicing of leaves of *Arabidopsis* resulted in increased expression of a proteasome subunit (Genschik *et al.*, 1992b). Contrary to this, no increase in the level of ubiquitin or the proteasome was observed at the cut surface in sunflower and mung bean cuttings (Ingvardsen *et al.*, 2001). When the proteasome is inhibited by lactacystin, wound inducible genes are not expressed. This indicates that an inhibitor exists that must be degraded by the proteasome before the wound response is initiated (Ito *et al.*, 1999).

The exposure of plants to heavy metals such as $HgCl_2$ increases the expression of ubiquitin genes in *Nicotiana* (Genschik *et al.*, 1992a). Also, a strong accumulation of a ubiquitin-conjugating enzyme (UBC1) was observed in tomato after exposure to $CdCl_2$ (Feussner *et al.*, 1997). Furthermore, darkness, starvation and an enhanced level of ozone alter the level of ubiquitin gene expression (Chevalier *et al.*, 1996; Sun and Callis, 1997; Wegener *et al.*, 1997).

6.3.3 Growth and differentiation

Programmed cell death occurs in insects as well as in animals and plants. During programmed cell death in the muscles of insects, poly-ubiquitin genes and genes coding for the enzymes involved in ubiquitin conjugation are induced. Also, the proteasome has been shown to be important for muscle degradation in insects as well as in rodents (Attaix *et al.*, 1999). When the proteasome is inhibited in human leukemic cells, programmed cell death is induced (Shinohara *et al.*, 1996; Drexler, 1997). Although the ubiquitin-dependent pathway seems to be involved in programmed cell death in animals and insects, the exact role has not been clarified. The reason might be that the proteasome dependent degradation is a secondary event that functions after other proteases have cleaved the muscle proteins (Hasselgren, 1999; Smith *et al.*, 2000).

In plants, programmed cell death occurs during senescence of leaves and flowers, during the development of vascular tissue, in somatic embryogenesis and during interactions with the environment such as aerenchyma formation and plant–pathogen interactions (Gray and Johal, 1998). Even though programmed cell death in plants and animals seems different in a lot of ways, similarities have also been found. The ubiquitin-dependent pathway seems to be involved in at least three examples of programmed cell death in plants; xylogenesis, sieve element differentiation and organ senescence.

6.3.3.1 *Xylogenesis*

Although ubiquitin and the proteasome is present in all plant tissue, histochemical investigations have shown specific and enhanced ubiquitin and proteasome levels in the procambium, and in maturing tracheary elements throughout the sunflower plant (Ingvardsen *et al.*, 2001). This very enhanced level of ubiquitin and the proteasome implies an important function in xylogenesis. This is confirmed by the high level of ubiquitin or ubiquitin gene expression found in the vascular tissue of rice, cotton and *Coleus* (Cornejo *et al.*, 1993; Stephenson *et al.*, 1996). Also, plants that harbor disturbances in the ubiquitin-dependent pathway develop abnormalities during vascular tissue formation (Bachmair *et al.*, 1990). The ubiquitin-conjugating enzymes (E2) from the *AtUBC1–3* gene family are also highly expressed in the immature vascular tissue (Thoma *et al.*, 1996).

The proteasome must be active for formation of tracheary element in cell cultures of *Zinnia* (Woffenden *et al.*, 1998), but the activity seems to be dangerous for the surrounding cells when the cytosolic contents are released as a consequence of the differentiation (Endo *et al.*, 2001). Auxin is also known to be necessary for xylem differentiation, which strengthens the proposal that the ubiquitin–proteasome pathway is involved in the regulation of the auxin response. Auxin has been suggested to activate genes by stimulating ubiquitylation of repressor proteins (Guilfoyle *et al.*, 1998; Walker and Estelle, 1998).

6.3.3.2 *Sieve element differentiation*

The ubiquitin-dependent pathway is also important for differentiation of phloem sieve elements. Ubiquitin has been found in the sieve tube exudate from *Ricinus communis* and a cDNA encoding a poly-ubiquitin protein has been found in phloem tissue from *Pinus sabiniana*. However, the exact role of ubiquitin and/or the ubiquitin-dependent pathway in the sieve element is not known (Carter *et al.*,

1995; Schobert *et al.*, 1998). It seems that the concentration of ubiquitin and the proteasome is even higher in the companion cells when compared with the sieve elements (Ingvardsen *et al.*, 2001), which presumably is due to a higher need for regulation of these cells.

6.3.3.3 Senescence

The expression of polyubiquitin genes has been found to increase during leaf senescence in several plants, including: *N. sylvestris*, potato, wheat and *A. thaliana* (Genschik *et al.*, 1994b; Garbarino *et al.*, 1995; Pinedo *et al.*, 1996; Park *et al.*, 1998), as have genes coding for some of the proteasome subunits during cotyledon senescence (Ito *et al.*, 1997). However, the level of polyubiquitin gene expression was found to be almost constant during senescence of *Pharbitis nil* cotyledons when the mRNA was analyzed according to leaf area—a decrease rather than an increase was observed (C. Ingvardsen, B. Veierskov and W. Laing, unpublished data). The ubiquitin-dependent pathway has also been shown to be involved in flower development and senescence, where an increase in the expression of poly-ubiquitin genes was found (Courtney *et al.*, 1994) and Genschik *et al.* (1994a) found that the expression level of the ubiquitin-conjugating enzyme *Ubd4* was increased during senescence of leaves in *N. sylvestris*. However, the proteasome does not seem to be induced by senescence, as Bahrami and Gray (1999) found a fall in the expression of a proteasome subunit during senescence. This is in agreement with our own observations showing that the expression of an $\alpha6$ subunit of the proteasome decreases during senescence in *Pharbitis nil* cotyledons (C. Ingvardsen and B. Veierskov, unpublished data).

In green tissue, the chloroplasts contain most of the cellular protein and so degradation of the chloroplast proteins is important for the senescence process. As it is known that proteins can be transported over membranes for degradation in the cytoplasm by the ubiquitin-dependent pathway, a role for this pathway in regulation of chloroplast proteins cannot be excluded. It has been debated whether or not the ubiquitylation of chloroplast proteins is due to cytosolic contamination or not (Hoffman *et al.*, 1991; Veierskov and Ferguson, 1991). Some ubiquitin was found in the chloroplast of unicellular alga *Chlamydomonas reinhardii*, but immunohistological investigations from higher plants have not shown any ubiquitin in the chloroplasts (Wettern *et al.*, 1990; Beers *et al.*, 1992; Ehlers *et al.*, 1996).

6.4   ATP-dependent Clp protease

6.4.1   Physical properties

X-Ray diffraction has shown that Clp proteases (casinolytic protease) consist of two functionally distinct parts: a central core and a regulatory ATPase unit.

The central core consists of two identical heptameric rings. Two types have been identified, ClpP and ClpQ (also called HsvV). The central core has a chamber that is 50 Å in diameter with axial openings of 10 Å. An opening of this size limits the ability of folded polypeptides to enter (Wang *et al.*, 1997) and so the substrates for the ClpP and ClpQ protease must be unfolded before entry into the proteolytic chamber in common with the 20S proteasome. The ClpP associates with either of two hexameric ATPases ClpA or ClpX (Katayama *et al.*, 1988; Gottesman *et al.*, 1993). The proteolytic active ClpAP (see figure 6.1) appears to have three separate compartments: the proteolytic chamber inside the ClpP, a cavity between ClpA and ClpP, and finally one within ClpA (Grimaud *et al.*, 1998). The secondary structure of ClpA and ClpX are very similar in the C-terminal and ATPase domains, placing them within the same AAA superfamily (Schweder *et al.*, 1996). The structure of the holo-ClpAP protease from *E. coli* resembles that of the 26S proteasome. ClpP resembles the β-rings of the 20S core of the proteasome and ClpA resambeling the ATPase ring of the 19S regulatory complex as seen in Figure 1 (Grimaud *et al.*, 1998).

ClpA has twice the affinity of ClpX for binding to ClpP (Grimaud *et al.*, 1998). Alone ClpP is incapable of degrading polypeptides longer than six amino acids, whereas ClpAP is able to degrades large proteins to short peptides of 7–10 amino acids without any apparent sequence specificity (Thompson *et al.*, 1994). ClpQ associates with the ClpY ATPase, also called HslU (Kessel *et al.*, 1996; Rohrwild *et al.*, 1996).

6.4.2   Subunits and active sites

The proteolytic mechanisms of the two Clp proteases differ; the ClpP is a serine-type protease while ClpQ has a threonine active site. The subunits of ClpP are 21 kDa, and the size of the ATPase subunits are 83 kDa for ClpA and 46 kDa for ClpX. ClpA has two non-homolog ATPase domains. The first ATP binding domain is required for ClpA oligomerization, while ATP hydrolysis by the other ATP domain is necessary for proteolysis (Hoskins *et al.*, 2000). Degradation of target proteins is facilitated by rotation of the ClpA hexamer around the common axis shared with ClpP, an effect that is intensified by the misalignment of the ClpA/ClpP complex (Beuron *et al.*, 1998).

Both ClpA and ClpX are members of a molecular chaperon family known as Clp/Hsp100, and are able to function as autonomous chaperons (Gottesman *et al.*, 1997). The Clp/Hsp100 family consists of two groups that are separated into types based on sequence similarity (Schirmer *et al.*, 1996). The first group contains proteins between 85 and 105 kDa with two distinct ATP-binding domains. These proteins are divided into five subtypes designated ClpA to ClpE. Of these, ClpA is present in all eukaryotes, and ClpC and ClpD have been identified in plants. The second group only consists of ClpX and ClpY, both having a single ATP-binding domain. Only ClpX has been found in all eukaryotes (Schirmer *et al.*, 1996).

6.4.3 *Target identification*

How target proteins are identified is still not completely understood. The Clp proteases do not use the ubiquitin targeting system, in common with the 26S proteasome, or any similar system. It is the ATPases of the Clp proteases that facilitates not only unfolding, but also substrate recognition. The different ATPases (ClpA–E) seem to determine substrate specificity of the ClpP complex (Gottesman, 1999). The recognition of target proteins may be based on signals located at their *C*- and *N*-termini. The ClpP protease *N*-end rule has been shown to degrade β-galactosidase fusion proteins depending on the residue at the *N*-terminus (Tobias *et al.*, 1991), whereas the *C*-terminal peptide sequence ANDENYALAA caused degradation by ClpXP as well as ClpAP. The ClpXP has different substrate specificity than the ClpAP complex, although ClpP has the ability to form a mixed complex with ClpA and ClpX, thereby broadening substrate specificity. It has been shown recently that ClpXPs are able to distinguish between a herero- and homodimer of the UmuD (D′) subunit of the error-prone DNA polymerase polV (Gonzalez *et al.*, 2000). Here it appears that each heterodimer provides a portion of the degradation signal, ensuring that neither of the two forms is degraded when present as homodimers.

6.4.4 *Localization and biological function*

The level of ClpP is normally low in eubacteria, but is inducible under many types of stress such as heat shock, starvation, salt and oxidative stress. In plants, nuclear coded isomers of ClpP have been identified as well as one plastid encoded isomer (pClpP) (Sokolenko *et al.*, 1998).

The pClpP complex is localized in the chloroplast stroma whereas only little is known of the nuclear-coded isoforms. The pClpP is synthesized constitutively in all plant tissue ranging from etiolated to green leaves and in roots. The protein does not seems to be induced by either heat shock or dehydration (Jabben *et al.*, 1989). However pClpP appears to be essential for the normal phototropic reaction. The four additional ClpP proteases identified from the *Arabidopsis* genome do not have chloroplastic transit peptide, but two of them do have extended *N*-terminal sequences indicating that they are targeted to other organelles.

6.5 Lon proteases

The Lon or La protease is another highly conserved ATP-dependent protease with serine in the active site. It is found in Archaea, Eubacteria and eukaryotic mitochondria, including plants, but has not been found in chloroplasts. In common with the Clp proteases, the Lon is not dependent on a targeting system for recognition (as ubiquitin is for the proteasome). Sequence motifs that call for degradation correspond with binding sites on the ATPases, but little is known as to what determines recognition.

6.5.1   Physical properties

The Lon proteases are homomeric, as both the ATPase and the proteolytic activity are located within the same polypeptide chain. The *Saccharomyces cerevisiae* mitochondrial Lon is a heptameric complex of 800 kDa (Stahlberg *et al.*, 1999). The outer ring diameter is about 11.5 nm with an inner hole of 2.5 nm and the total length of the complex is 17 nm (Stahlberg *et al.*, 1999). Lon thus distinguishes itself from other ATP-dependent proteases by consisting of identical subunits, and having ATPase activity and proteolytic activity within the same subunits.

6.5.2   Target identification

The Lon protease degrades abnormal proteins and exerts chaperone-like activity and is part of the heat shock response. Lon has also been found to stabilize the mitochondrial genome and regulate the expression of genes within the mitochondria. The Lon protease is also involved in the regulation of cell division by controlling turnover of short-lived regulatory proteins such as the cell division inhibitors SulA and RcsA (van Melderen and Gottesman, 1999).

6.6   Conclusions

Self-compartmentalizing proteolytic complexes exist in all known living organisms and all show a high degree of similarity in their physical properties. The proteolytic activity is located inside a cylinder-shaped complex formed by four rings each consisting of seven subunits. Access to the complexes is introduced by gating the terminal ends of the cylinder. This gating may be performed either by another ring consisting of ATPases, as is found in the Clp proteases, or by a ring consisting of ATPases upon which a lid is attached, as is observed in the proteasome. The substrate proteins to be degraded all have a recognition site that is recognized by the regulating gate. The recognition sites may be located on the target protein itself as is used by the Lon and Clp proteases, or may be performed by attachment of a poly-ubiquitin chain to the target protein (as in the proteasome).

Although these proteolytic systems are known to regulate the majority of protein turnover in the cells, we only have a sparse knowledge of processes that are regulated in plants. Emerging results indicate a function of the proteasome in auxin and jasmonic acid signaling pathways as well as flower induction. This proteolytic pathway is very conserved among species, and many similar processes are regulated in mammalian cells and plants. It is, however, important to remember that plant development is unique, and thus specific plant-related processes regulated by this proteolytic pathway are to be expected.

References

Aberle, H., Bauer, A., Stappert, J., Kispert, A. and Kemler, R. (1997) β-catenin is a target for the ubiquitin–proteasome pathway. *EMBO J.*, **16**, 3797-3804.

Adams, G.M., Falke, S., Goldberg, A.L., Slaughter, C.A., DeMartino, G.N. and Gogol, E.P. (1997) Structural and functional effects of PA700 and modulator protein on proteasomes. *J. Mol. Biol.*, **273**, 646-657.

Akopian, T.N., Kisselev, A.F. and Goldberg, A.L. (1997) Processive degradation of proteins and other catalytic properties of the proteasome from *Thermoplasma acidophilum*. *J. Biol. Chem.*, **272**, 1791-1798.

Altuvia, Y. and Margalit, H. (2000) Sequence signals for generation of antigenic peptides by the proteasome: implications for proteasomal cleavage mechanism. *J. Mol. Biol.*, **295**, 879-890.

Amerik, A.Y., Li, S.J. and Hochstrasser, M. (2000) Analysis of the deubiquitinating enzymes of the yeast *Saccharomyces cerevisiae*. *Biol. Chem.*, **381**, 981-992.

Aranda, M.A., Escaler, M., Wang, D. and Maule, A.J. (1996) Induction of HSP70 and polyubiquitin expression associated with plant virus replication. *Proc. Natl Acad. Sci. USA*, **93**, 15,289-15,293.

Arendt, C.S. and Hochstrasser, M. (1999) Eukaryotic 20S proteasome catalytic subunit propeptides prevent active site inactivation by *N*-terminal acetylation and promote particle assembly. *EMBO J.*, **18**, 3575-3585.

Aso, T., Lane, W.S., Conaway, J.W. and Conaway, R.C. (1995) Elongin (SIII)—a multisubunit regulator of elongation by RNA-polymerase-II. *Science*, **269**, 1439-1443.

Attaix, D., Combaret, L., Tilignac, T. and Taillandier, D. (1999) Adaptation of the ubiquitin–proteasome proteolytic pathway in cancer cachexia. *Mol. Biol. Rep.*, **26**, 77-82.

Bachmair, A., Becker, F., Masterson, R.V. and Schell, J. (1990) Perturbation of the ubiquitin system causes leaf curling, vascular tissue alterations and necrotic lesions in a higher plant. *EMBO J.*, **9**, 4543-4549.

Bahrami, A.R. and Gray, J.E. (1999) Expression of a proteasome α-type subunit gene during tobacco development and senescence. *Plant Mol. Biol.*, **39**, 325-333.

Banerjee, A., Deshaies, R.J. and Chau, V. (1995) Characterization of a dominant-negative mutant of the cell cycle ubiquitin-conjugating enzyme Cdc34. *J. Biol. Chem.*, **270**, 26,209-26,215.

Becker, J., Kempf, R., Jeblick, W. and Kauss, H. (2000) Induction of competence for elicitation of defense responses in cucumber hypocotyls requires proteasome activity. *Plant J.*, **21**, 311-316.

Beers, E.P., Moreno, T.N. and Callis, J. (1992) Subcellular localization of ubiquitin and ubiquitinated proteins in *Arabidopsis thaliana*. *J. Biol. Chem.*, **267**, 15,432-15,439.

Bercovich, B., Stancovski, I., Mayer, A. *et al.* (1997) Ubiquitin-dependent degradation of certain protein substrates *in vitro* requires the molecular chaperone Hsc70. *J. Biol. Chem.*, **272**, 9002-9010.

Beuron, F., Maurizi, M.R., Belnap, D.M. *et al.* (1998) At sixes and sevens: characterization of the symmetry mismatch of the ClpAP chaperone-assisted protease. *J. Struct. Biol.*, **123**, 248-259.

Binet, M.N., Weil, J.H. and Tessier, L.H. (1991) Structure and expression of sunflower ubiquitin genes. *Plant Mol. Biol.*, **17**, 395-407.

Bogyo, M., McMaster, J.S., Gaczynska, M., Tortorella, D., Goldberg, A.L. and Ploegh, H. (1997) Covalent modification of the active site threonine of proteasomal β subunits and the *Escherichia coli* homolog HslV by a new class of inhibitors. *Proc. Natl Acad. Sci. USA*, **94**, 6629-6634.

Braun, B.C., Glickman, M., Kraft, R. *et al.* (1999) The base of the proteasome regulatory particle exhibits chaperone-like activity. *Nat. Cell Biol.*, **1**, 221-226.

Byrd, C., Turner, G.C. and Varshavsky, A. (1998) The N-end rule pathway controls the import of peptides through degradation of a transcriptional repressor. *EMBO J.*, **17**, 269-277.

Callis, J. and Vierstra, R.D. (2000) Protein degradation in signaling. *Curr. Opin. Plant Biol.*, **3**, 381-386.

Carter, M.C.A., Kulikauskas, R.M. and Park, R.B. (1995) Structure and expression of ubiquitin gene transcripts in pine. *Canad. J. Forest Res.*, **25**, 1-7.

Chevalier, C., LeQuerrec, F. and Raymond, P. (1996) Sugar levels regulate the expression of ribosomal protein genes encoding protein S28 and ubiquitin-fused protein S27a in maize primary root tips. *Plant Sci.*, **117**, 95-105.

CohenFix, O., Peters, J.M., Kirschner, M.W. and Koshland, D. (1996) Anaphase initiation in *Saccharomyces cerevisiae* is controlled by the APC-dependent degradation of the anaphase inhibitor Pds1p. *Genes Devel.*, **10**, 3081-3093.

Conconi, M. and Friguet, B. (1997) Proteasome inactivation upon aging and on oxidation-effect of HSP 90. *Mol. Biol. Rep.*, **24**, 45-50.

Confalonieri, F. and Duguet, M. (1995) A 200-amino acid ATPase module in search of a basic function. *Bioessays*, **17**, 639-650.

Conrath, U., Klessig, D.F. and Bachmair, A. (1998) Tobacco plants perturbed in the ubiquitin-dependent protein degradation system accumulate callose, salicylic acid, and pathogenesis-related protein 1. *Plant Cell Rep.*, **17**, 876-880.

Cornejo, M.J., Luth, D., Blankenship, K.M., Anderson, O.D. and Blechl, A.E. (1993) Activity of a maize ubiquitin promoter in transgenic rice. *Plant Mol. Biol.*, **23**, 567-581.

Courtney, S.E., Rider, C.C. and Stead, A.D. (1994) Changes in protein ubiquitination and the expression of ubiquitin-encoding transcripts in daylily petals during floral development and senescence. *Physiol. Plant.*, **91**, 196-204.

Dahlmann, B., Kopp, F., Kristensen, P. and Hendil, K.B. (1999) Identical subunit topographies of human and yeast 20S proteasomes. *Arch. Biochem. Biophys.*, **363**, 296-300.

del Pozo, J.C. and Estelle, M. (2000) F-box proteins and protein degradation: an emerging theme in cellular regulation. *Plant Mol. Biol.*, **44**, 123-128.

DeMartino, G.N. and Slaughter, C.A. (1999) The proteasome, a novel protease regulated by multiple mechanisms. *J. Biol. Chem.*, **274**, 22,123-22,126.

Deshaies, R.J. (1997) Phosphorylation and proteolysis: partners in the regulation of cell division in budding yeast. *Curr. Opin. Genet. Devel.*, **7**, 7-16.

Deshaies, R.J. (1999) SCF and cullin/RING H2-based ubiquitin ligases. *Annu. Rev. Cell. Devel. Biol.*, **15**, 435-467.

Deveraux, Q., Ustrell, V., Pickart, C. and Rechsteiner, M. (1994) A 26-s protease subunit that binds ubiquitin conjugates. *J. Biol. Chem.*, **269**, 7059-7061.

Drexler, H.C.A. (1997) Activation of the cell death program by inhibition of proteasome function. *Proc. Natl Acad. Sci. USA*, **94**, 855-860.

Duggleby, H.J., Tolley, S.P., Hill, C.P., Dodson, E.J., Dodson, G. and Moody, P.C.E. (1995) Penicillin acylase has a single-amino-acid catalytic centre. *Nature*, **373**, 264-268.

Ehlers, K., Schulz, M. and Kollmann, R. (1996) Subcellular localization of ubiquitin in plant protoplasts and the function of ubiquitin in selective degradation of outer-wall plasmodesmata in regenerating protoplasts. *Planta*, **199**, 139-151.

Eleuteri, A.M., Kohanski, R.A., Cardozo, C. and Orlowski, M. (1997) Bovine spleen multicatalytic proteinase complex (proteasome)—replacement of X, Y, and Z subunits by LMP7, LMP2, and MECL1 and changes in properties and specificity. *J. Biol. Chem.*, **272**, 11,824-11,831.

Endo, S., Demura, T. and Fukuda, H. (2001) Inhibition of proteasome activity by the TED4 protein in extracellular space: a novel mechanism for protection of living cells from injury caused by dying cells. *Plant Cell Physiol.*, **42**, 9-19.

Etienne, P., Petitot, A.S., Houot, V., Blein, J.P. and Suty, L. (2000) Induction of tcI 7, a gene encoding a β-subunit of proteasome, in tobacco plants treated with elicitins, salicylic acid or hydrogen peroxide. *FEBS Lett.*, **466**, 213-218.

Ferguson, D.L., Guikema, J.A. and Paulsen, G.M. (1990) Ubiquitin pool modulation and protein-degradation in wheat roots during high temperature stress. *Plant Physiol.*, **92**, 740-746.

Feussner, K., Feussner, I., Leopold, I. and Wasternack, C. (1997) Isolation of a cDNA coding for an ubiquitin-conjugating enzyme UBC1 of tomato—the first stress-induced UBC of higher plants. *FEBS Lett.*, **409**, 211-215.

Figueiredo-Pereira, M.E. and Cohen, G. (1999) The ubiquitin/proteasome pathway: friend or foe in zinc-, cadmium-, and H_2O_2-induced neuronal oxidative stress. *Mol. Biol. Rep.*, **26**, 65-69.

Flinn, E.M., Busch, C.M.C. and Wright, A.P.H. (1998) myc Boxes, which are conserved in myc family proteins, are signals for protein degradation via the proteasome. *Mol. Cell. Biol.*, **18**, 5961-5969.

Fu, H.Y., Doelling, J.H., Arendt, C.S., Hochstrasser, M. and Vierstra, R.D. (1998) Molecular organization of the 20S proteasome gene family from *Arabidopsis thaliana*. *Genetics*, **149**, 677-692.

Fu, H.Y., Girod, P.A., Doelling, J.H. *et al.* (1999) Structure and functional analyses of the 26S proteasome subunits from plants - plant 26S proteasome. *Mol. Biol. Rep.*, **26**, 137-146.

Fujimuro, M., Sawada, H. and Yokosawa, H. (1997) Dynamics of ubiquitin conjugation during heat-shock response revealed by using a monoclonal antibody specific to multi-ubiquitin chains. *Eur. J. Biochem.*, **249**, 427-433.

Garbarino, J.E., Rockhold, D.R. and Belknap, W.R. (1992) Expression of stress-responsive ubiquitin genes in potato tubers. *Plant Mol. Biol.*, **20**, 235-244.

Garbarino, J.E., Oosumi, T. and Belknap, W.R. (1995) Isolation of a polyubiquitin promoter and its expression in transgenic potato plants. *Plant Physiol.*, **109**, 1371-1378.

Genschik, P., Parmentier, Y., Durr, A. *et al.* (1992a) Ubiquitin genes are differentially regulated in protoplast-derived cultures of *Nicotiana sylvestris* and in response to various stresses. *Plant Mol. Biol.*, **20**, 897-910.

Genschik, P., Philipps, G., Gigot, C. and Fleck, J. (1992b) Cloning and sequence analysis of a cDNA clone from *Arabidopsis thaliana* homologous to a proteasome α subunit from *Drosophila*. *FEBS Lett.*, **309**, 311-315.

Genschik, P., Durr, A. and Fleck, J. (1994a) Differential expression of several E2-type ubiquitin carrier protein genes at different developmental stages in *Arabidopsis thaliana* and *Nicotiana sylvestris*. *Mol. Gen. Genet.*, **244**, 548-556.

Genschik, P., Marbach, J., Uze, M., Feuerman, M., Plesse, B. and Fleck, J. (1994b) Structure and promoter activity of a stress and developmentally regulated polyubiquitin-encoding gene of *Nicotiana tabacum*. *Gene*, **148**, 195-202.

Gerards, W.L.H., de Jong, W.W., Boelens, W. and Bloemendal, H. (1998) Structure and assembly of the 20S proteasome. *Cell. Mol. Life Sci.*, **54**, 253-262.

Gilon, T., Chomsky, O. and Kulka, R.G. (1998) Degradation signals for ubiquitin system proteolysis in *Saccharomyces cerevisiae*. *EMBO J.*, **17**, 2759-2766.

Girod, P.A. and Vierstra, R.D. (1993) A major ubiquitin conjugation system in wheat-germ extracts involves a 15 kDa ubiquitin-conjugating enzyme (E2) homologous to the yeast UBC4/UBC5 gene-products. *J. Biol. Chem.*, **268**, 955-960.

Glickman, M.H., Rubin, D.M., Coux O. *et al.* (1998a) A subcomplex of the proteasome regulatory particle required for ubiquitin-conjugate degradation and related to the COP9-signalosome and eIF3. *Cell*, **94**, 615-623.

Glickman, M.H., Rubin, D.M., Fried, V.A. and Finley, D. (1998b) The regulatory particle of the *Saccharomyces cerevisiae* proteasome. *Mol. Cell. Biol.*, **18**, 3149-3162.

Glickman, M.H., Rubin, D.M., Fu, H.Y. *et al.* (1999) Functional analysis of the proteasome regulatory particle. *Mol. Biol. Rep.*, **26**, 21-28.

Glotzer, M., Murray, A.W. and Kirschner, M.W. (1991) Cyclin is degraded by the ubiquitin pathway. *Nature*, **349**, 132-138.

Gmachl, M., Gieffers, C., Podtelejnikov, A.V., Mann, M. and Peters, J.M. (2000) The RING-H2 finger protein APC11 and the E2 enzyme UBC4 are sufficient to ubiquitinate substrates of the anaphase-promoting complex. *Proc. Natl Acad. Sci. USA*, **97**, 8973-8978.

Gonzalez, M., Rasulova, F., Maurizi, M.R. and Woodgate, R. (2000) Subunit-specific degradation of the UmuD/D' heterodimer by the ClpXP protease: the role of *trans* recognition in UmuD' stability. *EMBO J.*, **19**, 5251-5258.

Gottesman, S. (1999) Regulation by proteolysis: developmental switches. *Curr. Opin. Microbiol.*, **2**, 142-147.

Gottesman, S., Clark, W.P., Decrecylagard, V. and Maurizi, M.R. (1993) Clpx, an alternative subunit for the ATP-dependent Clp protease of *Escherichia coli*. Sequence and *in vivo* activities. *J. Biol. Chem.*, **268**, 22,618-22,626.

Gottesman, S., Maurizi, M.R. and Wickner, S. (1997) Regulatory subunits of energy-dependent proteases. *Cell*, **91**, 435-438.

Gray, J. and Johal, G.S. (1998) Programmed cell death in plants, in *Arabidopsis* (eds M. Anderson and J. Roberts), Sheffield Academic Press, Sheffield, UK, pp. 360-394.

Gray, W.M., del Pozo, J.C., Walker, L. *et al.* (1999) Identification of an SCF ubiquitin–ligase complex required for auxin response in *Arabidopsis thaliana*. *Genes Dev.*, **13**, 1678-1691.

Grimaud, R., Kessel, M., Beuron, F., Steven, A.C. and Maurizi, M.R. (1998) Enzymatic and structural similarities between the *Escherichia coli* ATP-dependent proteases, ClpXP and ClpAP. *J. Biol. Chem.*, **273**, 12,476-12,481.

Groll, M., Ditzel, L., Löwe, J. *et al.* (1997) Structure of 20S proteasome from yeast at 2.4 Å resolution. *Nature*, **386**, 463-471.

Groll, M., Bajorek, M., Kohler, A. *et al.* (2000) A gated channel into the proteasome core particle. *Nat. Struct. Biol.*, **7**, 1062-1067.

Grossberger, R., Gieffers, C., Zachariae, W. *et al.* (1999) Characterization of the DOC1/APC10 subunit of the yeast and the human anaphase-promoting complex. *J. Biol. Chem.*, **274**, 14,500-14,507.

Grune, T., Reinheckel, T., Joshi, M. and Davies, K.J.A. (1995) Proteolysis in cultured liver epithelial cells during oxidative stress. Role of the multicatalytic proteinase complex, proteasome. *J. Biol. Chem.*, **270**, 2344-2351.

Grune, T., Reinheckel, T. and Davies, K.J.A. (1996) Degradation of oxidized proteins in K562 human hematopoietic cells by proteasome. *J. Biol. Chem.*, **271**, 15,504-15,509.

Grune, T., Reinheckel, T. and Davies, K.J.A. (1997) Degradation of oxidized proteins in mammalian cells. *FASEB J.*, **11**, 526-534.

Guilfoyle, T., Hagen, G., Ulmasov, T. and Murfett, J. (1998) How does auxin turn on genes? *Plant Physiol.*, **118**, 341-347.

Haas, A.L. and Siepmann, T.J. (1997) Pathways of ubiquitin conjugation. *FASEB J.*, **11**, 1257-1268.

Handke, C., Boyle, C. and Wettern, M. (1993) Effect of aging, abiotic and biotic stress upon ubiquitination in young barley plants. *Angewandte Botanik*, **67**, 120-123.

Haracska, L. and Udvardy, A. (1995) Cloning and sequencing a non-ATPase subunit of the regulatory complex of the *Drosophila* 26S-protease. *Eur. J. Biochem.*, **231**, 720-725.

Hasselgren, P.O. (1999) Role of the ubiquitin-proteasome pathway in sepsis-induced muscle catabolism. *Mol. Biol. Rep.*, **26**, 71-76.

Hatfield, P.M., Gosink, M.M., Carpenter, T.B. and Vierstra, R.D. (1997) The ubiquitin-activating enzyme (El) gene family in *Arabidopsis thaliana*. *Plant J.*, **11**, 213-226.

Hicke, L. and Riezman, H. (1996) Ubiquitination of a yeast plasma membrane receptor signals its ligand-stimulated endocytosis. *Cell*, **84**, 277-287.

Hoffman, N.E., Ko, K., Milkowski, D. and Pichersky, E. (1991) Isolation and characterization of tomato cDNA and genomic clones encoding the ubiquitin gene *ubi3*. *Plant Mol. Biol.*, **17**, 1189-1201.

Hoskins, J.R., Singh, S.K., Maurizi, M.R. and Wickner, S. (2000) Protein binding and unfolding by the chaperone ClpA and degradation by the protease ClpAP. *Proc. Natl Acad. Sci. USA*, **97**, 8892-8897.

Huibregtse, J.M., Scheffner, M. and Howley, P.M. (1991) A cellular protein mediates association of p53 with the E6 oncoprotein of human papillomavirus types 16 or 18. *EMBO J.*, **10**, 4129-4135.

Ingvardsen, C. and Veierskov, B. (2001) Ubiquitin- and proteasome-dependent proteolysis in plants. *Physiol. Plant.*, **112**, 451-459.

Ingvardsen, C., Vierskov, B. and Joshi P.A. (2001) Immunohistochemical localisation of ubiquitin and the proteasome in sunflower (*Helianthus annuus* cv. Giganteus). *Planta*. **213**, 333-341.

Ito, N., Tomizawa, K., Tanaka, K. *et al.* (1997) Characterization of 26S proteasome α- and β-type and ATPase subunits from spinach and their expression during early stages of seedling development. *Plant Mol. Biol.*, **34**, 307-316.

Ito, N., Seo, S., Ohtsubo, N., Nakagawa, H. and Ohashi, Y. (1999) Involvement of proteasome-ubiquitin system in wound-signaling in tobacco plants. *Plant Cell Physiol.*, **40**, 355-360.

Jabben, M., Shanklin, J. and Vierstra, R.D. (1989) Ubiquitin-phytochrome conjugates. Pool dynamics during *in vivo* phytochrome degradation. *J. Biol. Chem.*, **264**, 4998-5005.

Jackson, P.K., Eldridge, A.G., Freed, E. *et al.* (2000) The lore of the RINGs: substrate recognition and catalysis by ubiquitin ligases. *Trends Cell Biol.*, **10**, 429-439.

Jarrousse, A.S., Petit, F., Kreutzer-Schmid, C., Gaedigk, R. and Schmid, H.P. (1999) Possible involvement of proteasomes (prosomes) in AUUUA-mediated mRNA decay. *J. Biol. Chem.*, **274**, 5925-5930.

Johnson, P.R., Swanson, R., Rakhilina, L. and Hochstrasser, M. (1998) Degradation signal masking by heterodimerization of MATα2 and MATa1 blocks their mutual destruction by the ubiquitin-proteasome pathway. *Cell*, **94**, 217-227.

Johnston, S.C., Riddle, S.M., Cohen, R.E. and Hill, C.P. (1999) Structural basis for the specificity of ubiquitin C-terminal hydrolases. *EMBO J.*, **18**, 3877-3887.

Jørgensen, L. and Hendil, K.B. (1999) Proteasome subunit zeta, a putative ribonuclease, is also found as a free monomer. *Mol. Biol. Rep.*, **26**, 119-123.

Kaelin, W.G., Iliopoulos, O., Lonergan, K.M. and Ohh, M. (1998) Functions of the von Hippel-Lindau tumour suppressor protein. *J. Intern. Med.*, **243**, 535-539.

Katayama, Y., Gottesman, S., Pumphrey, J., Rudikoff, S., Clark, W.P. and Maurizi, M.R. (1988) The two-component, ATP-dependent Clp protease of *Escherichia coli*. Purification, cloning, and mutational analysis of the ATP-binding component. *J. Biol. Chem.*, **263**, 15,226-15,236.

Kessel, M., Wu, W.F., Gottesman, S., Kocsis, E., Steven, A.C. and Maurizi, M.R. (1996) Six-fold rotational symmetry of ClpQ, the *E. coli* homolog of the 20S proteasome, and its ATP-dependent activator, ClpY. *FEBS Lett.*, **398**, 274-278.

King, R.W., Glotzer, M. and Kirschner, M.W. (1996) Mutagenic analysis of the destruction signal of mitotic cyclins and structural characterization of ubiquitinated intermediates. *Mol. Biol. Cell*, **7**, 1343-1357.

Kisselev, A.F., Akopian, T.N. and Goldberg, A.L. (1998) Range of sizes of peptide products generated during degradation of different proteins by archaeal proteasomes. *J. Biol. Chem.*, **273**, 1982-1989.

Kitashiba, H. and Toriyama, K. (1997) Expression of a gene for a protein similar to HIV-1 Tat binding protein 1 (TBP1) in floral organs of *Brassica rapa*. *Plant Cell Physiol.*, **38**, 966-969.

Koegl, M., Hoppe, T., Schlenker, S., Ulrich, H.D., Mayer, T.U. and Jentsch, S. (1999) A novel ubiquitination factor, E4, is involved in multiubiquitin chain assembly. *Cell*, **96**, 635-644.

Kurasawa, Y. and Todokoro, K. (1999) Identification of human APC10/Doc1 as a subunit of anaphase promoting complex. *Oncogene*, **18**, 5131-5137.

Kwon, Y.T., Kashina, A.S. and Varshavsky, A. (1999) Alternative splicing results in differential expression, activity, and localization of the two forms of arginyl-tRNA-protein transferase, a component of the *N*-end rule pathway. *Mol. Cell. Biol.*, **19**, 182-193.

Lam, Y.A., DeMartino, G.N., Pickart, C.M. and Cohen, R.E. (1997) Specificity of the ubiquitin isopeptidase in the PA700 regulatory complex of 26 S proteasomes. *J. Biol. Chem.*, **272**, 28,438-28,446.

Lammer, D., Mathias, N., Laplaza, J.M. *et al.* (1998) Modification of yeast Cdc53p by the ubiquitin-related protein Rub1p affects function of the SCF^{Cdc4} complex. *Genes Dev.*, **12**, 914-926.

Laney, J.D. and Hochstrasser, M. (1999) Substrate targeting in the ubiquitin system. *Cell*, **97**, 427-430.

Lee, D.H., Sherman, M.Y. and Goldberg, A.L. (1996) Involvement of the molecular chaperone Ydj1 in the ubiquitin-dependent degradation of short-lived and abnormal proteins in *Saccharomyces cerevisiae*. *Mol. Cell. Biol.*, **16**, 4773-4781.

Li, X.Q., Stebbins, B., Hoffman, L., Pratt, G., Rechsteiner, M. and Coffino, P. (1996) The *N* terminus of antizyme promotes degradation of heterologous proteins. *J. Biol. Chem.*, **271**, 4441-4446.

Löwe, J., Stock, D., Jap, R., Zwickl, P., Baumeister, W. and Huber, R. (1995) Crystal structure of the 20S proteasome from the Archaeon *T. acidophilum* at 3.4 Å resolution. *Science*, **268**, 533-539.

Lupas, A., Zwickl, P. and Baumeister, W. (1994) Proteasome sequences in eubacteria. *Trends Biochem. Sci.*, **19**, 533-534.

Maupin-Furlow, J.A., Aldrich, H.C. and Ferry, J.G. (1998) Biochemical characterization of the 20S proteasome from the methanoarchaeon *Methanosarcina thermophila*. *J. Bacteriol.*, **180**, 1480-1487.

Mazeyrat, F., Mouzeyar, S., Courbou, I. *et al.* (1999) Accumulation of defense related transcripts in sunflower hypocotyls (*Helianthus annuus* L.) infected with *Plasmopara halstedii*. *Eur. J. Plant Pathol.*, **105**, 333-340.

McCafferty, H.R.K. and Talbot, N.J. (1998) Identification of three ubiquitin genes of the rice blast fungus *Magnaporthe grisea*, one of which is highly expressed during initial stages of plant colonisation. *Current Genet.*, **33**, 352-361.

McGrath, J.P., Jentsch, S. and Varshavsky, A. (1991) *UBA1*: an essential yeast gene encoding ubiquitin-activating enzyme. *EMBO J.*, **10**, 227-236.

Mehta, R.A., Warmbardt, R.D. and Mattoo, A.K. (1996) Tomato (*Lycopersicon esculentum* cv. Pik-Red) leaf carboxypeptidase: identification, *N*-terminal sequence, stress-regulation, and specific localization in the paraveinal mesophyll vacuoles. *Plant Cell Physiol.*, **37**, 806-815.

Murakami, Y., Matsufuji, S., Hayashi, S., Tanahashi, N. and Tanaka, K. (2000) Degradation of ornithine decarboxylase by the 26S proteasome. *Biochem. Biophys. Res. Commun.*, **267**, 1-6.

Murray, P.F., Giordano, C.V., Passeron, S. and Barneix, A.J. (1997) Purification and characterization of 20 S proteasome from wheat leaves. *Plant Sci.*, **125**, 127-136.

Murzin, A.G. (1998) How far divergent evolution goes in proteins. *Curr. Opin. Struct. Biol.*, **8**, 380-387.

Musti, A.M., Treier, M., Peverali, F.A. and Bohmann, D. (1996) Differential regulation of c-Jun and JunD by ubiquitin-dependent protein degradation. *Biological Chem.*, **377**, 619-624.

Orlowski, M. and Wilk, S. (2000) Catalytic activities of the 20 S proteasome, a multicatalytic proteinase complex. *Arch. Biochem. Biophys.*, **383**, 1-16.

Osmulski, P.K. and Gaczynska, M. (2000) Atomic force microscopy reveals two conformations of the 20 S proteasome from fission yeast. *J. Biol. Chem.*, **275**, 13,171-13,174.

Pamnani, V., Haas, B., Puhler, G., Sanger, H.L. and Baumeister, W. (1994) Proteasome-associated RNAs are nonspecific. *Eur. J. Biochem.*, **225**, 511-519.

Park, J.H., Oh, S.A., Kim, Y.H., Woo, H.R. and Nam, H.G. (1998) Differential expression of senescence-associated mRNAs during leaf senescence induced by different senescence-inducing factors in *Arabidopsis*. *Plant Mol. Biol.*, **37**, 445-454.

Patton, E.E., Willems, A.R., Sa, D. *et al.* (1998) Cdc53 is a scaffold protein for multiple Cdc34/Skp1/F box protein complexes that regulate cell division and methionine biosynthesis in yeast. *Genes Dev.*, **12**, 3144-3144.

Peters, J.M., King, R.W., Höög, C. and Kirschner, M.W. (1996) Identification of BIME as a subunit of the anaphase-promoting complex. *Science*, **274**, 1199-1201.

Pfleger, C.M. and Kirschner, M.W. (2000) The KEN box: an APC recognition signal distinct from the D box targeted by Cdh1. *Genes Dev.*, **14**, 655-665.

Pickart, C.M. and VanDemark, A.P. (2000) Opening doors into the proteasome. *Nat. Struct. Biol.*, **7**, 999-1001.

Pieterse, C.M.J., Risseeuw, E.P. and Davidse, L.C. (1991) An *in planta* induced gene of *Phytophthora infestans* codes for ubiquitin. *Plant Mol. Biol.*, **17**, 799-811.

Pinedo, M.L., Goicoechea, S.M., Lamattina, L. and Conde, R.D. (1996) Estimation of ubiquitin and ubiquitin mRNA content in dark senescing wheat leaves. *Biol. Plant.*, **38**, 321-328.

Potuschak, T., Stary, S., Schlogelhofer, P., Becker, F., Nejinskaia, V. and Bachmair, A. (1998) *PRT1* of *Arabidopsis thaliana* encodes a component of the plant *N*-end rule pathway. *Proc. Natl Acad. Sci. USA*, **95**, 7904-7908.

Pouch, M.N., Petit, F., Buri, J., Briand, Y. and Schmid, H.P. (1995) Identification and initial characterization of a specific proteasome (prosome) associated RNase activity. *J. Biol. Chem.*, **270**, 22,023-22,028.

Prombona, A., Tabler, M., Providaki, M. and Tsagris, M. (1995) Structure and expression of LeMA-1, a tomato protein belonging to the SEC18-PAS1-CDC48-TBP-1 protein family of putative Mg^{2+}-dependent ATPases. *Plant Mol. Biol.*, **27**, 1109-1118.

Pühler, G., Pitzer, F., Zwickl, P. and Baumeister, W. (1994) Proteasomes: multisubunit proteinases common to *Thermoplasma* and eukaryotes. *Syst. Appl. Microbiol.*, **16**, 734-741.

Ramos, P.C., Hockendorff, J., Johnson, E.S., Varshavsky, A. and Dohmen, R.J. (1998) Ump1p is required for proper maturation of the 20S proteasome and becomes its substrate upon completion of the assembly. *Cell*, **92**, 489-499.

Rechsteiner, M. and Rogers, S.W. (1996) PEST sequences and regulation by proteolysis. *Trends Biochem. Sci.*, **21**, 267-271.

Rechsteiner, M., Realini, C. and Ustrell, V. (2000) The proteasome activator 11 S REG (PA28) and Class I antigen presentation. *Biochem. J.*, **345**, 1-15.

Reichel, C. and Beachy, R.N. (2000) Degradation of tobacco mosaic virus movement protein by the 26S proteasome. *J. Virol.*, **74**, 3330-3337.

Reinheckel, T., Sitte, N., Ullrich, O., Kuckelkorn, U., Davies, K.J.A. and Grune, T. (1998) Comparative resistance of the 20S and 26S proteasome to oxidative stress. *Biochemical J.*, **335**, 637-642.

Richmond, C., Gorbea, C. and Rechsteiner, M. (1997) Specific interactions between ATPase subunits of the 26 S protease. *J. Biol. Chem.*, **272**, 13,403-13,411.

Rickey, T.M. and Belknap, W.R. (1991) Comparison of the expression of several stress-responsive genes in potato tubers. *Plant Mol. Biol.*, **16**, 1009-1018.

Rohrwild, M., Coux, O., Huang, H.C. *et al.* (1996) HslV-HslU: a novel ATP-dependent protease complex in *Escherichia coli* related to the eukaryotic proteasome. *Proc. Natl Acad. Sci. USA*, **93**, 5808-5813.

Roth, A.F., Sullivan, D.M. and Davis, N.G. (1998) A large PEST-like sequence directs the ubiquitination, endocytosis, and vacuolar degradation of the yeast a-factor receptor. *J. Cell Biol.*, **142**, 949-961.

Rotin, D. (1998) WW (WWP) domains: from Current topics in microbiology and immunology structure to function, in *Protein Modules in Signal Transduction* (ed. A.J. Pawson), Springer-Verlag, Berlin, pp. 115-133.

Rubin, D.M., Glickman, M.H., Larsen, C.N., Dhruvakumar, S. and Finley, D. (1998) Active site mutants in the six regulatory particle ATPases reveal multiple roles for ATP in the proteasome. *EMBO J.*, **17**, 4909-4919.

Salama, S.R., Hendricks, K.B. and Thorner, J. (1994) G_1 cyclin degradation: the PEST motif of yeast Cln2 is necessary, but not sufficient, for rapid protein-turnover. *Mol. Cell. Biol.*, **14**, 7953-7966.

Schirmer, E.C., Glover, J.R., Singer, M.A. and Lindquist, S. (1996) HSP100/Clp proteins: a common mechanism explains diverse functions. *Trends Biochem. Sci.*, **21**, 289-296.

Schobert, C., Baker, L., Szederkényi, J. *et al.* (1998) Identification of immunologically related proteins in sieve-tube exudate collected from monocotyledonous and dicotyledonous plants. *Planta*, **206**, 245-252.

Schweder, T., Lee, K.H., Lomovskaya, O. and Matin, A. (1996) Regulation of *Escherichia coli* starvation sigma factor (sigma(s)) by ClpXP protease. *J. Bacteriol.*, **178**, 470-476.

Seemüller, E., Lupas, A., Zuhl, F., Zwickl, P. and Baumeister, W. (1995) The proteasome from *Thermoplasma acidophilum* is neither a cysteine nor a serine-protease. *FEBS Lett.*, **359**, 173-178.

Seufert, W. and Jentsch, S. (1990) Ubiquitin-conjugating enzymes UBC4 and UBC5 mediate selective degradation of short-lived and abnormal proteins. *EMBO J.*, **9**, 543-550.

Shang, F., Gong, X. and Taylor, A. (1997) Activity of ubiquitin-dependent pathway in response to oxidative stress. Ubiquitin-activating enzyme is transiently up-regulated. *J. Biol. Chem.*, **272**, 23,086-23,093.

Sheaff, R.J., Singer, J.D., Swanger, J., Smitherman, M., Roberts, J.M. and Clurman, B.E. (2000) Proteasomal turnover of p21^{Cip1} does not require p21^{Cip1} ubiquitination. *Mol. Cell*, **5**, 403-410.

Shinohara, K., Tomioka, M., Nakano, H., Tone, S., Ito, H. and Kawashima, S. (1996) Apoptosis induction resulting from proteasome inhibition. *Biochem. J.*, **317**, 385-388.

Skowyra, D., Craig, K.L., Tyers, M., Elledge, S.J. and Harper, W. (1997) F-box proteins are receptors that recruit phosphorylated substrates to the SCF ubiquitin-ligase complex. *Cell*, **91**, 209-219.

Smith, C.K., Baker, T.A. and Sauer, R.T. (1999) Lon and Clp family proteases and chaperones share homologous substrate-recognition domains. *Proc. Natl Acad. Sci. USA*, **96**, 6678-6682.

Smith, L., Chen, L., Reyland, M.E. *et al.* (2000) Activation of atypical protein kinase C zeta by caspase processing and degradation by the ubiquitin-proteasome system. *J. Biol. Chem.*, **275**, 40,620-40,627.

Sokolenko, A., Lerbs-Mache, S., Altschmied, L. and Herrmann, R.G. (1998) Clp protease complexes and their diversity in chloroplasts. *Planta*, **207**, 286-295.

Stahlberg, H., Kutejova, E., Suda, K. *et al.* (1999) Mitochondrial Lon of *Saccharomyces cerevisiae* is a ring-shaped protease with seven flexible subunits. *Proc. Natl Acad. Sci. USA*, **96**, 6787-6790.

Stephenson, P., Collins, B.A., Reid, P.D. and Rubinstein, B. (1996) Localization of ubiquitin to differentiating vascular tissues. *Am. J. Bot.*, **83**, 140-147.

Sun, C.W. and Callis, J. (1997) Independent modulation of *Arabidopsis thaliana* polyubiquitn mRNAs in different organs and in response to environmental changes. *Plant J.*, **11**, 1017-1027.

Suzuka, I., Kogaban, Y., Sasaki, T., Minobe, Y. and Hashimoto, J. (1994) Identification of cDNA clones for rice homologs of the human immunodeficiency virus-1 Tat-binding protein and subunit-4 of human 26S-protease (proteasome). *Plant Sci.*, **103**, 33-40.

Tamura, T., Nagy, I., Lupas, A. *et al.* (1995) The first characterization of a eubacterial proteasome: the 20S complex of rhodococcus. *Curr. Biol.*, **5**, 766-774.

Tanaka, K. and Tsurumi, C. (1997) The 26S proteasome: subunits and functions. *Mol. Biol. Rep.*, **24**, 3-11.

Terrell, J., Shih, S., Dunn, R. and Hicke, L. (1998) A function for monoubiquitination in the internalization of a G protein-coupled receptor. *Mol. Cell*, **1**, 193-202.

Thoma, S., Sullivan, M.L. and Vierstra, R.D. (1996) Members of two gene families encoding ubiquitin-conjugating enzymes, AtUBC1-3 and AtUBC4-6, from *Arabidopsis thaliana* are differentially expressed. *Plant Mol. Biol.*, **31**, 493-505.

Thompson, M.W., Singh, S.K. and Maurizi, M.R. (1994) Processive degradation of proteins by the ATP-dependent Clp protease from *Escherichia coli*. Requirement for the multiple array of active-sites in ClpP but not ATP hydrolysis. *J. Biol. Chem.*, **269**, 18,209-18,215.

Tobias, J.W., Shrader, T.E., Rocap, G. and Varshavsky, A. (1991) The *N*-end rule in bacteria. *Science*, **254**, 1374-1377.

van Melderen, L. and Gottesman, S. (1999) Substrate sequestration by a proteolytically inactive Lon mutant. *Proc. Natl Acad. Sci. USA*, **96**, 6064-6071.

van Nocker, S., Sadis, S., Rubin, D.M. *et al.* (1996) The multiubiquitin-chain-binding protein Mcb1 is a component of the 26S proteasome in *Saccharomyces cerevisiae* and plays a nonessential, substrate-specific role in protein turnover. *Mol. Cell. Biol.*, **16**, 6020-6028.

Varshavsky, A., Turner, G., Du, F.Y. and Xie, Y.M. (2000) The ubiquitin system and the *N*-end rule pathway. *Biol. Chem.*, **381**, 779-789.

Veierskov, B. and Ferguson, I.B. (1991) Ubiquitin conjugating activity in leaves and isolated- chloroplasts from *Avena sativa* L. during senescence. *J. Plant Physiol.*, **138**, 608-613.

Walker, L. and Estelle, M. (1998) Molecular mechanisms of auxin action. *Curr. Opin. Plant Biol.*, **1**, 434-439.

Walz, J., Erdmann, A., Kania, M., Typke, D., Koster, A.J. and Baumeister, W. (1998) 26S proteasome structure revealed by three-dimensional electron microscopy. *J. Struct. Biol.*, **121**, 19-29.

Wang, J.M., Hartling, J.A. and Flanagan, J.M. (1997) The structure of ClpP at 2.3 angstrom resolution suggests a model for ATP-dependent proteolysis. *Cell*, **91**, 447-456.

Wang, G.L., Yang, J. and Huibregtse, J.M. (1999) Functional domains of the Rsp5 ubiquitin–protein ligase. *Mol. Cell. Biol.*, **19**, 342-352.

Wegener, A., Gimbel, W., Werner, T., Hani, J., Ernst, D. and Sandermann, H. (1997) Sequence analysis and ozone-induced accumulation of polyubiquitin mRNA in *Pinus sylvestris. Can. J. Forest Res.*, **27**, 945-948.

Wenzel, T., Eckerskorn, C., Lottspeich, F. and Baumeister, W. (1994) Existence of a molecular ruler in proteasomes suggested by analysis of degradation products. *FEBS Lett.*, **349**, 205-209.

Wettern, M., Parag, H.A., Pollmann, L., Ohad, I. and Kulka, R.G. (1990) Ubiquitin in *Chlamydomonas reinhardii*. Distribution in the cell and effect of heat-shock and photoinhibition on its conjugate pattern. *Eur. J. Biochem.*, **191**, 571-576.

Wilkinson, K.D. and Hochstrasser, M. (1998) The deubiquitinating enzymes, in *Ubiquitin and the Biology of the Cell* (eds J. Peters, J. R. Harris and D. Finley), Plenum Press, New York, pp. 99-126.

Winston, J.T., Koepp, D.M., Zhu, C.H., Elledge, S.J. and Harper, J.W. (1999) A family of mammalian F-box proteins. *Curr. Biol.*, **9**, 1180-1182.

Woffenden, B.J., Freeman, T.B. and Beers, E.P. (1998) Proteasome inhibitors prevent tracheary element differentiation in Zinnia mesophyll cell cultures. *Plant Physiol.*, **118**, 419-429.

Worley, C.K., Ling, R. and Callis, J. (1998) Engineering in vivo instability of firefly luciferase and *Escherichia coli* β-glucuronidase in higher plants using recognition elements from the ubiquitin pathway. *Plant Mol. Biol.*, **37**, 337-347.

Yaffe, M.B. and Elia, A.E.H. (2001) Phosphoserine/threonine-binding domains. *Curr. Opin. Cell Biol.*, **13**, 131-138.

Yan, N., Doelling, J.H., Falbel, T.G., Durski, A.M. and Vierstra, R.D. (2000) The ubiquitin-specific protease family from *Arabidopsis*. AtUBP1 and 2 are required for the resistance to the amino acid analog canavanine. *Plant Physiol.*, **124**, 1828-1843.

Yaron, A., Gonen, H., Alkalay, I. *et al.* (1997) Inhibition of NF-κB cellular function via specific targeting of the IκB-ubiquitin ligase. *EMBO J.*, **16**, 6486-6494.

Yoshimura, T., Kameyama, K., Takagi, T. *et al.* (1993) Molecular characterization of the "26S" proteasome complex from rat-liver. *J. Struct. Biol.*, **111**, 200-211.

Zhou, P.B. and Howley, P.M. (1998) Ubiquitination and degradation of the substrate recognition subunits of SCF ubiquitin-protein ligases. *Mol. Cell*, **2**, 571-580.

Zwickl, P., Kleinz, J. and Baumeister, W. (1994) Critical elements in proteasome assembly. *Nat. Struct. Biol.*, **1**, 765-770.

Zwickl, P., Baumeister, W. and Steven, A. (2000) Dis-assembly lines: the proteasome and related ATPase-assisted proteases. *Curr. Opin. Struct. Biol.*, **10**, 242-250.

7 The higher plant chaperonins

Rajach Sharkia, Paul Viitanen, Galit Levy-Rimler,
Celeste Weiss, Adina Niv and Abdussalam Azem

7.1 Chaperonins

The primary sequence of a protein contains all the information that is needed
for the protein to fold precisely to its active conformation (Anfinsen, 1973).
However, in the crowded cellular environment, most proteins require assistance
of a group of proteins, called molecular chaperones, in order to fold into their
functional form (Ellis, 1987; Hartl, 1996; Bukau and Horwich, 1998). In addition
to mediating the folding of newly translated, imported and stress-denatured
proteins, molecular chaperones participate in many vital cellular processes,
including: translocation of proteins within the cell, protein degradation, antigen
presentation, signal transduction, developmental processes and regulation of
apoptosis (Schatz and Dobberstein, 1996; Edwards, 1998; Mayer and Bukau,
1999). The initial observation that the expression of many molecular chaperones
is induced at elevated temperatures has led to their being classified as heat-
shock proteins (HSPs). The major families of HSPs have molecular weights
of 100, 90, 70, 60 and ca. 15–25 kDa (small HSPs) (Polissi *et al.*, 1995). The
chaperonins (CPNs) is a term reserved for the 60 kDa HSP family (hsp60),
which is a sequence-related family of chaperone proteins that is found in all
cells from bacteria to humans.

The CPNs are divided into two groups, type I and type II (Hendrick and
Hartl, 1995). Type I chaperonins exhibit high sequence homology to each
other and are present in eubacteria and in those eukaryotic organelles that are
derived from prokaryotes through the process of endosymbiosis (e.g. chloroplast
and mitochondria). The type I chaperonin family has two distinct members,
chaperonin 60 (cpn60) and chaperonin 10 (cpn10), which together facilitate the
folding of other proteins. Both are essential for the viability of yeast and bacteria
under all conditions. Type II chaperonins comprise only a cpn60 member that
is much less homologous to type I chaperonins (Gutsche *et al.*, 1999; Willison,
1999; Leroux and Hartl, 2000; Carrascosa *et al.*, 2001). The members of this
family are found in eukaryotic cytosol and Archaebacteria (termed thermosome
and TF55).

Plants contain both type I (mitochondria and chloroplast) and cytosolic type
II chaperonins (Boston *et al.*, 1996). In this chapter, we will review the recent
progress made in understanding the structure and function of plant chaperonins.

7.2 Type I chaperonins

7.2.1 Bacterial chaperonins

Due to the simplicity of the bacterial system and the relative ease with which GroEL can be purified, most of our understanding of chaperonin structure and function comes from the *Escherichia coli* GroE chaperonin proteins (Hartl, 1996; Lorimer and Todd, 1996; Bukau and Horwich, 1998; Sigler *et al.*, 1998). Therefore, as a basis for understanding plant chaperonins, it is important to summarize some of the more important properties of the bacterial prototypes. The bacterial chaperonin system comprises two proteins: the 60 kDa HSP (also called GroEL or CPN60) and the 10 kDa HSP (also called GroES or CPN10). Studies using X-ray crystallography and electron microscopy have given us a very detailed picture of the chaperonin structure (figure 7.1). GroEL is a cylindrical molecule comprising two stacked rings of seven identical subunits (Langer *et al.*, 1992; Braig *et al.*, 1994; Lorimer and Todd, 1996; Roseman *et al.*, 1996; Sigler *et al.*, 1998). Each subunit contains three regions: an apical, an intermediate and an equatorial domain. The apical domain contains binding sites for both GroES and denatured substrate protein. Seven apical domains of one ring form a single such site. Thus, each GroEL molecule contains one GroES and one peptide-binding site at each end of the GroEL cylinder. Although both sites share the same amino acid residues, GroES and the unfolded substrate do bind to the same ring concomitantly (Weissman *et al.*, 1995; Mayhew *et al.*, 1996; Llorca *et al.*, 1997; Sparrer *et al.*, 1997; Chen and Sigler, 1999). The equatorial domain contains the ATP binding site (seven sites in each ring) and some of the intersubunit contacts that stabilize the $GroEL_{14}$ oligomer (Boisvert *et al.*, 1996; Chen and Sigler, 1999). The intermediate domain links and transfers allosteric information between the apical and equatorial domains.

Under non-permissive conditions (e.g. conditions where spontaneous refolding of denatured protein does not occur), the chaperonin-assisted folding reaction strictly requires ATP hydrolysis and the participation of GroES. Two types of hetero-oligomeric complexes are formed in the presence of nucleotides: one is an asymmetric, bullet-shaped complex, that comprises one $GroEL_{14}$ and one $GroES_7$, and the second a symmetric, football-shaped complex comprising one $GroEL_{14}$ and two $GroES_7$ molecules (Azem *et al.*, 1994; Harris *et al.*, 1994; Llorca *et al.*, 1994; Schmidt *et al.*, 1994; Behlke *et al.*, 1997). The currently accepted model for the GroEL/GroES mediated protein folding reaction is summarized in figure 7.1 (Lorimer and Todd, 1996; Martin and Hartl, 1997; Horovitz, 1998; Sigler *et al.*, 1998; Saibil, 2000). Protein that is exposed to stress undergoes denaturation and is at risk of aggregation. When $GroEL_{14}$ is present, the denatured protein binds in the central cavity of the chaperonin, where it is isolated from other proteins in solution (Ellis, 1994). The

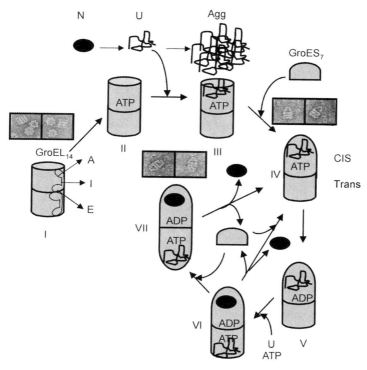

Figure 7.1 The mechanism of GroE-mediated protein folding. GroEL is a tetradecamer (GroEL₁₄, I) organized in two heptameric rings. Each GroEL subunit is composed of three domains: the apical (A) the intermediate (I) and the equatorial domain (Braig *et al.*, 1994; Boisvert *et al.*, 1996; Xu *et al.*, 1997; Chen and Sigler, 1999). Native protein (N) that is exposed to stress undergoes denaturation (U) and is at risk of aggregation (Agg). When GroEL₁₄ (I and II) is present, the denatured protein binds in GroEL's central cavity (forming complex III). The central cavity provides an isolated environment in which the substrate can fold, without the risk of unwanted interactions between unfolded proteins. GroES₇ binds to complex III, to the same ring where U and ATP are bound (IV, *cis* asymmetric complex), and initiates the productive chaperonin reaction cycle as follows: (i) hydrolysis of ATP (form V), (ii) folding of the enclosed substrate; and (iii) binding of a second unfolded substrate and ATP at the trans ring (complex VI) triggers the release of the folded protein from the *cis* ring (transition of form VI to form IV). It was suggested that the steps carried out by forms IV–V–VI constitute the basic obligatory cycle of the chaperonin-mediated protein folding reaction (Hartl, 1996; Ranson *et al.*, 1997; Rye *et al.*, 1999; Saibil, 2000; Grantcharova *et al.*, 2001). However, an increase in protein folding efficiency was observed (Azem *et al.*, 1995; Corrales and Fersht, 1996; Llorca *et al.*, 1996; Gorovits *et al.*, 1997; Sparrer *et al.*, 1997) under conditions where the reaction cycle proceeds via a symmetric GroEL₁₄(GroES₇)₂ hetero-oligomer (cycle IV–V–VI–VII). *Inset, form I*: electron micrograph of a GroEL₁₄ oligomer (two molecules viewed from the top and one molecule viewed from the side). *Inset, form IV*: electron micrograph of side views of the asymmetric GroEL₁₄GroES₇ hetero-oligomer. *Inset, form VII*: electron micrograph of side views of the symmetric GroEL₁₄(GroES₇)₂ hetero-oligomer.

GroEL$_{14}$- unfolded protein complex is very stable, so that refolding and subsequent release of the refolded protein can proceed only through the binding of ATP and GroES$_7$. Recent studies suggest that only when the unfolded substrate and GroES$_7$ are bound to the same GroEL$_{14}$ ring (*cis* ring) does efficient refolding of denatured substrate occur under the GroES$_7$ cap (Weissman *et al.*, 1995; Mayhew *et al.*, 1996). The binding of ATP and unfolded substrate to the opposite (*trans*) ring triggers the release of the refolded, sequestered substrate into solution. This important allosteric interaction between the two rings (Yifrach and Horovitz, 1995; Roseman *et al.*, 1996; Inbar and Horovitz, 1997; Rye *et al.*, 1997, Rye *et al.*, 1999; Saibil, 2000), cycling of the rings between high and low affinity for the substrate, and timing of the substrate release, are all regulated by the binding and hydrolysis of ATP by GroEL (Grantcharova *et al.*, 2001).

7.2.2 The chloroplast chaperonins

7.2.2.1 CH-CPN60, the Rubisco binding protein

Even before the concept of chaperone proteins existed in the literature, it was noticed that ribulose-1,5-bisphosphate carboxylase (Rubisco) large subunits transiently bound a very large multimeric protein, suitably named the 'Rubisco large subunit binding protein' (RBP), prior to assembly into their oligomeric form (Gutteridge and Gatenby, 1995; Ellis, 1996; Ellis and van der Vies, 1991). The fact that Rubisco bound to RBP before it was incorporated into an oligomer hinted that RBP might be required for the assembly of the plant Rubisco (Barraclough and Ellis, 1980; Roy *et al.*, 1982; Milos and Roy, 1984; Cannon *et al.*, 1986; Roy *et al.*, 1988; Roy, 1989; Hubbs and Roy, 1992; Hubbs and Roy, 1993). This theory was supported by the fact that MgATP caused Rubisco large subunits to dissociate from the chloroplast chaperonin and undergo incorporation into the Rubisco holoenzyme (Milos and Roy, 1984). Moreover, CH-CPN60 antibodies were shown to inhibit assembly of the Rubisco into its oligomeric form (Cannon *et al.*, 1986). Independent research during the same period investigated an *E. coli* protein, called GroEL, that seemed to be involved in assembly of a bacteriophage capsid protein. It was soon discovered that RBP and GroEL were highly homologous proteins and the term chaperonin was coined (Hemmingsen *et al.*, 1988). The following year, folding of prokaryotic Rubisco by GroEL was demonstrated both *in vivo* and *in vitro*, thereby solidifying the concept of a chaperone protein (Goloubinoff *et al.*, 1989a,b; Lorimer, 2001). Although the protein-folding capacity of RBP, the plant CH-CPN60, has since been reconstituted *in vitro* as well, it is interesting to note that the only substrate proteins refolded until now have been prokaryotic Rubisco and mammalian malate dehydrogenase, both heterologous proteins. Refolding of higher plant Rubisco *in vitro* by CH-CPN60 has yet to be demonstrated.

7.2.2.2 The CH-CPN60 monomer

While the GroEL/GroES model can be used as a framework within which the chloroplast chaperonin can be studied, a number of structural and functional properties distinguish the higher plant chloroplast homologs from the GroE model. Since CH-CPN60 is synthesized in the cytosol, it contains a presequence to facilitate import into the chloroplast (Hemmingsen and Ellis, 1986; Hemmingsen et al., 1988). Moreover, in contrast to the 14 identical subunits of the GroEL oligomer, two different subunit types were identified in CH-CPN60 purified from pea (Pisum sativum) (Hemmingsen and Ellis, 1986; Musgrove et al., 1987), barley (Hordeum vulgare) (Musgrove et al., 1987) and wheat (Triticum aestivum) (Musgrove et al., 1987). These subunit types, named α and β, were shown to be present in roughly equal amounts in the chloroplast. They exhibit different migration patterns when separated on SDS-PAGE (Hemmingsen and Ellis, 1986), are immunologically distinct, and produce different digestion patterns when treated with protease (Musgrove et al., 1987). Amino acid sequences of α- and β-subunits, based on cDNAs from Brassica napus and Arabidopsis thaliana, were shown to be only about 50% identical (Martel et al., 1990), no more similar to each other than they are to GroEL (table 7.1). This observation was shown to hold for α- and β-subunits of

Table 7.1 Identity[a] between the amino acid sequences of chaperonins from chloroplast, mitochondria and the α-subunit of TRiC/CCT from Arabidopsis thaliana

	Arabidopsis chloroplast-α	Arabidopsis mitochondria	Arabidopsis TriC/CCTα
Escherichia coli	47	56	14
Saccharomyces cerevisiae mitochondria	44	60	15
Arabidopsis thaliana chloroplast-α	—	44	16
Arabidopsis thaliana chloroplast-β	49	46	17
Pisum sativum chloroplast-α	86	44	18
Pisum sativum chloroplast-β	50	45	16
Arabidopsis thaliana mitochondria	44	—	16
Cucurbita maxima mitochondria-1	43	90	16
Cucurbita maxima mitochondria-2	44	92	16
Arabidopsis thaliana TRiC/CCTα	16	16	—

[a]The identities are expressed in %. Mitochondrial and chloroplast targeting signals were not included in the comparison.

chaperonin from pea as well (Dickson *et al.*, 2000). Consistent with its putative role in assembly of Rubisco, CH-CPN60 is constitutively expressed. Although the involvement of chloroplast chaperonins in protection from stress damage has yet to be shown definitively, expression does increases slightly during heat shock (Viitanen *et al.*, 1995; Hartman *et al.*, 1992).

7.2.2.3 The CH-CPN60 oligomer

In early studies it was shown that, similar to GroEL, CH-CPN60 is a tetrade-camer composed of ca. 60 kDa subunits organized in two heptameric rings (Pushkin *et al.*, 1982). However, in contrast to the highly stable GroEL oligomer, the CH-CPN60 oligomer is extremely unstable and dissociates to monomers under relatively mild conditions that are commonly used to study protein in general and the *E. coli* GroEL in particular (Roy *et al.*, 1988; Hemmingsen and Ellis, 1986; Musgrove *et al.*, 1987; Lissin, 1995; Viitanen *et al.*, 1998). This can be understood by studies over the past few years that have shown that CH-CPN60 exists in solution in dynamic equilibrium between monomers and tetradecamers (Dickson *et al.*, 2000). The monomer–oligomer equilibrium can easily be shifted toward dissociation by factors such as low temperature, ethylenediaminetetraacetic acid (EDTA), extreme dilution of the CH-CPN60 (μM range) and the presence of MgATP (Roy *et al.*, 1988; Musgrove *et al.*, 1987; Lissin 1995; Viitanen *et al.*, 1998). It is interesting to note that ATP is also required for the highly cooperative assembly of CH-CPN60$_{14}$ from its monomeric subunits *in vitro* (Lissin, 1995; Dickson *et al.*, 2000).

The observation that two isoforms of CPN60 are present in chloroplasts raises some important structural questions with mechanistic implications: do the two subunit types exist in a single oligomer and, if so, how are they organized within the tetradecamer? We recently showed that purified β-monomers of the pea CH-CPN60 self-assemble in the presence of adenine nucleotides into functional 14-mers that are active in protein folding (Dickson *et al.*, 2000). Similar to the results obtained with GroEL (Lissin *et al.*, 1990), the β-assembly reaction is highly cooperative and enhanced by GroES homologs. In contrast, the purified pea α-monomers can assemble into tetradecamers only in the presence of β-subunits, and in a nucleotide-dependent manner. Interestingly, α-monomers greatly stimulate the assembly of β-monomers and the particles that are formed contain similar amounts of both subunits (figure 7.2). Results from reconstitution *in vitro* confirm and extend previous studies *in vivo* in which the *Brassica napus* α- and β-subunits were expressed in *E. coli*, both together and individually (Cloney *et al.*, 1992a,b). While the studies *in vivo* were compromised by the fact that the recombinant plant proteins formed 'mixed' oligomers with the bacterial host's GroEL, it was nevertheless evident that the assembly of α-subunits into tetradecamers required the co-expression of β-subunits. In a second study, antibodies raised against α- and β-subunits were used to immunoprecipitate CH-CPN60 oligomers isolated from chloroplasts (Nishio *et al.*, 1999).

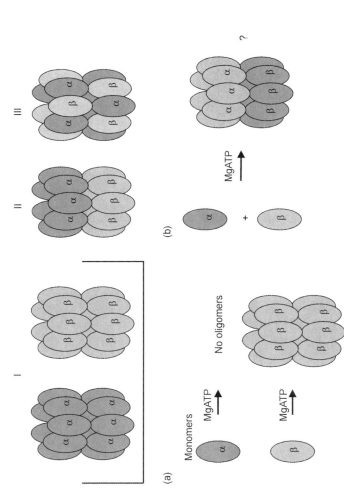

Figure 7.2 Organization of α- and β-subunits within the CH-CPN60 hetero-oligomer. *Top*: A number of possibilities exist for the oligomeric organization of chaperonin subunits in chloroplasts. The organization depicted in I is excluded by studies cited in the text and the experiment schematically outlined in (a) and (b) below. Still more studies are needed to demonstrate which organization, II or III, exists in the chloroplast. *Bottom*: Reconstitution of ch-cpn60 oligomers active in protein folding (Cloney *et al.*, 1992a,b; Dickson *et al.*, 2000). (a) Purified β-monomers can assemble to form β_{14}– oligomers while α-monomers alone do not oligomerize. (b) α-monomers assemble into tetradecamers only in the presence of β-monomers. The reconstituted oligomers may also be the form III presented in *top*.

Antibodies raised against either type of chaperonin subunit were able to co-immunoprecipitate the other chaperonin subunit from a native CH-CPN60 oligomer preparation. Thus the available evidence suggests that the divergent CH-CPN60 isoforms co-exist in the same macromolecular complex. However, it cannot be excluded that plastids might also contain a small population of pure β_{14} homo-oligomers. Although several organisms were shown to contain multiple genes for type I CPN60 (Tsugeki *et al.*, 1992; Kong *et al.*, 1993; Lehel *et al.*, 1993; Cloney *et al.*, 1994), the chloroplast chaperonin seems to be unique in that it contains equal amounts of two divergent isoforms.

While it is clear that the CH-CPN60 oligomers contain both α- and β-isotypes, their exact organization within the oligomers remains to be determined. Theoretical possibilities for the organization of α- and β-subunits within the CH-CPN60 oligomer are schematically presented in figure 7.2. The fact that the oligomers exhibit sevenfold symmetry, and that α-subunits do not oligomerize on their own, suggests that perhaps rings of α are formed on templates of β-7, to give a 14-mer composed of two homogeneous rings. However, this remains to be proven experimentally.

7.2.2.4 The CH-CPN10

In common with the GroE system, the chloroplast CPN60 functions with the aid of a helper protein that is homologous to GroES. In addition to its sequence homology to GroES, the CH-CPN10 can form a complex with GroEL in the presence of ATP. Taking advantage of this functional homology, CH-CPN10 was first purified as a complex with GroEL from crude pea extract (Bertsch *et al.*, 1992). This method subsequently served to identify GroES homologs from other sources as well (Bertsch *et al.*, 1992; Rospert *et al.*, 1993a; Lubben *et al.*, 1990; Burt and Leaver, 1994). In addition to forming a complex with GroEL, the purified pea CH-CPN10 was able to assist GroEL and CH-CPN60 in protein folding (Dickson *et al.*, 2000; Viitanen *et al.*, 1998). As expected, and in common with CH-CPN60, levels of CH-CPN10 are increased slightly as a result of thermal stress (Viitanen *et al.*, 1995; Hartman *et al.*, 1992).

Like other imported chloroplast proteins, the CH-CPN10 is synthesized in the cytosol with an *N*-terminal signal sequence that allows it to be targeted to the chloroplast. Following import, the signal sequence is cleaved from the protein (Bertsch *et al.*, 1992). In contrast to other known CPN10 homologs, the molecular mass of the mature spinach subunit was determined to be 21,385 kDa. This double CH-CPN10 gene effectively contains two domains, each coding for a GroES homolog, that are held together by a short linker chain to form one protein (Bertsch *et al.*, 1992; Baneyx *et al.*, 1995). Each domain of the gene is highly conserved at specific residues that are suggested to have functional significance (Bertsch *et al.*, 1992; Viitanen *et al.*, 1998). Among these regions is a polypeptide segment that is homologous to the 'mobile loop' region of GroES (Koonin and van der Vies, 1995; Landry *et al.*, 1993), which is thought to facilitate binding to GroEL during the protein folding cycle (Landry *et al.*, 1993). The spatial orientation and oligomeric state of this 'double' CH-CPN10 remain a mystery, although under the electron microscope it is seen to form rings that resemble GroES. While little is currently known about functional differences between the two halves of the CH-CPN10, one fascinating possibility would be that each half has adapted to interact with a different isoform of the CH-CPN60.

Synechocystis sp. PCC 6803, a photosynthetic cyanobacteria, also possesses more than one CPN60 gene (Lehel *et al.*, 1993). However, the co-chaperonin in this organism has a molecular weight of only 10 kDa. The fact that this chloroplast-containing phototroph contains only a single 10 kDa CPN10 suggests that the double CH-CPN10 occurred via a gene duplication event that took place following the endosymbiotic event that gave rise to chloroplasts, rather than a fusion of two distinct genes of differing endosymbiotic origin. Consistent with this hypothesis, the 21 kDa form of CPN10 has been detected in species that are evolutionarily separated by at least 4×10^8 years (Baneyx *et al.*, 1995).

It should be noted that, using antibodies to CH-CPN60 and CH-CPN10, the thylakoid lumen was also shown to contain chaperonins. In contrast to those of the chloroplast, the latter contain only one CPN60 isoform and a single 10 kDa CPN10 (Schlicher and Soll, 1996).

7.2.2.5 The structure and function of the CH-CPN60.CH-CPN10 hetero-oligomer

As observed previously, under non-permissive conditions, proper functioning of the bacterial CPN60 system requires formation of a transient hetero-oligomer composed of $GroEL_{14}$, $GroES_7$ and nucleotide. Similarly, in the presence of ADP, pea CH-CPN60 and CH-CPN10 form asymmetric heterocomplexes (Viitanen *et al.*, 1995). When examined using electron microscopy, spinach CH-CPN10 molecules can be seen as circular objects similar to the toroidal GroES from *E. coli* (Baneyx *et al.*, 1995). Therefore, CH-CPN60, CH-CPN10 and their mutual complexes are structurally similar to their respective GroEL/GroES counterparts. In contrast to the heptameric bacterial GroES and spinach CH-CPN10, it was recently shown that CH-CPN10 from *Arabidopsis thaliana* forms tetrameric structures that interact with GroEL (Koumoto *et al.*, 1999). It will be interesting to see whether this tetrameric CH-CPN10 is unique for *Arabidopsis thaliana*, or whether it holds for other plant species. Further studies are needed to determine how the tetrameric structure interacts with a heptameric ring of CPN60.

Two important functional assays, ATP hydrolysis and protein folding, are commonly used to investigate the function of chaperonins. GroEL exhibits weak ATPase activity, and each monomer cleaves about six molecules of ATP per minute (Todd *et al.*, 1994). ATP hydrolysis by GroEL is cooperative, with intra- and inter-ring positive and negative cooperativities, respectively (Gray and Fersht, 1991; Horovitz, 1998). In the presence of GroES, the GroEL ATPase activity is reduced by 50%. This is attributed to the effect of GroES binding on ATP hydrolysis by the *trans* ring, i.e. the ring not occupied with GoES (Horovitz, 1998). The extreme lability of CH-CPN60 *in vitro* has severely hindered the investigation of its enzymatic and chaperone properties in the past. After identifying conditions under which the oligomer is stable, we undertook a structure–function analysis of chloroplast chaperonin oligomers. Examination of ATP hydrolysis by CH-CPN60 oligomers shows that at low chaperonin concentrations, CH-CPN10 apparently enhances the CH-CPN60$_{14}$ ATPase activity (figure 7.3). However, at high chaperonin concentrations, similar to the bacterial GroE chaperonins, CH-CPN10 inhibits the CH-CPN60$_{14}$ ATPase activity by 50%. Since CH-CPN60 was shown to exist fully as an oligomer only at high protein concentrations (Dickson *et al.*, 2000), what is observed to be a stimulation of ATPase activity by CPN10 at low CPN60 concentration is probably due to enhanced oligomerization of CPN60 monomers by CPN10. Thus the inhibition observed at high chaperonin concentration reflects the physiological effect of CH-CPN10 on ATP hydrolysis by the CH-CPN60 oligomer.

Several lines of evidence suggest that the folding reaction mediated by GroEL and CH-CPN60 are very similar. Like GroEL, purified CH-CPN60 can form a complex with partially folded mitochondrial malate dehydrogenase and Rubisco (Dickson *et al.*, 2000; Viitanen *et al.*, 1995). In both cases, the chaperonin releases the bound substrate upon the binding of ATP and CPN10 (Goloubinoff *et al.*, 1989a; Viitanen *et al.*, 1995). The efficiency of Rubisco folding seems to be the same for both chaperonin types, with a half-time

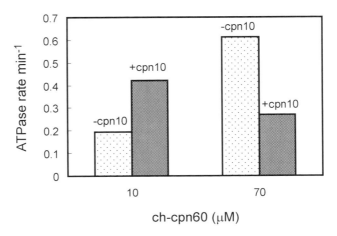

Figure 7.3 ATP hydrolysis by CH-($\alpha\beta$)CPN60$_{14}$. Initial rates of ATP hydrolysis (μmol ATP hydrolyzed/μmol ch-cpn60 monomer.min) were measured with the indicated CH-($\alpha\beta$)CPN60 concentrations (bottom of column), in the presence and absence of CH-CPN10 (indicated at the top of each column).

of about 3 min at 25°C. Furthermore, GroEL and purified pea CH-CPN60 were demonstrated to function equally well with bacterial, mitochondrial and chloroplast CPN10 (Dickson *et al.*, 2000; Viitanen *et al.*, 1995). Thus despite the structural differences mentioned previously, the plant chloroplast CPN10 is functionally similar to GroES by at least three criteria: (i) the ability to form a stable complex with GroEL in the presence of adenine nucleotides (Bertsch *et al.*, 1992; Baneyx *et al.*, 1995); (ii) the ability to inhibit GroEL's ATPase activity; and (iii) the ability to assist GroEL in the refolding of prokaryotic Rubisco (Bertsch *et al.*, 1992; Baneyx *et al.*, 1995).

It should be noted that, in contrast to GroEL and the mitochondrial CPN60 (mammalian and yeast) (Dubaquie *et al.*, 1997; Viitanen *et al.*, 1992), CH-CPN60 chaperone activity is not K^+-dependent (Viitanen *et al.*, 1995). This conclusion was based on the observation that it was not necessary to supplement CH-CPN60-assisted protein folding reactions with potassium ions, even when the purified protein was subjected to gel filtration to remove potentially contaminating monovalent cations. Similar results were obtained when the ATPase activity of CH-CPN60 was examined in the presence and absence of exogenously added monovalent cations (Viitanen *et al.*, 1995). While these results suggest that CH-CPN60 is unique in this regard, the possibility of very tightly bound potassium ions needs to be explored further.

7.2.2.6 Open questions concerning the chloroplast chaperonins

Despite the structural and functional similarities between CH-CPN60 and GroEL, many aspects of chloroplast chaperonin structure and function remain obscure. Some of the points yet to be clarified are as follows:

I. How do the two CH-CPN60 isoforms interact with CH-CPN10? Since the key amino acid residues in the binary CH-CPN10 are conserved, it was suggested that both halves of the molecule are active and may perform different functions (Bertsch *et al.*, 1992). Indeed, when expressed individually, both the *N*-terminal and *C*-terminal domains of the spinach CH-CPN10 are able to complement *E. coli* GroES-deficient mutants (Baneyx *et al.*, 1995). Unfortunately, neither domain is functionally active *in vitro* with either GroEL (Baneyx *et al.*, 1995; Bertsch *et al.*, 1995) or the CH-CPN60 (P.V. Viitanen, unpublished observations). Therefore, the autonomous function of both CH-CPN10 domains has yet to be demonstrated more conclusively. The possibility that the α- and β-subunits of CH-CPN60 may be organized in two separate rings in a single molecule (see figure 7.2) suggests an intriguing scenario in which the two tandemly-linked CPN10 domains may interact differentially with the two CH-CPN60 divergent subunits. However, organization of the two isoforms within the CH-CPN60 oligomer, and the precise interaction of CH-CPN10 with both isoforms, remains to be determined.

II. Do the unique structural features of the chloroplast chaperonin system reflect a special adaptation to their homologous substrates? The GroEL/GroES chaperonins are structurally adapted to assist in the folding of bacterial proteins. Although chaperonins were shown, both *in vivo* and *in vitro*, to refold substrates from heterologous systems, mounting evidence suggests that during evolution, the chaperonins may have acquired structural changes that allow them more efficiently to fold the unique substrates present in their native environment. Specifically:

(i)   As noted above, the *E. coli* chaperonin system performs its essential function of folding bacterial proteins with the help of its co-chaperonin, GroES. Interestingly, the bacteriophage T4-encoded Gp31 co-chaperonin and the bacteriophage RB49-encoded co-chaperonin CoCO proteins are capable of substituting for GroES in essential functions in *E. coli* (reviewed in Ang *et al.*, 2000). However, while both bacteriophage co-chaperonins can assist GroEL in the correct folding of Gp23, a bacteriophage capsid protein, GroES, was not able to carry out that function. A high-resolution structure of Gp31 shows that the bacteriophage co-chaperonin possesses structural differences that result in increased size and hydrophilicity of the enclosed cavity within the GroEL–Gp31 complex. The larger cavity enables GroEL to accommodate the bacteriophage Gp23 large protein within its folding chamber and subsequently to mediate its correct folding (Hunt *et al.*, 1997).

(ii) While the chloroplast and bacterial CPN60s are able to function with co-chaperonin from any source (e.g. bacterial, chloroplast or mitochondrial CPN10s), the mitochondrial (MT) CPN60 can mediate protein folding only with the help of the mitochondrial CPN10 (Rospert *et al.*, 1993b; Viitanen *et al.*, 1992). This strict dependence of MT-CPN60 on MT-CPN10 may reflect a structural adaptation, which allows for more efficient folding of specific proteins that are present in the mitochondria.

(iii) Transgenic tobacco plants, produced by expression of antisense β-subunits, show drastic phenotypic alterations including slow growth, delayed flowering, stunting, leaf chlorosis and, in extreme cases, lethality (Zabaleta *et al.*, 1994a). Interestingly, while the folding of numerous proteins was apparently affected, the β-CH-CPN60 antisense plants accumulated normal or slightly elevated levels of active Rubisco. Since a number of isotypes of the β-subunit form have been identified (Zabaleta *et al.*, 1994a), it is possible that the remaining β-genes substituted the functions of antisense-neutralized gene.

III. Does CH-CPN60 play a role in protein import into chloroplasts? The chloroplast imports the majority of its proteins from the cytoplasm. Similar to the mitochondrial system (Schatz and Dobberstein, 1996; Neupert, 1997; Glick, 1995; Matouschek *et al.*, 2000; Ryan *et al.*, 1997), the chloroplast proteins are imported mainly in an unfolded conformation and refolding takes place with the assistance of stromal chaperones (reviewed in Chen and Schnell, 1999; May and Soll, 1999). It is well established that CPN60 neither participates in mediating protein import of mitochondrial proteins nor interacts with the mitochondrial import channel (Neupert, 1997). In contrast, there are certain instances in which the interaction of CPN60 and the precursor protein appear to be part of the import process in chloroplasts (Lubben *et al.*, 1989; Mandueno *et al.*, 1993; Tsugeki *et al.*, 1993; Bonk *et al.*, 1997). For example, during the import of the two carotenoid biosynthetic enzymes, phytoene desaturase and lycopene cyclase, a high molecular weight soluble import-intermediate, containing CH-CPN60, is formed. The precursor bound to CH-CPN60 was shown to be competent for membrane association (Bonk *et al.*, 1997). Additionally, it was demonstrated that CH-CPN60 interacts with the chloroplast import machinery by association with Tic110, an integral membrane protein of the chloroplast envelope inner membrane (Chen and Schnell, 1999; May and Soll, 1999; Kessler and Blobel, 1996). It was further suggested that Tic110 serves to recruit CH-CPN60 for the folding of newly imported proteins (Kessler and Blobel, 1996). Thus some evidence does indicate that CH-CPN60 may participate in mediating protein import in chloroplasts.

IV. How is the function of CH-CPN60 regulated? Chaperonins in general are constitutively expressed proteins whose expression is increased by heat shock and other stress conditions. While little is known about how the expression levels and activity of CH-CPN60 are regulated, isolated studies over the past decade suggest that factors other than stress may regulate the expression levels and activity of CH-CPN60 as well. Such studies have indicated that:

(i)   While one isoform of the β-chaperonin subunit in *Arabidopsis* is not induced upon heat shock, its gene was found to be developmentally regulated and wound-repressible (Zabaleta *et al.*, 1994b);

(ii) The levels of CH-CPN60 mRNA were found to be induced in rye under a variety of stress conditions. Interestingly, the levels of the CH-CPN60 protein were increased only in etiolated and leached leaves (Zabaleta *et al.*, 1992; Schmitz *et al.*, 1996);

(iii) Recently, the isolation of the first chloroplast calmodulin binding protein, from *Arabidopsis thaliana*, has been reported. The protein was identified to be CH-CPN10. The authors suggested that a calcium signaling system might regulate protein folding in chloroplasts (Yang and Poovaiah, 2000).

7.2.3   Mitochondrial chaperonins (MT-CPN60)

Like chloroplast proteins, the majority of mitochondrial proteins are encoded by nuclear genes, synthesized in the cytoplasm and then imported into their correct mitochondrial subcompartment (Schatz and Dobberstien, 1996; Neupert, 1997;

Ryan *et al.*, 1997). Since mitochondria can import only unfolded proteins, one role of mitochondrial chaperones, in addition to folding of stress-denatured proteins, is to fold newly imported proteins. Until now, most of the studies on mitochondrial import and folding have been carried out with yeast mitochondria. Results from various *in vivo* and *in vitro* studies showed that MT-CPN60 is able to bind and refold denatured proteins in an ATP-dependent manner, with the assistance of MT-CPN10. The essential role played by MT-CPN60 for mito-chondrial function is demonstrated by the fact that mitochondrial chaperonins, MT-CPN60 and MT-CPN10, are two of the few matrix-localized proteins that are essential for viability of yeast under all conditions (Reading *et al.*, 1989; Rospert *et al.*, 1993a).

CPN60 and CPN10 homologs are also present in plant mitochondria. Mito-chondrial CPN60s were identified in *Arabidopsis thaliana* (Parasad and Stewart, 1992), *Zea mays* (Prasad and Hallberg, 1989), *Curcurbita maxima* (Tsugeki *et al.*, 1992) and *Brassisca napus* (Cloney *et al.*, 1994; Cole *et al.*, 1994). At the level of the amino acid sequence, they share high identity with the type I chaperonins found in bacteria and yeast mitochondria (ca. 60%; table 7.1) and lesser identity with the chloroplast chaperonins (ca. 45%). High conserva-tion is also observed among the plant mitochondrial chaperonins themselves. For example, the *Arabidopsis thaliana* MT-CPN60 is ca. 90% identical with *Curcubita maxima* MT-CPN60. Interestingly, plant mitochondria contain two isoforms of the CH-CPN60 subunits. However, in contrast to the two isoforms present in the chloroplast, the two plant MT-CPN60 isoforms are highly identical to each other (ca. 90%;). It is not known whether the two MT-CPN60 isoforms assemble in the mitochondria in same molecule (hetero-oligomer), or whether they assemble in two distinct homo-oligomers. However, it was suggested that the two MT-CPN60 isoforms might be differentially expressed in different tissues or at different stages of development (Tsugeki *et al.*, 1992). An early study with partially purified MT-CPN60 from *Zea mays* showed that, similar to other chaperonins, the plant mitochondrial chaperonins form tetradecamers (Prasad and Hallberg, 1989). Although, it was never demonstrated directly, the high conservation of the plant mitochondrial chaperonins (primary sequence and oligomeric state) with other type I chaperonins suggests that their primary function is to mediate the folding of newly imported and stress denatured mitochondrial proteins.

Plant mitochondria also contain a CPN10 homolog (Burt and Leaver, 1994; Hartman *et al.*, 1992; Koumoto *et al.*, 1996). The latter is of 'normal' size, 10 kDa, as opposed to the 21 kDa double homolog found in the plant chloro-plast. MT-CPN10 from *Arabidopsis thaliana* exhibits 50 and 30% identity to mammalian mitochondrial CPN10 and to GroES, respectively (Koumoto *et al.*, 1996).

Studies on the expression pattern of plant mitochondrial chaperonins have shown that these proteins are induced as a result of heat shock. Additionally, higher levels (2–4 times) of MT-CPN60s were detected in plants in early developmental stages than in more adult tissue (Prasad and Hallberg, 1989; Prasad and Stewart, 1992; Tsugeki *et al.*, 1992). Since plants in the early stages of development require very active mitochondria, MT-CPN60 may play a central role in the biogenesis of plant mitochondria (Tsugeki *et al.*, 1992).

7.3 Type II chaperonins

The eukaryotic cytosol also contains chaperonin proteins (called TRiC, CCT and TCP-1) that belong to the type II chaperonin family. Mammalian and yeast TRiC/CCT cytosolic chaperonins have been studied extensively (Willison, 1999; Gutsche *et al.*, 1999; Leroux and Hartl, 2000; Carrascosa *et al.*, 2001). The latter are composed of eight divergent but homologous subunits, with a molecular weight of ca. 60 kDa, designated α (alpha), β (beta), γ (gamma), δ (delta), ε (epsilon), η (eta), θ (theta) and ζ (zeta). It is commonly accepted that the various cytosolic chaperonin subunits arose during evolution by a process of gene duplication (Gupta, 1995; Archibald *et al.*, 2000). The first cytosolic chaperonin subunit from a plant source, the α-subunit, was cloned and sequenced from *Arabidopsis thaliana* (Mori *et al.*, 1992). At the level of the amino acid sequence, the *Arabidopsis* α-subunit exhibits low homology to type I chaperonins (ca. 16%; see table 7.1). The sequences of the different subunits of cytosolic (type II) chaperonins are also quite divergent. As presented in table 7.2, the identity between the α-subunit of the *Arabidopsis thaliana* cytosolic chaperonin and three other subunits γ, θ, and ζ is 25%, 27% and 28%, respectively. However, the individual subunits from different eukaryotes share great sequence identity. For example, the α-subunit from *Arabidopsis thaliana* shares about 65% identity with α-subunits from mouse and human sources. There is no plant for which the sequences of all the cytosolic chaperonin subunits have been determined (Mori *et al.*, 1992; Ehmann *et al.*, 1993; Ahnert *et al.*, 1996). However, the *Arabidopsis thaliana* complete genome sequence contains, in addition to the above mentioned subunits, additional putative TRiC/CCT subunits.

The structure of type II chaperonins from several sources, but not from plants, was determined at high resolution. Similar to GroEL, these oligomers

Table 7.2 Identity between the subunits of *Arabidopsis thaliana* cytosolic chaperonins and their human and mouse homologs

	Arabidopsis thaliana subunit	α *Homo sapiens*	Mouse
α	—	64	65
γ	28	48	49
θ	27	48	49
ζ	25	56	62

assemble to form a double toroidal cylinder encompassing a large central cavity (Klumpp *et al.*, 1997; Ditzel *et al.*, 1998). Each toroid is composed of eight different but homologous subunits (α–ζ). Thus, in contrast to the sevenfold symmetry of GroEL, the symmetry within the cytosolic chaperonin ring is eightfold (Carrascosa *et al.*, 2001; Llorca *et al.*, 1999; Leroux and Hartl, 2000; Gutsche *et al.*, 1999). While type I chaperonins function with the assistance of a co-chaperonin, CPN10, cytosolic chaperonins do not function with a CPN10 homolog. Instead, the apical domain of cytosolic chaperonins contains a built-in lid that replaces the functions performed by GroES (e.g. capping the substrate within the central cavity). The closure of the chaperonin lid is induced by ATP hydrolysis and not by ATP binding, as is the case for GroEL (Carrascosa *et al.*, 2001; Leroux and Hartl, 2000; Gutsche *et al.*, 2000) A well-studied function of mammalian and yeast cytosolic chaperonins is their ability to fold the cytoskeletal proteins actin and tubulin (Gao *et al.*, 1992; Yaffe *et al.*, 1992). In addition, overwhelming evidence suggests that cytosolic chaperonins may also fold a wide array of cytosolic proteins (Leroux and Hartl, 2000; Thulasiraman, 1999).

Very little data are available on the function of plant cytosolic chaperonins. However, recent evidence suggests that cytosolic chaperonins may also play an important role in the biogenesis of the microtubular network of the plant cytoskeleton (Moser *et al.*, 2000; Himmelspach *et al.*, 1997; Nick *et al.*, 2000). The role of cytosolic chaperonins in chaperoning plant microtubules seems to be important during cell division (Moser *et al.*, 2000; Himmelspach *et al.*, 1997; Nick *et al.*, 2000).

7.4 Concluding remarks

Although the plant chaperonin was one of the first members of this class to be discovered, most of the existing mechanistic knowledge on chaperonin proteins to this day comes from studies of the bacterial GroEL and GroES proteins. The ease of propagation of *E. coli*, the relative simplicity of the bacterial model and the extreme stability of the bacterial chaperonins have allowed for intense investigation of these proteins. It is now becoming apparent that while the general principle of chaperoning exists in plant cells, there is significant variation in the structure and function of the plant homologs. Not only do plant cells house distinct types of chaperonins in their various organelles, these proteins are more complex than their bacterial counterparts.

Due to their relative complexity, great technical difficulties are encountered when working with the chloroplast chaperonins. Under conditions *in vitro* that have been commonly used to investigate GroEL and many other proteins, the CH-CPN60 dissociates into monomers. Such conditions include low temperatures, micromolar protein concentrations and the presence of EDTA.

Recent data that provides us with a better grasp of the CH-CPN60 structure and stability lays the groundwork for future studies, which promise to yield a more detailed understanding of plant chaperonin structure and function. In contrast to bacterial and mitochondrial CPN60s, chloroplasts express several isoforms of chaperonin subunits. One possible explanation for this variability could be differential expression. As previously noted, some plants express specific chaperonins during different stages of development or in different tissues. It is also possible that the different subunit types have evolved to serve specific chloroplast substrates. Adding to this puzzle is the existence of the 'double' chloroplast CH-CPN10. More studies are required to clarify how the expression of type I chaperonins is regulated, and how these proteins are structurally adapted to function in chloroplasts.

Even less is known about the function of mitochondrial and cytosolic plant chaperonins in the folding of their respective compartmental proteins. Recent studies have only just begun to identify the various subunits that form the TRiC/CCT complex in plants. Identification of all the plant TRiC/CCT complex constitutes the first step in studying the function and regulation of plant cytosolic chaperonins. Concerning the mitochondrial homologs, initial studies that were carried out have begun to look at their expression patterns. Research on mammalian TRiC/CCT and mitochondrial CPN60 can serve as a basis for this work, but, as seen for type I chaperonins, significant variations on the theme may be encountered.

Acknowledgments

A.A. is supported by The Israel Science Foundation founded by The Israel Academy of Science. R.S. is supported by a fellowship from the Israeli Ministry of Art and Science.

References

Ahnert, V., May, C., Gerke, R. and Kindl, H. (1996) Cucumber T-complex protein: molecular cloning, bacterial expression and characterization within a 22-S cytosolic complex in cotyledons and hypocotyls. *Eur. J. Biochem.*, **235**, 114-119.

Anfinsen, C.B. (1973) Principles that govern the folding of protein chains. *Science*, **181**, 223-230.

Ang, D., Keppel, F., Klein, G., Richardson, A. and Georgopoulos, C. (2000) Genetic analysis of bacteriophage-encoded cochaperonins. *Annu. Rev. Genet*, **34**, 439-456.

Archibald, J.M., Logsdon, J.M. and Doolittle, W.F. (2000) Origin and evolution of eukaryotic chaperonins: phylogenetic evidence for ancient duplication in CCT genes. *Mol. Biol. Evol*, **17**, 1456-1466.

Azem, A., Kessel, M. and Goloubinoff P. (1994) Characterization of a functional $GroEL_{14}(GroES_7)_2$ chaperonin hetero-oligomer. *Science*, **265**, 653-656.

Azem, A., Diamant, S., Kessel, M., Weiss, C. and Goloubinoff, P. (1995) The protein-folding activity of chaperonins correlates with the symmetric GroEL$_{14}$(GroES$_7$)$_2$ hetero-oligomer. *Proc. Natl Acad. Sci. USA*, **92**, 12021-12025.

Baneyx, F., Bertsch, U., Kalbach, C.E., van der Vies, S.M., Soll, J. and Gatenby, A.A. (1995) Spinach chloroplast cpn21 cochaperonin possesses two functional domains fused together in a toroidal structure, and exhibits nucleotide-dependent binding to plastid chaperonin 60. *J. Biol. Chem*, **270**, 10,695-10,702.

Barraclough, R. and Ellis, R.J. (1980) Protein sunthesis in chloroplasts. IX. Assembly of newly-synthesized large subunits into ribulose bisphosphate carboxylase in isolated intact pea chloroplasts. *Biochim. Biophys. Acta*, **608**, 50-53.

Behlke, J., Risatu, O. and Schonfeld, H.J. (1997) Nucleotide-dependent complex formation between the *Escherichia coli* chaperonins GroEL and GroES studied under equilibrium conditions. *Biochemistry*, **36**, 5149-5156.

Bertsch, U. and Soll, J. (1995) Functional analysis of isolated cpn10 domains and conserved amino acid residues in spinach chloroplast co-chaperonin by site-directed mutagenesis. *Plant Mol. Biol.*, **29**, 1039-1055.

Bertsch, U., Soll, J., Seetharam, R. and Viitanen, P.V. (1992) Identification, characterization, and DNA sequence of a functional 'double' groES-like chaperonin from chloroplasts of higher plants. *Proc. Natl Acad. Sci USA*, **89**, 8696-8700.

Boisvert, D.C., Wang, J., Otwinowski, Z., Horwich, A.L. and Sigler, P.B. (1996) The 2.4 Å crystal structure of the bacterial chaperonin GroEL complexed with ATP gamma S. *Nat. Struct. Biol.*, **3**, 170-177.

Bonk, M., Hoffmann, B., Von Lintig, J. *et al.* (1997) Chloroplast import of four carotenoid biosynthetic enzymes *in vitro* reveals differential fates prior to membrane binding and oligomeric assembly. *Eur. J. Biochem.*, **247**, 942-950.

Boston, R.S., Viitanen, P.V. and Vierling, E. (1996) Molecular chaperones and protein folding in plants. *Plant Mol. Biol.*, **32**, 191-222.

Braig, K., Otwinowski, Z., Hegde, R. *et al.* (1994) The crystal structure of the bacterial chaperonin GroEL at 2.8 Å. *Nature*, **371**, 578-586.

Bukau, B. and Horwich, A.L. (1998) The Hsp70 and Hsp60 chaperone machines. *Cell*, **92**, 351-366.

Burt, W.J.E. and Leaver, C.J. (1994) Identification of a chaperonin-10 homologue in plant mitochondria. *FEBS Lett.*, **339**, 139-141.

Cannon, S., Wang, P. and Roy, H. (1986) Inhibition of ribulose bisphosphate carboxylase assembly by antibody to a binding protein. *J. Cell Biol.*, **103**, 1327-1335.

Carrascosa, J.L., Llorca, O. and Valpuesta, J.M. (2001) Structural comparison of prokaryotic and eukaryotic chaperonins. *Micron*, **32**, 43-50.

Chen, L. and Sigler, P.B. (1999) The crystal structure of GroEL/peptide complex: plasticity as a basis for substrate diversity. *Cell*, **99**, 757-768.

Chen, X. and Schnell, D.J. (1999) Protein import into chloroplasts. *Trends Cell Biol.*, **9**, 222-227.

Cloney, L.P., Wu, H.B. and Hemmingsen, S.M. (1992a) Expression of plant chaperonin-60 genes in *Escherichia coli*. *J. Biol. Chem.*, **267**, 23,327-23,332.

Cloney, L.P., Bekkaoui, D.R., Wood, M.G. and Hemmingsen, S.M. (1992b) Assessment of plant chaperonin-60 gene function in *Escherichia coli*. *J. Biol. Chem.*, **267**, 23,333-23,336.

Cloney, L.P., Bekkaoui, D.R., Feist, G.L., Lane, W.S. and Hemmingsen, S.M. (1994) *Brassica napus* plastid and mitochondrial chaperonin-60 proteins contain multiple distinct polypeptides. *Plant Physiol.*, **105**, 233-241.

Cole, K.P., Blakely, S.D. and Dennis, D.T. (1994) Isolation of a full-length cDNA encoding *Brassica napus* mitochondrial chaperonin 60. *Plant Physiol.*, **105**, 451.

Corrales, F.J. and Fersht, A.R. (1996) Kinetic significance of GroEL$_{14}$(GroES$_7$)$_2$ complexes in molecular chaperone activity. *Fold. Des.*, **1**, 265-273.

Dickson, R., Weiss, C., Howard, R.J. *et al.* (2000) Reconstitution of higher plant chloroplast chaperonin 60 tetradecamers active in protein folding. *J. Biol. Chem.*, **275**, 11,829-11,835.

Ditzel, L., Löwe, J., Stock, D., Stetter, K.O., Hubber, H., Huber, R. and Steinbacher, S. (1998) Crystal structure of thermosome, the archaeal chaperonin and homolog of CCT. *Cell*, **93**, 125-138.

Dubaquie, Y., Looser, R. and Rospert, S. (1997) Significance of chaperonin 10-mediated inhibition of ATP hydrolysis by chaperonin 60. *Proc. Natl Acad. Sci. USA.*, **94**, 9011-9016.

Edwards, M.J. (1998) Apoptosis, the heat shock response, hyperthermia, birth defects, disease and cancer. Where are the common links? *Cell Stress Chaperones*, **4**, 213-220.

Ehmann, B., Krenz, M., Mummert, E. and Schafer, E. (1993) Two Tcp-1-related but highly divergent gene families exist in oat encoding proteins of assumed chaperone function. *FEBS Lett.*, **336**, 313-316.

Ellis, R.J. (1987) Proteins as molecular chaperones. *Nature*, **328**, 378-379.

Ellis, RJ. (1994) Molecular chaperones: opening and closing the Anfinsen cage. *Curr. Biol.*, **4**, 633-635.

Ellis, R.J. (1996) Discovery of molecular chaperones. *Cell Stress Chaperones*, **1**, 155-160.

Ellis, R.J. and van der Vies, S.M. (1991) Molecular chaperones. *Annu. Rev. Biochem.*, **60**, 321-347.

Gao, Y., Thomas, J.O., Chow, R.L., Lee, G.H. and Cowan, N.J. (1992) A cytoplasmic chaperonin that catalyzes beta-actin folding. *Cell*, **69**, 1043-1050.

Glick, B.S. (1995) Can Hsp70 proteins act as force-generating motors? *Cell*, **80**, 11-14.

Goloubinoff, P., Christeller, J.T., Gatenby, A.A. and Lorimer, G.H. (1989a) Reconstitution of active dimeric ribulose bisphosphate carboxylase from an unfolded state depends on two chaperonin proteins and Mg–ATP. *Nature*, **342**, 884-889.

Goloubinoff, P., Gatenby, A.A. and Lorimer, G.H. (1989b) GroE heat-shock proteins promote assembly of foreign prokaryotic ribulose bisphosphate carboxylase oligomers in *Escherichia coli*. *Nature*, **337**, 44-47.

Gorovits, B.M., Ybarra, J., Seale, J.W. and Horowitz, P.M. (1997) Conditions for nucleotide-dependent GroES-GroEL interactions. $GroEL_{14}(GroES_7)_2$ is favored by an asymmetric distribution of nucleotides. *J. Biol. Chem.*, **272**, 26,999-27,004.

Grantcharova, V., Alm, E.J., Backer, D. and Horwich, A.L. (2001) Mechanism of protein folding. *Curr. Opin. Struct. Biol.*, **11**, 70-82.

Gray, T.E. and Fersht, A.R. (1991) Cooperativity in ATP hydrolysis by GroEL is increased by GroES. *FEBS Lett.*, **292**, 254-258.

Gupta, R.S. (1995) Evolution of the chaperonin families (Hsp60, Hsp10 and Tcp-1) of proteins and origin of eukaryotic cells. *Mol. Microbiol.*, **15**, 1-11.

Gutsche, I., Essen, L.O. and Baumeister, W. (1999) Group II chaperonins: new TRiC(k)s and turns of a protein folding machine. *J. Mol. Biol.*, **293**, 295-312.

Gutsche, I., Holzinger, J., Rössle, M., Heumann, H., Baumeister, W. and May, P.R. (2000) Conformational rearrangements of an archaeal chaperonin upon ATPase cycle. *Curr. Biol.*, **10**, 405-408.

Gutteridge, S. and Gatenby, A.A. (1995) Rubisco synthesis, assembly, mechanism and regulation. *Plant Cell*, **7**, 808-819.

Harris, J.R., Pluncktun, A. and Zahn, R. (1994) Transmission electron microscopy of GroEL, GroES, and the symmetrical GroEL/GroES complex. *J. Struc. Biol.*, **112**, 216-230.

Hartl, F.U. (1996) Molecular chaperones in cellular protein folding. *Nature*, **381**, 571-579.

Hartman, D.J., Dougan, D., Hoogenraad, N.J. and Hoj, P.B. (1992) Heat shock proteins of barley mitochondria and chloroplasts. Identification of oraganellar hsp10 and 12: putative chaperonin 10 homologues. *FEBS Lett.*, **305**, 147-150.

Hemmingsen, S.M. and Ellis, R.J. (1986) Purification and properties of ribulosebisphosphate carboxylase large-subunit binding protein. *Plant Physiol.*, **80**, 269-276.

Hemmingsen, S.M., Woolford, C., van der Vies, S.M. *et al.* (1988) Homologous plant and bacterial proteins chaperone oligomeric protein assembly. *Nature*, **333**, 330-334.

Hendrick, J.P. and Hartl, F.U. (1995) The role of molecular chaperones in protein folding. *FASEB J.*, **15**, 1559-1569.

Himmelspach, R., Nick, P., Schäfer, E. and Ehman, B. (1997) Developmental and light-dependent changes of the cytosolic chaperonin containing TCP-1 (CCT) subunits in maize seedlings, and localization in coleoptiles. *Plant J.*, **12**, 1299-1310.

Horovitz, A. (1998) Structural aspects of GroEL function. *Curr. Opin. Struct. Biol.*, **8**, 93-100.

Hubbs, A.E. and Roy, H. (1992) Synthesis and assembly of large subunits into ribulose bisphosphate carboxylase/oxygenase in chloroplast extracts. *Plant Physiol.*, **100**, 272-281.

Hubbs, A.E. and Roy, H. (1993) Assembly of *in vitro* synthesized large subunits into ribulose bisphosphate carboxylase/ogygenase is sensitive to Cl⁻, requires ATP, and does not proceed when large subunits are synthesized at temperature >32°C. *Plant Physiol.*, **101**, 523-533.

Hunt, J.F., van der Vies, S.M., Henry, L. and Deisenhofer, J. (1997) Structural adaptations in the specialized bacteriophage T4 cochaperonin Gp31 expand the size of the Anfinsen cage. *Cell*, **90**, 361-371.

Inbar, E. and Horovitz, A. (1997) GroES promotes the T to R transition of the GroEL ring distal to GroES in the GroEL-GroES complex. *Biochemistry*, **36**, 12,276-12,281.

Kessler, F. and Blobel, G. (1996) Interaction of the protein import machineries in the chloroplast. *Proc. Natl Acad. Sci. USA*, **93**, 7684-7689.

Kong, T.H., Coates, A.R.M., Butcher, P.D., Hickman, C.J. and Shinnick, T.M. (1993) Mycobacterium tuberculosis expresses two chaperonin 60 homologs. *Proc. Natl Acad. Sci. USA*, **90**, 2608-2612.

Koonin, E.V. and van der Vies, S.M. (1995) Conserved sequence motifs in bacterial and bacteriophage chaperonins.*Trends Biochem. Sci.*, **20**, 14-15.

Klumpp, M., Baumeister, W. and Essen, L.O. (1997) Structure of the substrate binding domain of the thermosome, an archaeal group II chaperonin. *Cell*, **91**, 263-270.

Koumoto, Y., Tsugeki, R., Shimada, T. *et al.* (1996) Isolation and characterization of cDNA encoding mitochondrial chaperonin 10 from *Arabidopsis thaliana* by functional complementation of *Escherichia coli* groES mutant. *Plant J.*, **10**, 1119-1125.

Koumoto, Y., Shimada, T., Kondo, M. *et al.* (1999) Chloroplast cpn20 forms a tetrameric structure in *Arabidopsis thaliana. Plant J.*, **17**, 467-477.

Landry, S.J., Zeilstra-Ryalls, J., Fayet, O., Georgopoulos, C. and Gierasch, L.M. (1993) Characterization of a functionally important mobile domain of GroES. *Nature*, **364**, 255-258.

Langer, T., Pfeifer, G., Martin, J., Baumeister, W. and Hartl, F.U. (1992) Chaperonin-mediated protein folding: GroES binds to one end of the GroEL cylinder, which accommodates the protein substrate within its central cavity. *EMBO J.*, **11**, 4757-4765.

Lehel, C., Los, D., Wada, H. *et al.* (1993) A second *groEL*-like gene, organized in a *groESL* operon is present in the genome of *Synechocystis* sp. PCC 6803. *J. Biol. Chem.*, **268**, 1799-1804.

Leroux, M.R. and Hartl, F.U. (2000) Protein folding: versatility of the cytosolic chaperonin TRiC/CCT. *Curr. Biol.*, **10**, 260-264.

Lissin, N.M. (1995) *In vitro* dissociation and self-assembly of three chaperonin 60s: the role of ATP. *FEBS lett.*, **361**, 55-60.

Lissin, N.M., Venyaminov, S.Yu. and Girshovich, A.S. (1990) (Mg-ATP)-dependent self assembly of molecular chaperone GroEL. *Nature*, **348**, 339-342.

Llorca, O., Marco, S., Carrascosa, J.L. and Valpuesta, J.M. (1994) The formation of symmetrical GroEL-GroES complexes in the presence of ATP. *FEBS Lett.*, **345**, 181-186.

Llorca, O., Carrascosa, J.L. and Valpuesta, J.M. (1996) Biochemical characterization of symmetric GroEL-GroES complexes. Evidence for a role in protein folding. *J. Biol. Chem.*, **271**, 68-76.

Llorca, O., Marco, S., Carrascosa, J.L. and Valpuesta, J.M. (1997) Symmetric GroEL-GroES complexes can contain substrate simultaneously in both GroEL rings. *FEBS Lett.*, **405**, 195-199.

Llorca, O., Smyth, M.G., Carrascosa, J.L. *et al.* (1999) 3D reconstitution of the ATP-bound form of CCT reveals the asymmetric folding conformation of a type II chaperonin. *Nat. Struct. Biol.*, **6**, 639-642.

Lorimer, G.H. (2001) A personal account of chaperonin history. *Plant Physiol.*, **125**, 38-41.

Lorimer, G.H. and Todd, M.S. (1996) GroE structures galore. *Nat. Struc. Biol.*, **3**, 116-121.

Lubben, T.H., Donaldson, G.K., Viitanen, P.V. and Gatenby, A.A. (1989) Several proteins imported into chloroplasts form stable complexes with the groEL-related chloroplast molecular chaperone. *Plant Cell*, **1**, 1223-1230.

Lubben, T.H., Gatenby, A.A., Donaldson, G.K., Lorimer, G.H. and Viitanen, P.V. (1990) Identification of a GroES-like chaperonin in mitochondria that facilitate protein folding. *Proc. Natl Acad. Sci. USA*, **87**, 7683-7687.

Mandueno, F., Napier, J.A. and Gray, J.C. (1993) Newly imported Rieske iron-sulfer protein associates with both cpn60 and hsp70 in the chloroplast stroma. *Plant Cell*, **5**, 1865-1876.

Martel, R., Cloney, L.P., Pelcher, L.E. and Hemmingsen, S.M. (1990) Unique composition of plastid chaperonin-60 α and β polypeptide-encoding genes are highly divergent. *Gene*, **94**, 181-187.

Martin, J. and Hartl, F.U. (1997) Chaperone-assisted protein folding. *Curr. Opin. Struct. Biol.*, **7**, 41-52.

Matouschek, A., Pfanner, N. and Voos, W. (2000) Protein unfolding by mitochondria: the hsp70 import motor. *EMBO Rep.*, **1**, 404-410.

May, T. and Soll, J. (1999) Chloroplast precursor protein translocation. *FEBS Lett.*, **452**, 52-56.

Mayer, M.P. and Bukau, B. (1999) Molecular chaperones: the busy life of Hsp90. *Curr. Biol.*, **9**, 322-325.

Mayhew, M., da Silva, A.C., Martin, J., Erdjument-Bromage, H., Tempest, P. and Hartl, F.U. (1996) Protein folding in the central cavity of the GroEL-GroES chaperonin complex. *Nature*, **379**, 420-426.

Milos, P. and Roy, H. (1984) ATP-released large subunits participate in the assembly of ribulose bisphosphate carboxylase. *J. Cell Biochem.*, **24**, 153-162.

Mori, M., Murata, K., Kubota, H., Yamamoto, A., Matsushiro, A. and Takashi, M. (1992) Cloning of a cDNA encoding the Tcp-1 (t complex polypeptide 1) homologue of *Arabidopsis thaliana*. *Gene*, **122**, 381-382.

Moser, M., Schäfer, E. and Ehmann, B. (2000) Characterization of protein and transcript levels of the chaperonin containing tailless complex protein-1 and tubulin during light-regulated growth of oat seedlings. *Plant Physiol.*, **124**, 313-321.

Musgrove, J.E., Johnson, R.A. and Ellis, R.J. (1987) Dissociation of the ribulosebisphosphate-carboxylase large-subunit binding protein into dissimilar subunits. *Eur. J. Biochem.*, **163**, 529-534.

Neupert, W. (1997) Protein import into mitochondria. *Annu. Rev. Biochem.*, **66**, 863-917.

Nick, P., Heuing, A. and Ehman, B. (2000) Plant chaperonins: a role in microtubule-dependent wall formation? *Protoplasma*, **211**, 234-244.

Nishio, K., Hirohashi, T. and Nakai, M. (1999) Chloroplast chaperonins: evidence for heterogeneous assembly of α and β cpn60 polypeptides into a chaperonin oligomer. *Biochem. Biophys. Res. Commun.*, **266**, 584-587.

Polissi A., Goffin L. and Georgopoulos, C. (1995) The *Escherichia coli* heat shock response and bacteriophage lambda development. *FEMS Microbiol Rev.*, **17**, 159-169.

Prasad, T.K. and Hallberg, R.L. (1989) Identification and metabolic characterization of the *Zea mays* mitochondrial homolog of the *Escherichia coli* GroEL protein. *Plant Mol. Biol.*, **12**, 609-618.

Prasad, T.K. and Stewart, C.R. (1992) cDNA clones encoding *Arabidopsis thaliana* and *Zea mays* mitochondrial chaperonin hsp60 and gene expression during seed germination and heat shock. *Plant Mol. Biol.*, **18**, 873-885.

Pushkin, A.V., Tsuprun, V.L., Solovjea, N.A., Shubin, V.V., Evstigneeva, Z.G. and Kretovich, W. (1982) High molecular weight pea leaf protein similar to the GroE protein of *Escherichia coli*. *Biochim. Biophys. Acta*, **704**, 379-384.

Ranson, N.A., Burston, S.G. and Clarke, A.R. (1997) Binding, encapsulation and ejection: substrate dynamics during a chaperonin-assisted folding reaction. *J. Mol. Biol.*, **266**, 656-664.

Reading, D.S., Hallberg, R.L. and Mayers, A.M. (1989) Characterization of the yeast hsp60 gene coding for mitochondrial assembly factor. *Nature*, **337**, 655-659.

Roseman, A.M., Chen, S., White, H., Braig, K. and Saibil, H.R. (1996) The chaperonin ATPase cycle: mechanism of allosteric switching and movements of substrate-binding domains in GroEL. *Cell*, **87**, 241-251.

Rospert, S., Junne, T., Glick, B.S. and Schatz, G. (1993a) Cloning and disruption of the gene encoding yeast mitochondrial chaperonin 10, the homolog of *E. coli* GroES. *FEBS Lett.*, **335**, 358-360.

Rospert, S., Glick, BS., Jeno, P. *et al.* (1993b) Identification and functional analysis of chaperonin 10, the GroES homolog from yeast mitochondria. *Proc. Natl Acad. Sci. USA*, **90**, 10,967-10,971.

Roy, H. (1989) Rubisco assembly: a model system for studying the mechanism of chaperonin action. *Plant Cell*, **1**, 1035-1042.

Roy, H., Bloom, M., Milos, P. and Manroe, M. (1982) Studies on the assembly of large subunits of ribulose bisphosphate carboxylase in isolated pea chloroplasts. *J. Cell Biol.*, **94**, 20-27.

Roy, H., Hubbs, A. and Cannon, S. (1988) Stability and dissociation of the large subunit RuBisCO binding protein complex *in vitro* and *in organello*. *Plant Physiol.*, **86**, 50-53.

Ryan, M.T., Naylor, D.J., Hoj, P.B., Clark, M.S. and Hoogenraad, N.J. (1997) The role of molecular chaperones in mitochondrial protein import and folding. *Int. Rev. Cytol.*, **174**, 127-193.

Rye, H.S., Burston, S.G., Fenton, W.A. *et al.* (1997) Distinct actions of *cis* and *trans* ATP within the double ring of the chaperonin GroEL. *Nature*, **388**, 792-798.

Rye, H.S., Roseman, A.M., Chen, S. *et al.* (1999) GroEL-GroES cycling: ATP and nonnative polypeptide direct alternations of folding-active rings. *Cell*, **97**, 325-338.

Saibil, H.R. (2000) Molecular chaperones: containers and surfaces for folding, stabilizing or unfolding proteins. *Curr. Opin. Struct. Biol.*, **10**, 251-258.

Schatz, G. and Dobberstein, B. (1996) Common principles of protein translocation across membranes. *Science*, **271**, 1519-1526.

Schlicher, T. and Soll, J. (1996) Molecular chaperones are present in the thylakoid lumen of pea chloroplasts. *FEBS Lett.*, **379**, 302-304.

Schmidt, M., Rutkat, K., Rachel, R. *et al.* (1994) Symmetric complexes of GroE chaperonins as part of functional cycle. *Science*, **265**, 656-659.

Schmitz, G., Schmidt, J. and Feierabend, J. (1996) Comparison of the expression levels of a plastidic chaperonin 60 in different plant tissues and under photosynthetic and non-photosynthetic conditions. *Planta*, **200**, 326-334.

Sigler, P.B., Xu, Z., Rye, HS., Burston, S.G., Fenton, W.A. and Horwich, A.L. (1998) Structure and function in GroEL-mediated protein folding. *Annu. Rev. Biochem.*, **67**, 581-608.

Sparrer, H., Rutkat, K. and Buchner, J. (1997) Catalysis of protein folding by symmetric chaperone complexes. *Proc. Natl Acad. Sci. USA*, **94**, 1096-1100.

Thulasiraman, V., Yang, C.F. and Frydman, J. (1999) *In vivo* newly translated polypeptides are sequestered in a protected folding environment. *EMBO J.*, **18**, 85-95.

Todd, M.J., Viitanen, P.V. and Lorimer, G.H. (1994) Dynamics of the chaperonin ATPase cycle: implications for facilitated protein folding. *Science*, **265**, 659-666.

Tsugeki, R. and Nishimura, M. (1993) Interaction of homologues of Hsp70 and Cpn60 with ferradoxin-NADP$^+$ reductase upon its import into chloroplasts. *FEBS Lett.*, **320**, 198-202.

Tsugeki, R., Mori, H. and Nishimura, M. (1992) Purification, cDNA cloning and northern-blot analysis of mitochondrial chaperonin 60 from pumpkin cotyledons. *Eur. J. Biochem.*, **209**, 453-458.

Viitanen, P.V., Lubben, T.H., Reed, J., Goloubinoff, P. O'Keefe, D.P. and Lorimer, G.H. (1990) Chaperonin-facilitated refolding of ribulosebisphosphate carboxylase and ATP hydrolysis by chaperonin 60 (GroEL) are K$^+$ dependent. *Biochemistry*, **29**, 5665-5671.

Viitanen, P.V., Lorimer, G.H., Seetharam, R. *et al.* (1992) Mammalian mitochondrial chaperonin 60 functions as a single toroidal ring. *J. Biol. Chem.*, **267**, 695-698.

Viitanen, P.V., Schmidt, M., Buchner, J. *et al.* (1995) Functional characterization of the higher plant chloroplast chaperonins. *J. Biol. Chem.*, **270**, 18,158-18,164.

Viitanen, P.V., Bacot, K., Dickson, R. and Webb, T. (1998) Purification of recombinant plant and animal GroES homologs: chloroplast and mitochondrial chaperonin 10. *Methods Enzymol.*, **290**, 218-230.

Weissman, J.S., Hohl, C.M., Kovalenko, O. *et al.* (1995) Mechanism of GroEL action: productive release of polypeptide from a sequestered position under GroES. *Cell*, **83**, 577-587.

Willison, K.R. (1999) Composition and function of the eukaryotic cytosolic chaperonin-containing TCP-1, in *Molecular chaperones and folding catalysis* (ed. B. Bukau), Harwood Academic, Amsterdam, pp. 555-571.

Xu, Z., Horwich, A.L. and Sigler, P.B. (1997) The crystal structure of the asymmetric GroEL-GroES(ADP)7 chaperonin complex. *Nature*, **388**, 741-750.

Yaffe, M.B., Farr, G.W., Miklos, D., Horwich, A.L., Sternlicht, M.L. and Sternlicht, H. (1992) TCP1 complex is a molecular chaperone in tubulin biogenesis. *Nature*, **358**, 245-248.

Yang, T. and Poovaiah, W. (2000) *Arabidopsis thaliana* chaperonin 10 is a calmodulin-binding protein. *Biochem. Biophys. Res. Commun.*, **275**, 601-607.

Yifrach, O. and Horovitz, A. (1995) Nested cooperativity in the ATPase activity of the oligomeric chaperonin GroEL. *Biochemistry*, **34**, 5303-5308.

Zabaleta, E., Oropenza, A., Jimenez, B., Salerno, G., Crespi, M. and Herrera-Estrella, L. (1992) Isolation and characterization of genes encoding chaperonin beta from *Arabidopsis thaliana*. *Gene*, **111**, 175-181.

Zabaleta, E., Assad, N., Oropeza, A., Salerno, G. and Herrera-Estrella, L. (1994a) Expression of one of the members of the *Arabidopsis* chaperonin 60β gene family is developmentally regulated and wound-repressible. *Plant Mol. Biol.*, **24**, 195-202.

Zabaleta, E., Oropeza, A., Assad, N., Mandel, A., Salerno, G. and Herrera-Estrella, L. (1994b) Antisense expression of chaperonin 60β in transgenic tobacco plants leads to abnormal phenotypes and altered distribution of photoassimilates. *Plant J.*, **6**, 425-432.

8 Receptor kinases

Andrew C. Allan and Keith Hudson

8.1 Introduction

In eukaryotes, membrane-bound receptor kinases play an important part in perception of extracellular signals. These proteins have features that allow both perception of an appropriate signal and also the ability directly to initiate kinase phosphorylation cascades by their own kinase activity, to transmit the signal through the cell. Protein phosphorylation via kinases is a key mechanism for signal transduction in both eukaryotes and prokaryotes. Pathways involving kinase activity direct development, growth and death, as well as cellular responses to changes in the extracellular environment.

In plants, receptor kinases are usually of the Ser-Thr type, i.e. they autophosphorylate on a Ser or a Thr residue. There are several receptor histidine kinases (rHKs) and apparently no receptor tyrosine kinases in plants (The *Arabidopsis* Genome Initiative, 2000). By far the most abundant plant receptor kinase class, the Ser-Thr receptor kinases, are composed of an extracellular domain, a single pass transmembrane region and a cytoplasmic kinase domain. Isolation of mutations in plant receptor kinase genes has indicated functions in processes as diverse as hormone perception, developmental regulation and pathogen defense. However, in plants no natural ligand has been fully characterized. So, despite the abundance of these proteins and their structural similarity to mammalian receptor kinases (The *Arabidopsis* Genome Initiative, 2000), they are more accurately termed receptor-like kinases (RLKs). Biophysical characterization of several RLKs has been performed and several have been shown to localize to the cell surface. Similarly, Ser–Thr kinase activity has been shown with recombinant *Escherichia coli* expressed RLKs.

Advances in the characterization of ligand interactions with plant RLKs have recently been made. For example, the binding of brassinosteroids to the *Arabidopsis* RLK BRI1, the interaction of CLV3 with CLV1, a receptor responsible for cell fate in the meristem, and the likely interaction of pollen proteins with the stigma receptors that elicit self-incompatibility in *Brassica*. The ligands for the rHKs are also beginning to emerge; the rHK ETR1 has been shown to bind ethylene (Hall *et al.*, 1999). Expressing the *ETR1* in yeast generates saturable ethylene binding sites (Schaller and Bleecker, 1995). The receptor for another classical plant growth regulator, cytokinin, has been demonstrated to be the rHK CRE1 (Inoue *et al.*, 2001).

In a recent review (Hardie, 1999), the structure and function of 18 *Arabidopsis* RLKs was discussed. However, there are over 340 predicted RLKs in the *Arabidopsis* genome (The *Arabidopsis* Genome Initiative, 2000). Over 330 of these have been analyzed for sequence similarity (K. Hudson, S. Lund and R. Onrust, 2000, unpublished data). Many of these RLKs may be functionally interchangeable; of the entire *Arabidopsis* genome 35% of genes are not members of gene families, while a staggering 37% lie within families of five or more (Initiative, 2000). Alternatively, many may not be expressed as proteins. However, these 340 RLKs and 11 rHKs present us with the upper limit of potential cell surface perception-kinase proteins that can respond to stress and developmental cues in plants.

Although the plant receptor kinases are mainly RLKs and rHKs, other plant receptor complexes can act as functional receptor kinases. Furthermore, plants appear to have several light receptors which have kinase activity, e.g. the phytochromes (Yeh and Lagarias, 1998) and phototropin (Salomon *et al.*, 2000).

8.2 A survey of the mechanistic classes of receptor kinases in plants

8.2.1 RLKs

There are at least 340 predicted RLKs in the *Arabidopsis* genome (The *Arabidopsis* Genome Initiative, 2000). As mentioned previously, these proteins are composed of an extracellular domain, a transmembrane region and a cytoplasmic kinase domain (see figure 8.1). A comparison of all *Arabidopsis* RLK extracellular domains indicates eight or possibly nine major types of extracellular domain and a small number of others that do not cluster neatly by this analysis. Nearly all of these extracellular domain types have been described in the literature previously (Herve *et al.*, 1999; The *Arabidopsis* Genome Initiative, 2000). Surprisingly, cluster analysis of the *Arabidopsis* RLK kinase domains gives a phylogenetic tree that more-or-less reflects the assignments made according to extracellular domain (see figure 8.2). For example, the kinase domains of all the lectin-type of RLKs (see following sections for discussion on extracellular domain structure) are more related to each other than other RLK kinase domains. This suggests that the RLK kinase domain has co-evolved with its extracellular partner. Since it is the extracellular domain that determines ligand binding, we will describe further the nine groups of known extracellular domain.

8.2.1.1 RLKs with the leucine-rich repeat motif

RLKs with leucine-rich repeat (LRR) type modules in their extracellular domain represent the largest class in *Arabidopsis* with at least 82 members. The LRR motif is also frequently present in other plant extracellular proteins. For example, soluble proteins like polygalacturonase inhibitor protein. Other membrane

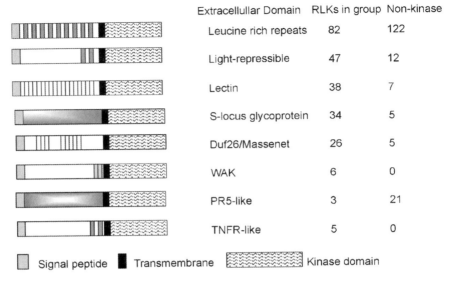

	Extracellullar Domain	RLKs in group	Non-kinase
	Leucine rich repeats	82	122
	Light-repressible	47	12
	Lectin	38	7
	S-locus glycoprotein	34	5
	Duf26/Massenet	26	5
	WAK	6	0
	PR5-like	3	21
	TNFR-like	5	0

☐ Signal peptide   ■ Transmembrane   ▨ Kinase domain

Figure 8.1 Schematic of the general molecular structure of *Arabidopsis* RLKs. Eight of the classes, based on consensus motifs within their extracellular domains, are shown. The numbers of family members identified in *Arabidopsis* (The *Arabidopsis* Genome Initiative, 2000) belonging to each class is given, as are non-kinase relatives (i.e. not RLKs but having strong homology to the extracellular domain). Shaded regions in the extracellular domain (apart from the signal peptide) represent group specific sequence consensus.

Figure 8.2 Phylogenic analysis of the kinase domains of 125 predicted *Arabidopsis* RLKs. Sequences were downloaded from PubMed, whilst the MATDB database was used to confirm new gene nomenclature. After alignment of the entire predicted protein using ClustalX, kinase domains were extracted, re-aligned and used for generation of a nearest neighbor tree. As can be seen, kinase domains still generally cluster according to what each extracellular domain would have been. Predicted proteins that are exceptions to this trend are marked (∗∗), while those that RLKs which are difficult to assign to one of the eight major classes of RLK are also marked (∗).

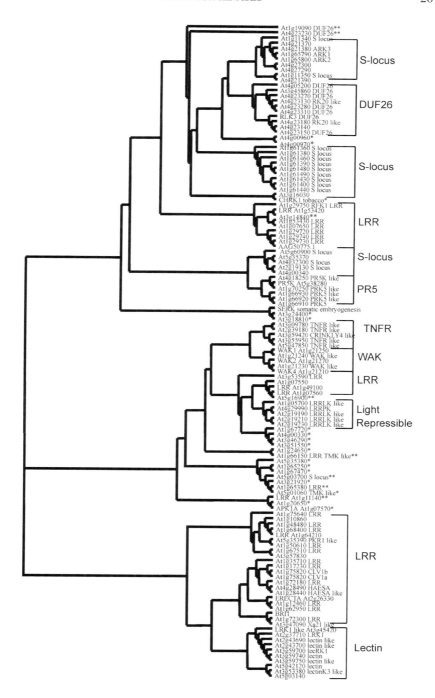

attached receptor-like proteins (RLPs; with a similar structure to RLKs, but lacking the kinase domain) also contain LRRs and account for over 122 genes (The *Arabidopsis* Genome Initiative, 2000).

The abundance of LRR proteins predicted to occur outside the cell is mirrored inside the cell with at least 80 LRR proteins in the Toll/interleukin-1 receptor (TIR) or nucleotide-binding site class. General surveys of the LRR motif abundance indicate its presence in both eukaryotes and prokaryotes, with a variety of cell locations and functions. The common feature amongst these proteins is the involvement of the LRRs in protein–protein interactions. A well-known receptor with LRRs in its extracellular domain is the *Drosophila* Toll receptor. This protein is involved in both development and defense, utilizing LRRs in the extracellular domain to bind protein ligands.

The extracellular domain of LRR-type RLKs has multiple tandem copies of a sequence of 24 residues in length with the following consensus: Leu-X-X-Leu-X-X-Leu-X-X-Leu-X-Leu-X-X-Asn-X-Leu-Gly-X-Ileu-Pro-X-X (X = any residue). Individual families have subtle variations on this pattern, usually with one of the Leu residues being replaced by another aliphatic residue. These LRRs are bounded by pairs of Cys residues. This capping is common in LRR proteins and many mammalian extracellular LRR proteins also have multiple Cys residues surrounding the LRRs. The length of the extracellular domain, and hence the number of LRRs, varies considerably in this group. The extreme varies from one *Arabidopsis* RLK (AT4g20140), with 862 residues and 32 LRRs in the extracellular domain, to the family with the smallest extracellular domain containing only 4 LRRs (e.g. AT1g34210 with 243 extracellular residues). Although there are smaller predicted extracellular domains in RLKs related to AT1g34210, these are likely mispredictions. There are at least 20 families in the plant LRR RLK class, with each family having a very similar number of residues and LRRs in the extracellular domain. Some receptors, like the *Arabidopsis* BRI1, have a 70 amino acid insertion between the 18th and 19th LRRs, which is conserved in a homologous rice RLK (Li and Chory, 1997). Data from mutants has suggested this region is critical for function, although its exact role is unknown.

One LRR domain has been subjected to structural analysis; the 3D structure of ribonuclease inhibitor, a protein built entirely of 16 LRRs, has been determined (Kobe and Deisenhofer, 1993). This reveals that its LRRs form a horseshoe-shaped structure with a 16-strand parallel β-sheet on the inside and 16 short helices on the outside. It is proposed that the exposed face of the parallel β-sheets are used to achieve strong protein–protein interactions (Kobe and Deisenhofer, 1993). The crystal structure of porcine ribonuclease inhibitor and its complex with ribonuclease A (Kobe and Deisenhofer, 1995) can be found at http://www.rcsb.org/pdb/ and viewed using Swiss-Pdb Viewer, downloaded from http://www.expasy.ch/spdbv/mainpage.html.

Several plant receptors in the LRR class have been studied for cell location and activity of the kinase domain. The *Arabidopsis* RLKs BRI1 and HAESA have been shown to be plasma-membrane localized. It has been demonstrated that numerous RLKs, including CLV1, BRI1 and HAESA, have *in vitro* kinase activity. Genetic studies have revealed that LRR-type RLKs have an array of functions including roles in pollen development and pollination (Mu *et al.*, 1994), disease resistance (Song *et al.*, 1995) and morphogenesis (Torii *et al.*, 1996). The two RLKs with predicted ligands (CLV1 and BRI1) belong in this LRR class. BRI1 has been shown to be plasma membrane localized, and *in vitro* kinase assays indicate that BRI1 is a functional Ser-Thr kinase (Friedrichsen *et al.*, 2000).

8.2.1.2 The light-repressible RLK

The light-repressible RLK family has 47 members and 12 non-kinase relatives in *Arabidopsis* (The *Arabidopsis* Genome Initiative, 2000). However, at least ten of these have either no signal sequence or transmembrane domain that, along with a preponderance of multiple exons, may represent mispredicted open reading frames. So far only one of these has been described experimentally; *lrrpk* was found to be expressed at elevated levels in etiolated *Arabidopsis* cotyledons, and expression was not detected elsewhere in the plant (Deeken and Kaldenhoff, 1997). The group has been named light-repressible because *lrrpk* is predominantly expressed in the absence of light. This may be misleading; recently another RLK in this group was found to be exclusively expressed during leaf senescence in *Phaseolus vulgaris* (Hajouj *et al.*, 2000).

Elements in *lrrpk's* promoter were similar to ones found in that of phytochrome A, suggesting that factors downstream of this phytoreceptor modulate *lrrpk's* expression. The extracellular domain of *lrrpk* also contains an LRR proximal to the transmembrane domain. These repeats are also bounded by pairs of Cys residues, which implies interaction with peptide molecules. This domain of 138 amino acids lies next to the transmembrane domain, leaving over 370 extracellular amino acids in the *N*-terminal sequence without apparent consensus to other RLKs (see figure 8.3). The extracellular domain is of very consistent size in this group, all members having approximately 500 amino acid residues. That there are so many *Arabidopsis* RLKs with homology to *lrrpk*, but such a lack of functional knowledge, makes this class of receptor kinase exciting for future study.

8.2.1.3 The lectin type

The *Arabidopsis* genome has 38 lectin RLKs and a further seven non-kinase relatives (The *Arabidopsis* Genome Initiative, 2000). The lectin class of RLK is so-called because its extracellular domain shares homology to legume lectins. Lectins are found throughout the plant kingdom and are unusual in their ability to bind oligosaccharides. Despite this common feature, there is great diversity of lectin structure, including differences in the number of carbohydrate binding sites per subunit of protein (Kennedy *et al.*, 1995).

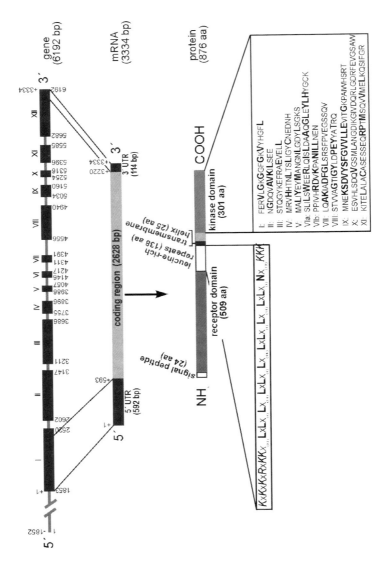

Figure 8.3 The structural organization of the light repressible RLK *lrrpk* gene and its encoded protein (reproduced with permission from Deeken and Kaldenhoff, 1997). The 12 exons which encode the *lrrpk* gene are symbolized by black boxes. Regions in the extracellular domain and the 11 conserved kinase domains are shown, with highly conserved amino acids shown in bold letters.

Arabidopsis lecRK1 was the first lectin RLK examined (Herve *et al.*, 1996) and has been shown to be present in the plasma membrane (Herve *et al.*, 1999). These authors also showed, through a polymerase chain reaction (PCR) survey, that lectin RLK genes are conserved across dicot species, suggesting they perform an important function. At present none of the ligands for a lectin RLK, nor any of the biological roles for these receptors, is known. Current thought suggests these RLKs may be detectors of the ever increasing range of carbohydrate-based elicitors of plant response (e.g. Klarzynski *et al.*, 2000). Such elicitors are sometimes derived from the breakdown of the cell wall (either of the host or of the pathogen) during fungal infection. In a recent review (Herve *et al.*, 1999) ten lectin RLK genes of five families were analyzed by sequence comparison and tertiary model building. One implication of this analysis was that the RLK lectin domains are unlikely to bind monosaccharides. This conclusion was based on a model of the lectin domain of the RLKs constructed from known lectin structure. The predicted structure shows key residues in the oligosacharide-binding site were substituted, and two residues required for calcium and manganese binding were absent. These metal chelating residues are used to coordinate oligosacharide binding with the metal ions. However, other hydrophilic residues present in an extended oligosacharide binding pocket were conserved so the prediction was made that more complex glucans may be ligands for these lectin RLKs. Alternatively, other compounds such as plant hormones and specific peptides, which have been found to bind legume lectins (John *et al.*, 1997), could be ligands for these RLKs.

8.2.1.4   *The S-locus glycoprotein type*

The S-glycoprotein type of RLK shares homology with the well-studied extra-cellular domains of the S-locus glycoproteins (SLGs) and S-locus receptor kinases (SRKs) of *Brassica* species. These proteins form the female recognition system present on the stigma that detects incompatible pollen, i.e. pollen of the same haplotype as the stigmatic surface. The SLG is a secreted protein, while the SRK has a SLG extracellular domain, a transmembrane region and a kinase domain. The SLG domain is characterized by a sequence of approximately 400 residues in length, ten conserved cysteines in the membrane proximal third of the extracellular domain, and other conserved residues (Walker, 1994). The male determinant in this system is a cysteine-rich protein (SCR), present in the pollen coat. Genetic evidence suggests that the SCR is the ligand for the SRK receptor. These three genes (*SRK, SLG* and *SCR*) cluster together and act as a single allele in *Brassica*. The locus is multi-allelic and each gene is highly polymorphic in the population, suggesting the three genes co-evolved at each locus.

SRK-like genes have been cloned and characterized from both monocot and dicot species. In *Arabidopsis* there are 34 *SRK*-like genes and five non-kinase relatives, with the *SRK*-like genes forming at least three families when analyzed by extracellular (data not shown) or kinase domains (see figure 8.2).

However, *Arabidopsis* does not have self-incompatibility, so these RLKs must have divergent functions. Further evidence that the function of *SRK*-like genes is not restricted to self-incompatibility is provided by the observation that most are expressed in parts of the plant other than the stigma and pollen. For example, expression of the *Arabidopsis SRK*-like gene *ARK1* (the first *Arabidopsis* S-glycoprotein homolog studied, Tobias *et al.*, 1992), is predominantly in leaf tissue. *ARK2* activity is exclusively in above-ground tissues, while *ARK3* promoter activity was detected in roots as well as above-ground tissues (Dwyer *et al.*, 1994). Futhermore, overexpression of *ARK1* results in severe stunting and disrupted cellular expansion (Tobias *et al.*, 1992).

In wheat, the recently cloned resistance gene to leaf rust, *LRK10*, was found to be an RLK containing several conserved amino acids of the S-domain glycoprotein. LRK10 and S-domain proteins appear to belong to the same superfamily of specific recognition proteins (Feuillet *et al.*, 1997). Some of these cysteine groups are also present in the DUF26 domain (see next section), suggesting these two RLK types are related (see figure 8.2). No *LRK10*-like genes exist in *Arabidopsis*, but searches of the public databases indicate homologs in rice, raising the possibility that this is a monocot specific family. In addition no homologs of the SCR genes, the candidate ligands in *Brassica*, have been reported in the *Arabidopsis* genome.

8.2.1.5 The Massenet/DUF26 type

The Massenet/DUF26 RLK family in *Arabidopsis* has 26 members and a further five non-kinase related genes (The *Arabidopsis* Genome Initiative, 2000). These genes form at least three families. The extracellular DUF26 (domain of unknown function 26, IPR002902, see figure 8.4) domain is characterized by a region of around 280 amino acids and the presence of cysteine residues in the pattern Cys-X8-Cys-X-X-Cys. The structure of this domain or the nature of potential ligands is unknown.

This family appears to be associated with defense responses, since all studied members have increased expression levels upon pathogen attack. One well-studied member of this group is the bean (*Phaseolus vulgaris*) RK20-1. This RLK was found to be differentially regulated in response to pathogens, symbionts and nodulation factors (Lange *et al.*, 1999). *Arabidopsis* RLK3 is a strong homolog of RK20-1. This RLK is also upregulated by pathogen attack, whilst an oxidative burst has been implicated in the signal cascade switching on *RLK3* expression (Czernic *et al.*, 1999). RLK3 is both rich in leucines (20 residues, but not arranged so they can form a LRR; Czernic *et al.*, 1999) and cysteines. Other *Arabidopsis* genes such as *RKC1* and close relatives have also been shown to be upregulated by the phytohormone salicylic acid (SA), which modulates systemic acquired resistance (Ohtake *et al.*, 2000). The clustering together of the kinase domains of Massenet/DUF26 RLK family and S-locus RLKs (see figure 8.2) suggests that they may be related.

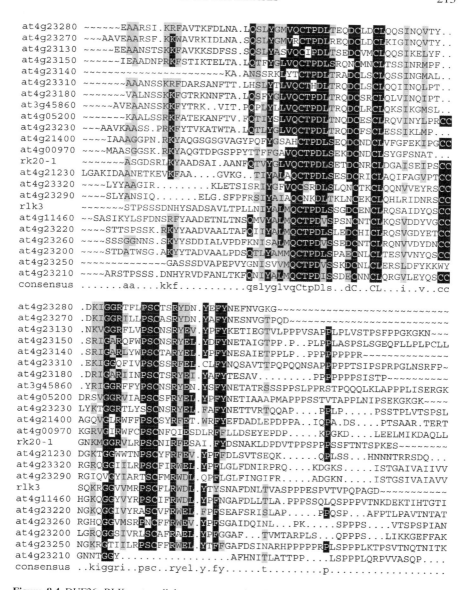

```
at4g23280  ~~~~~~EAARSI.KRFAVTKFDLNA.LQSLYGMVQCTPDLTEQDCLDCLQQSINQVTY..
at4g23270  ~~~AAVEAARSF.KKWAVRKIDLNA.SQSLYGMVRCTPDLREQDCLDCLKIGINQVTY..
at4g23130  ~~~~~EEAANSTSKKFAVKKSDFSS.SQSLYASVQCIPDLTSEDCVMCLQQSIKELYF..
at4g23150  ~~~~~~~IEAADNPRKFSTIKTELTA.LQTFYGLVQCTPDLSRQNCMNCLTSSINRMPF..
at4g23140  ~~~~~~~~~~~~~~~~~~~~~~~~KA.ANSSRKLYTCTPDLTRADCLSCLQSSINGMAL..
at4g23310  ~~~~~~~AAANSSKRFDARSANFTT.LHSLYTLVQCTHDLTRQDCLSCLQQIINQLPT..
at4g23180  ~~~~~~~~VALNSSKKFGTRKNNFTA.LQSFYGLVQCTPDLTRQDCSRCLQLVINQIPT..
at3g45860  ~~~~~AVEAANSSKKFYTRK..VIT.PQPLYLLVQCTPDLTRQDCLRCLQKSIKGMSL..
at4g05200  ~~~~~~~KAALSSRKFATEKANFTV.FQTIYSLVQCTPDLTNQDCESCLRQVINYLPRCC
at4g23230  ~~~AAVKAASS.PRKFYTVKATWTA.LQTLYGLVQCTPDLTRQDCFSCLESSIKLMP...
at4g21400  ~~~~IAAAGGPN.RKYAQGSGSGVAGYPQFYGSAHCTPDLSEQDCNDCLVFGFEKIPGCC
at4g00970  ~~~~MAASGGSK.RKYAQGTDPGSPPYTTFFGAVQCTPDLSEKDCNDCLSYGFSNAT...
rk20-1     ~~~~~~~ASGDSRLKYAADSAI.AANFQTVYGLVQCTPDLSETDCNRCLDGAISEIPSCC
at4g21230  LGAKIDAANETKEVKFAA....GVKG..TIYALAQCTPDLSESDCRICLAQIFAGVPTCC
at4g23320  ~~~~LYYAAGIR.........KLETSISRIYGFVQCSRDLSLQNCTKCLQQNVVEYRSCC
at4g23290  ~~~~SLYANSIQ.......ELG.SFPFRSIYAIAQCNKDLTKLNCEKCLQHLRIDNRSCC
rlk3       ~~~~~~~~STPSSSDNHYSADSAVLTPLLNIYALMQCTPDLSSGDCENCLRQSAIDYQSCC
at4g11460  ~~SASIKYLSFDNSRFYAADETNLTNSQMVYALMQCTPDVSPSNCNTCLKQSVDDYVGCC
at4g23220  ~~~~STTSPSSK.RKYYAADVAALTAFQIIYALMQCTPDLSLEDCHICLRQSVGDYETCC
at4g23260  ~~~~SSSGGNNS.SKYYSDDIALVPDFKNISALMQCTPDLSVSSEDCNTCLRQNVVYDNCC
at4g23200  ~~~~STDATWSG.AKYYTADVAALPDSQTLYAMMQCTPDLSPAECNLCLTESVVNYQSCC
at4g23250  ~~~~~~~~~~~~~~GASSSDVAPEPVYGNISVVMQCTPDVSSKDCNLCLERSLDFYKKWY
at4g23210  ~~~~ARSTPSSS.DNHYRVDFANLTKFQNIYALMQCTPDISSDECNNCLQRGVLEYQSCC
consensus  .......aa....kkf...........qslyglvqCtpDls..dC..CL...i..v..cc

at4g23280  .DKIGGRTFLPSCTSRYDN.YEFYNEFNVGKG~~~~~~~~~~~~~~~~~~~~~~~~~~~~~
at4g23270  .DKIGGRILLPSCASRYDN.YAFYNESNVGTPQD~~~~~~~~~~~~~~~~~~~~~~~~~~~
at4g23130  .NKVGGRFLVPSCNSRYEV.YPFYKETIEGTVLPPPVSAPPLPLVSTPSFPPGKGKN~~~
at4g23150  .SRIGARQFWPSCNSRYEL.YDFYNETAIGTPP.P..PLPPLASPSLSGEQFLLPLPCLL
at4g23140  .SRIGARLYWPSCTARYEL.YPFYNESAIETPPLP..PPPPPPPR~~~~~~~~~~~~~~~
at4g23310  .EKIGGQFIVPSCSSRPEL.CLFYNQSAVTTPQPQQNSAPPPPPTSIPSPRPGLNSRFP~
at4g23180  .DRIGARIINPSCTSRYEI.YAFYTESAV........PPPPPPPSISTP~~~~~~~~~~~
at3g45860  .YRIGGRFFYPSCNSRYEN.YSFYNETATRSSSPPSLPPRSTPQQQLKLAPPPLISERGK
at4g05200  DRSVGGRVIAPSCNSRYEL.YPFYNETIAAAPMAPPPSSTVTAPPLNIPSEKGKGK~~~~
at4g23230  LYKTGGRTLYSSCNSRYEL.FAFYNETTVRTQQAP....PPLP.....PSSTPLVTSPSL
at4g21400  AGQVGLRWFFPSCSYRPET.WRFYEFDADLEPDPPA..IQPA.DS....PTSAAR.TERT
at4g00970  KGRVGIRWFCPSCNFQIESDLRPFLLDSEYEPDP.....KPGKD....LEELMIKDAQLL
rk20-1     GNKMGGRVLRPSCNIRPESAI.FYDSNAKLDPDVTPPSPPPSSFTNTSPKES~~~~~~~~
at4g21230  DGKTGGWWTNPSCYFRPEV.YPPFDLSVTSEQK......QPLSS...HNNNTRRSDQ...
at4g23320  RGRQGGIILRPSCFIRWEL.YPPLGLFDNIRPRQ....KDGKS.....ISTGAIVAIIVV
at4g23290  RGIQVGYIARTSCFMRWDL.QPPLGLFINGIFR.....ADGKN.....ISTGSIVAIAVV
rlk3       SQKRGGVVMRPSCFLRWDL.YTYSNAFDNLTVASPPPESPVTVPQPAGD~~~~~~~~~~~
at4g11460  HGKQGGYVYRPSCIFRWDL.YPPNGAFDLLTLA.PPPSSQLQSPPPVTNKDEKTIHTGTI
at4g23220  NGKQGGIVYRASCVFRWEL.FPPSEAFSRISLAP.....PPQSP...AFPTLPAVTNTAT
at4g23260  RGHQGGVMSRPNCFRWEV.YPPSGAIDQINL...PK....SPPPS....VTSPSPIAN
at4g23200  LGRQGGSIVRLSCAFRAEL.YPPGGAF...TVMTARPLS...QPPPS...LIKKGEFFAK
at4g23250  NGKRGTIILRPSCFFRWEL.YTPFGAFDSINARHPPPPPRPLSPPPLKTPSVTNQTNITK
at4g23210  GNNTGGY.................AFHNITLATTPP....LSPPPLQRPVVASQP....
consensus  ..kiggri..psc..ryel.y.fy......t.........p.................
```

Figure 8.4 DUF26 RLK extracellular consensus domains. 18 DUF26 domains from InterPro (http://www.ebi.ac.uk/interpro/) were downloaded using the alignment function into FASTA format and then exported to the Unix-based SeqLab. A pile up of these domains and the full sequence of RLK3 and RK20-1 was made and BoxShade then used to generate the above consensus sequence.

8.2.1.6   The wall-associated type

According to the *Arabidopsis* sequencing initiative there are six members of the wall-associated RLK family (WAKs). These RLKs have a novel structure in that they are covalently connected to the extracellular matrix (ECM) of the cell wall, and also have a typical transmembrane and cytoplasmic kinase domain (He *et al.*, 1999). The WAK protein therefore has the potential to transduce signals in the cytoplasm (via its Ser-Thr kinase domain) to actions in the cell wall. WAKs have an *N*-terminal region that is only released from the cell wall by boiling in sodium lauryl sulfate (SDS) or cleaving the wall with pectinases. However, even after pectinase digestion the WAK protein is still attached to a pectin fragment (Wagner and Kohorn, 2001). The *WAK1* gene was mapped and found to lie in a cluster of five highly similar genes (*WAK1–5*) within a 30 kb region in the genome (He *et al.*, 1999). The extracellular domains of WAKs are 40–64% identical to each other (their kinases are 86% identical, He *et al.*, 1999; Wagner and Kohorn, 2001).

Early interest in WAKs was prompted by the presence of several epidermal growth factor (EGF) regions in their extracellular domain (He *et al.*, 1996, 1999). EGF domains are a sequence of around 40 to 50 amino acids, which includes six cysteine residues that form three disulfide bonds (see figure 8.5). These domains are common in secreted or extracellular domains of a large number of animal proteins, including EGF (Savage *et al.*, 1972). A subset of EGF-like domains are calcium binding (Rao *et al.*, 1995), but these also have a similar structure to the standard domain; calcium binding is thought to facilitate protein–protein interactions by this domain. For example, the transmembrane neurogenic *Drosophila* proteins, Notch and Delta, interact at the cell surface in a calcium-dependent manner. Notch has 36 EGF domains while Delta has nine,

GNQT**C**EQVGSTSI**C**GGNST**C**LDSTPRNGYI**C**R**C**NEGFDGNPYLSAG**C**
DVNE**C**TTSSTIHRHN**C**SDPKT**C**RNKVGGFY**C**K**C**QSGYRLDTTTMS**C**

nxnnC-x(3,14)-C-x(3,7)-Cxxbxxxxaxc-x(1,6)-C-x(8,13)-Cx

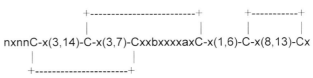

'n': negatively charged or polar residue [DEQN]
'b': possibly beta-hydroxylated residue [DN]
'a': aromatic amino acid
'C': cysteine, involved in disulfide bond
'x': any amino acid.

Figure 8.5 The two EGF-like domains found in WAK1 (He *et al.*, 1999) and the consensus sequence of the mammalian EGF domain (data collected from InterPro:IPR001881: http://www.ebi.ac.uk/interpro/) showing the linking of cysteines via intramolecular disulphides.

and a subset of both are calcium binding (Rao *et al.*, 1995). WAK1 (AT1g21250), WAK2 (AT1g21270) and WAK4 (AT1g21210) have a calcium-binding EGF-like region proximal to the membrane in the extracellular domain.

In animals and fungi a physical connection between the ECM and the cell permits communication that is essential for cell growth and division. The WAKs may be at least partly responsible for this role in plants. Analysis of *WAK* gene expression and the requirement of *WAK* expression for cell expansion support this hypothesis (Wagner and Kohorn, 2001). Therefore the WAK RLKs may not have soluble ligands, but rather respond to forces in the cell wall. However, the presence of extracellular domains in WAK RLKs that are known to interact frequently with other proteins may suggest interactions with arabinogalactans, or glycine and proline rich proteins that have been shown to have important roles in plant development.

8.2.1.7   The pathogenesis-related type

There are three *Arabidopsis* genes that encode RLKs whose extracellular domains have strong similarity to the pathogenesis-related protein 5 (PR5) and 21 non-kinase relatives (The *Arabidopsis* Genome Initiative, 2000). The first studied example in *Arabidopsis* was PR5K (Wang *et al.*, 1996). Acidic pathogenesis-related (PR) proteins, of which PR5 is a member, accumulate in the apoplast of plants under pathogen attack. However, treatments that induce PR proteins have no effect on the level of the RLK PR5K transcript (Wang *et al.*, 1996), suggesting that this RLK may be part of a constitutive surveillance system. The structural similarity between the extracellular domain of PR5K and the antimicrobial PR5-protein (50% identity at the amino acid level, Wang *et al.*, 1996) suggests a possible interaction with a common pathogen-related ligand. Perhaps induced expression of PR5 complements the action of PR5K RLK in detecting and reacting to the pathogen (e.g. via dimerization of the soluble extracellular PR5 with its RLK homolog).

In tobacco there is a pathogenesis-related RLK termed the 'chitinase-related RLK' (CHRK1, Kim *et al.*, 2000). Although this protein is unrelated to PR5K, it has an extracellular domain with 41% identity to tobacco class V chitinases, which are another class of pathogenesis-related proteins (Melchers *et al.*, 1994). Analysis of this domain predicts that it will not have chitinase activity but may well bind chitin oligomers (Kim *et al.*, 2000), suggesting CHRK1 has a role in detecting chitin fragments during the plants defense response. However, using a BLAST search of the *Arabidopsis* protein database, no *Arabidopsis* CHRK1 homologs can be found (Sakai *et al.*, 2000), although the kinase domain of CHRK1 is similar to the ARK3-like At3g16030 (see figure 8.2).

8.2.1.8   The tumor necrosis factor-like type

The tumor necrosis factor (TNFR)-like type of RLK was first observed after molecular characterization of the CRINKLY4 mutant in maize (Becraft *et al.*,

1996). The TNFR-like domain is found in a number of mammalian proteins, some of which are known to be receptors for growth factors (e.g. tumor necrosis factor type I and type II receptors and the receptors for CD30, CD40 and CD27 cytokines). The gene that encodes for the CRINKLY4 mutant, *CR4*, appears to regulate a number of developmental responses including cell proliferation, pattern and differentiation, suggesting a function analogous to growth factor responses in animals (Jin *et al.*, 2000). TNFR-like domains are *N*-terminal, cysteine-rich and can be subdivided into 3–4 repeats. Usually within each repeat are six conserved cysteines, all of which are involved in intra-chain disulfide bonds (Banner *et al.*, 1993). *Arabidopsis* has five genes that share some homology to the maize CRINKLY4 RLK, with AT3g59420 appearing to share the greatest identity both overall (61%) and within the proposed TNFR-like domain (see figure 8.6). However, only three of these predicted RLK proteins have the characteristic cysteine-rich TNFR repeats first discussed for CRINKLY4 (see figure 8.6 and Becraft *et al.*, 1996). All five *Arabidopsis* genes contain several 39 amino acid motifs, which are of unknown function and repeated seven times in CRINKLY4 (Becraft, 1998).

8.2.1.9   *The extensin type*

The *Arabiopsis* genome has provided evidence that another type of RLK may exist. Eight predicted genes (e.g. AT4g34440, AT1g12040, AT4g32710) are present in *Arabidopsis* that have an extracellular domain with similarity to cell wall extensin proteins, a transmembrane domain and a kinase domain. Extensins are cell-wall located proteins that are extremely proline rich (e.g. *Arabidopsis* extensin 4 is 40% proline), and are often stress inducible (Dubreucq *et al.*, 2000) and implicated in wall morphogenesis (Baumberger *et al.*, 2001). The prolines of extensins are usually within Ser(Pro)$_4$ repeats and have a high content of Tyr and Lys residues. The extracellular domains of extensin-like RLKs have a similar structure; AT4g32710 has 105 prolines in its extracellular domain (40% of total). However, none of these genes is predicted to have a signal sequence that would allow secretion, unlike extensin proteins. No other evidence to support expression of these predicted genes has been reported, so this category of RLK remains an enigma.

8.2.2   *Receptor histidine kinases*

Receptor histidine kinases (rHKs) have been well studied in prokaryotic cells, where they are involved in detecting a number of extracellular changes. These kinases were first identified in plants (Chang *et al.*, 1993) as the putative receptor for the gaseous plant growth regulator ethylene. Since then much work has focused on these kinases, implicating them in the detection of ethylene, cytokinin, osmolarity and light. Because of their relatedness to bacterial receptors

```
at3g09780  ~~~~~~~~~PGMCR.AGPCNEKEFAFNASILNEPDLTSLC......VRKELMVCSPCGS
at2g39180  ~~~~~~~CSPGMCSPRGNCGDWFAFNASILKESELTSLC.....SFHNLNICIRCGI
crinkly4   ~~~~~~~GICVPT.ACSHGYYEYVNH..GEVGSIKVC......KPANSRLCLPCST
at3g59420  ~~~~~~~GLCIDT.PCPPGTHELSNQ.EN....SPC.....KFTGSHICLPCST
tnfr       KCGGHDYEKDGLC..CASCHPGFYASRLCGPGSNTVCSPCEDGTFTASTNHAPACVSCRG
consensus  ..........GmC...g.C..gwfaf......e.l.slC.......vClpC.s

at3g09780  DCSHGFFLSSSCTANSDRICT.PCSLCQNSSCSDICKLHNSNFPDKH~~~
at2g39180  SCLEGYFPSSTCNPNADRVCT.PCSLCQNSSCYGICKIRATKSKEHEQKE~~~
crinkly4   GCPEGLYESSPCNATADRVCQFDCLKCVTDECLSFCLSQKRTKSRKL~~~~~~~~
at3g59420  SCPPGMYQKSVCTERSDQVCVYNCSSCSSHDCSSNCSSSATSGGKEKGKF~~~~~~~~
tnfr       PCTGHLSESQPCDRTHDRVC..NCST.......GNYCLLKGQNGSRICAPQTKCPAGYGVS
consensus  .C..glf.ss.C....DrvC...Cs.c...c...C.l.as...r........~
```

Figure 8.6 Tumor necrosis factor-like domains of *Arabidopsis* RLKs. The maize CRINKLY4 protein (accession T04108) was used in a BLAST search of the MATDB database (http://mips.gsf.de/proj/thal/db/) and five strong homologs (expect value of less than 1.9e-69) used in a SeqLab-based pile-up. Of these two were discarded (at5g47850 and at3g55950) as not having any homology in the core TNFR domain discussed by Becraft *et al.* (1996). This domain was then aligned with the mammalian TNFR1 and displayed using BOXSHADE.

and (apparently) their unique, non-proteinaceous ligands, there are some striking structural features of receptor histidine kinases.

rHKs function within a histidine to aspartate (His-to-Asp) network (thoroughly reviewed by Urao *et al.*, 2000). Detection of the ligand is on a histidine sensory kinase (with an *N*-terminal input domain, and a *C*-terminal kinase domain) where a histidine residue autophosphorylates itself. This phosphate is then passed to a response regulator (with an *N*-terminal receiver domain having an aspartate residue, which receives the phosphate from the rHK and a *C*-terminal output domain). This phospho-relay has been termed 'the two component system'. What complicates this simple protein to protein transfer of phosphate from His-to-Asp is that in some rHKs the sensor kinase and response regulator are on the same protein (a so called hybrid histidine kinase, e.g. ETR1). In other instances there is more than one transfer step. This second Asp-to-His-to-Asp transfer is mediated by an intermediate phospho-relay protein.

In *Arabidopsis* there appears to be 11 rHKs (which are a mixture of straight sensor kinases and hybrid rHKs, see table 8.1), 16 response regulators, eight pseudoresponse regulators (The *Arabidopsis* Genome Initiative, 2000), and at least three phospho-relay proteins (Suzuki *et al.*, 1998, Urao *et al.*, 2000). The His-to-Asp phospho-relay in turn interacts with downstream kinases; in the case of ethylene this protein is CTR1. CTR1 is a Raf/MAPKKK-related protein that has Ser-Thr kinase activity (Kieber *et al.*, 1993). It therefore appears that this bacterial-type two-component receptor system eventually feeds into a more eukaryotic kinase signal transduction cascade.

The ethylene receptor, both the first isolated plant rHK and the most extensively studied, provides an ideal illustration of the complexity of rHK function. ETR1 is a hybrid rHK with an *N*-terminus that is embedded in the membrane, via three membrane spanning domains. A homodimer of ETR1 binds ethylene via disulfide bonds between intramolecular cysteines in the transmembrane region. The intracellular portion of the protein has a histidine kinase domain and a *C*-terminal response regulator domain. However, ETR1 is not the only ethylene–related rHK. ETR2 and EIN4 are also hybrid rHKs with the transmembrane-kinase domain-receiver domain (TM-KD-RD) structure (see table 8.1), while ERS1 and ERS2 have the ethylene binding domain but lack a receiver domain. It has been shown that ETR1 is a functional rHK; autophosphorylation of the purified fusion protein was observed in yeast (Gamble *et al.*, 1998). However, many of the other ethylene receptors do not have these domains (see table 8.1) and some do not even have the histidine (e.g. a change to glutamate in ETR2). The deletion of the histidine does not eliminate the ability of the dominant mutant, when reintroduced into wild-type plants, to produce insensitivity to ethylene (Chang and Meyerowitz, 1995). A dominant *etr1-1* mutant without the histidine, still makes plants ethylene insensitive as does the tomato LeETR5, which lacks the histidine within the kinase domain, when transformed into

Table 8.1 Components of the plant histidine kinase signaling system

2-Component protein	Plant type	Input 'ligand'	Structure[a]	Reference
Histidine kinase				
ETR1	*Arabidopsis*	Ethylene	TM–KD–RD	1
ETR2	*Arabidopsis*	Ethylene	TM–KD(H/E)–RD	2
EIN4	*Arabidopsis*	Ethylene	TM–KD–RD	3
ERS1	*Arabidopsis*	Ethylene	TM–KD	4
ERS2	*Arabidopsis*	Ethylene	TM–KD(H/D)	5
CKI1	*Arabidopsis*	Cytokinin	TM–KD—RD	6
CRE1	*Arabidopsis*	Cytokinin	TM–KD—RD	7
ATHK1	*Arabidopsis*	Osmolarity?	TM–KD—RD	8
NR	Tomato	Ethylene	TM–KD	9
LeETR1(=eTAE1)	Tomato	Ethylene?	TM–KD–RD	10, 11
LeETR2(=TFE27)	Tomato	Ethylene?	TM–KD–RD	11
Cm-ETR1	*Cucumis melo*	Ethylene?	TM–KD–RD	12
Cm-ERS1	*Cucumis melo*	Ethylene?	TM–KD	12
RP-ERS1	*Rumex palustris*	Ethylene?	TM–KD	13
DC-ERS1	*Dianthus caryophyllus*	Ethylene?	TM–KD	14
DC-ERS2	*Dianthus caryophyllus*	Ethylene?	TM–KD	15
Phosphorelay intermediates				
ATHP1=APH2	*Arabidopsis*		HPt	16
ATHP2=APH3	*Arabidopsis*		HPt	16
ATHP3=APH1	*Arabidopsis*		HPt	16, 17
ATHP5	*Arabidopsis*		HPt	AB041767
AHP4	*Arabidopsis*		HPt	BAB01275
ZmHP2	Maize		HPt	18
ZmHP1	Maize		HPt	AB024293
Response regulators **Type A**				
ARR3	*Arabidopsis*		RD–	19
ARR4(=ATRR1, IBC7)	*Arabidopsis*		RD–	19, 20, 21
ARR5(=ATRR2, IBC6)	*Arabidopsis*		RD–	19, 20, 21
ARR6	*Arabidopsis*		RD–	19
ARR7	*Arabidopsis*		RD–	19
ARR8(=ATRR3)	*Arabidopsis*		RD–	22, 21
ARR9(=ATRR4)	*Arabidopsis*		RD–	22
ZmRR1	Maize		RD–	23
ZmRR2	Maize		RD–	24
Type B				
ARR1	*Arabidopsis*		RD—	25
ARR2	*Arabidopsis*		RD—	25
ARR10	*Arabidopsis*		RD—	22
ARR11	*Arabidopsis*		RD—	22

Table 8.1 (continued)

2-Component protein	Plant type	Input 'ligand'	Structure[a]	Reference
ARR12	*Arabidopsis*		RD—	22
ARR13	*Arabidopsis*		RD—	22
ARR14	*Arabidopsis*		RD—	22

[a]The structure column uses the abbreviations of transmembrane (TM), kinase domain (KD), receiver domain (RD), histidine containing phosphotransfer domain (HPt) to summarize each proteins basic structure. One dash corresponds to between 50–100 amino acids. Substitutions of amino acids from histidine to aspartate is shown as (H/D), and histidine to glutamate as (H/E).

References (unless a direct submission to a database, in which case an accession number is shown): 1 (Chang *et al.*, 1993), 2 (Sakai *et al.*, 1998b), 3 (Hua *et al.*, 1998), 4 (Hua *et al.*, 1995), 5 (Hua *et al.*, 1998), 6 (Kakimoto, 1996), 7 (Inoue *et al.*, 2001), 8 (Urao *et al.*, 1999), 9 (Wilkinson *et al.*, 1995), 10 (Zhou *et al.*, 1996), 11 (Lashbrook *et al.*, 1998), 12 (Sato-Nara *et al.*, 1999), 13 (Vriezen *et al.*, 1997), 14 (Charng *et al.*, 1997), 15 (Shibuya *et al.*, 1998), 16 (Miyata *et al.*, 1998) 17 (Suzuki *et al.*, 1998), 18 (Sakakibara *et al.*, 1999), 19 (Imamura *et al.*, 1998), 20 (Brandstatter and Kieber, 1998), 21 (Urao *et al.*, 1998), 22 (Imamura *et al.*, 1999), 23 (Sakakibara *et al.*, 1998), 24 (Sakakibara *et al.*, 1999), 25 (Sakai *et al.*, 1998a).

Re-drawn with permission from Urao *et al.* (2000), with added sequences from the MIPS database.

Arabidopsis (Tieman and Klee, 1999). A tomato receptor missing most of the kinase catalytic domain is also capable of transducing the ethylene signal (H. Klee, personal communication). Signaling via heterodimers of the various ethylene-rHKs could explain these observations.

A novel hybrid-type histidine kinase, ATHK1, has been isolated from dehydrated *Arabidopsis* plants (Urao *et al.*, 1999). This histidine kinase has structural similarity to the yeast osmosensor rHK, SLN1. The ATHK1 transcript accumulates under conditions of high or low osmolarity. Overexpression of the ATHK1 in yeast complements the yeast mutant *sln1-ts*, implying that ATHK1 functions as an osmosensor. If the cDNAs conserved His or Asp residues in ATHK1 are replaced then ATHK1 fails to complement the *sln1-ts* mutant indicating this protein functions as a histidine kinase (Urao *et al.*, 1999).

Response regulators, the proteins which possess an *N*-terminal receiver domain with an aspartate residue for receiving phosphate from the rHK, have been divided into two types, as judged from their structural designs, biochemical properties, and expression profiles (Imamura *et al.*, 1999). Type A response regulators have a receiver domain and short *C*-terminal extensions, whereas type B have long enough *C*-terminals to function as output domains (Urao *et al.*, 2000). ARR1 and ARR2, two type-B response regulators, have been shown to have a large *C*-terminus with similarity to the MYB (so named as it was first characterized as a proto-oncogene from *my*elo*b*lastosis virus) class of transcription factor (Imamura *et al.*, 1999; Sakai *et al.*, 1998a). This raises the intriguing possibility that transcription is regulated directly by these components of the phospho-relay.

8.3   Towards an understanding of receptor like kinase function(s)

8.3.1   Expression patterns

All RLKs studied so far appear to be plasma membrane located (for examples, see Delorme *et al.*, 1995; Feuillet *et al.*, 1998; Friedrichsen *et al.*, 2000; Herve *et al.*, 1999). Despite the lack of isolated ligands for RLKs, there is no doubt from genetic studies that plant RLKs are integral to the first steps of many plant responses. Some clues as to function of uncharacterized RLKs are provided by expression patterns.

Certain RLKs appear to show tissue or developmental-based regulation. For example, the mRNA of SbRLK1, a leucine-rich repeat RLK from sorghum, accumulates to much higher levels in mesophyll cells than in the bundle-sheath and is almost undetectable in roots. This expression pattern indicates that SbRLK1 might be involved in the regulation of specific processes in mesophyll cells (Annen and Stockhaus, 1999). The maize CRINKLY4 (CR4) RLK mediates a growth factor-like differentiation response and is required for the normal differentiation of leaf epidermis (Becraft *et al.*, 1996). CR4 appears to regulate an array of developmental responses, including cell proliferation, fate, pattern and differentiation, suggesting a function analogous to growth factor responses in animals. In carrot, cells having the competence to form embryos express an LRR-type RLK called somatic embryogenesis receptor-like kinase (SERK, Schmidt *et al.*, 1997). Somatic embryos formed from these cells were monitored using a SERK promoter–luciferase reporter gene, leading to the conclusion that SERK expression ceases after the globular stage of somatic embryogenesis. In whole plants, SERK mRNA could only be detected in the zygotic embryo. These results suggest that a highly specific signal transduction chain exists in both the zygotic embryo shortly after fertilization and somatic cells competent to form embryos (Schmidt *et al.*, 1997). The *Arabidopsis* RLK HAESA is expressed in the abscission zones of sepals, petals and stamens, and at the base of petioles where leaves attach to the stem. Plants having antisense HAESA show delayed abscission of floral organs suggesting that this RLK functions in floral organ abscission (Jinn *et al.*, 2000). Also during senescence in bean (*Phaseolus vulgaris*) an LRR-containing RLK called SARK is upregulated (Hajouj *et al.*, 2000). An S-type RLK is upregulated in the tobacco style during pollination, apparently via ethylene (Li and Gray, 1997). We have studied the expression of 92 RLKs in *Arabidopsis* using a reverse transcriptase (RT) PCR approach. A summary of results is shown in table 8.2. As shown, there are a number unique to a certain tissue, which may well direct development or be induced by developmental changes. Nevertheless, RLKs that are present as 'house-keepers' (i.e. present at all stages of the plants life) may well be of more interest when their ligands are isolated.

Table 8.2 The expression profile of *Arabidopsis* RLKs

Tissue	No. of RLKs expressed[a]	No. unique to tissue
Seeds	20	0
Siliques	19	0
Roots	56	5
Leaves	48	1
Bolt stems	37	0
Flowers (open + buds)	66	10
Flower buds	54	6
Open flowers	66	4
Whole seedlings	65	0
Cell suspensions	51	0

[a]All primers gave expected band size by PCR on genomic DNA and where counted as positive also on cDNA. Expression profiles were compatible with previously published data for BRI1 and HAESA.

8.3.2 Transcriptional and post-transcriptional regulation

An interesting level of regulation occurs with the LRR-type RLK inrpk1, isolated from Japanese morning glory, *Ipomoea nil* (Bassett *et al.*, 2000). Inrpk1 mRNA increases 20-fold in cotyledons in response to a single floral-inducing short day (SD). This pattern of expression and differential processing suggests a role for inrpk1 in short-day-induced flowering. Moreover, two transcripts appear to be made from the single gene of *INRPK1*. The largest (4.4 kb) transcript encodes the predicted full length polypeptide (INRPK1), whereas a 1.6 kb transcript apparently originates from a secondary transcription initiation site within the gene and potentially encodes a protein kinase identical to INRPK1 but lacking 21 of the LRRs. This smaller transcript predominates in vegetative roots. Alternative 3′-splicing of a large cryptic intron in the LRR region, creates one transcript (2.6 kb) potentially encoding a small, secretable polypeptide.

Many studies have observed an induction of certain RLK mRNAs under certain stress conditions. The *RPK1* gene of *Arabidopsis* is rapidly induced by stresses such as dehydration, high salt and low temperature, suggesting that the gene is involved in a general stress response (Hong *et al.*, 1997). Some RLKs are stimulated by wounding or pathogens (e.g. the tobacco NtTMK1, Cho and Pai, 2000). In cultured *Arabidopsis* cells, the *At-RLK3* gene is activated upon oxidative stress and salicylic acid (SA) treatment (Czernic *et al.*, 1999), and RK20-1, the bean homolog (*Phaseolus vulgaris*) of RLK3, is found to be induced by pathogens, symbionts and nodulation factors in roots (Lange *et al.*, 1999). Other pathogenesis-related RLKs (e.g. *Lrk10* gene in wheat) are constitutively expressed in the aerial parts of the plant and not in the roots (Feuillet *et al.*, 1998). The rice *Xa21* gene confers race-specific resistance to *Xanthomonas oryzae* pv *oryzae*, and turns out to be an LRR RLK (Ronald *et al.*,

1996). This is a classical gene-for-gene based resistance, with *Xa21* encoding the *R* gene allowing cell surface recognition of a pathogen ligand and subsequent activation of intracellular defense responses (Song *et al.*, 1995).

Promoter analysis of the RLK genes suggests the method of transcript induction. The 5′ upstream regions of the light-repressible RLK *lrrpk* gene have sequence elements that are similar to those identified in promoters of phytochrome A genes (Deeken and Kaldenhoff, 1997). A number of RLK genes of the LRR class and those of the S-locus domain class are induced by SA (Ohtake *et al.*, 2000; Pastuglia *et al.*, 1997), a phytohormone that modulates pathogen responses. Analysis of the upstream regions of these SA-responsive RLK genes shows that they contain the TTGAC sequence (the SA-responsive element). The wall-associated RLKs (WAKs) appear to be induced by pathogen infection and SA, or 2,2-dichloroisonicotinic acid (He *et al.*, 1998). In addition, overexpression of *Wak1*, or the kinase domain alone, can provide resistance to otherwise lethal SA levels (He *et al.*, 1998). A subset of RLKs therefore appear to be involved in SA-mediated defense responses.

8.3.3 Self-incompatibility

Perhaps the best functional study of an RLK system in plants is self-incompatibility in *Brassica*, in which the stigma rejects self-pollen but accepts non-self-pollen for fertilization. This self-incompatibility is sporophytically controlled by the polymorphic S locus. Two tightly linked genes map to the S locus, which encode a plasma membrane-anchored S-receptor kinase (SRK) and cell wall-localized S-locus glycoprotein (SLG). The extracellular domain of SRK (the S domain) exhibits high homology to that of SLG, but is not identical (Kusaba and Nishio, 1999). The male determined side of self-incompatibility is an independent pollen S factor found in the genome outside the cloned S locus in *Brassica napus* (Cui *et al.*, 2000). The S-receptor glycoprotein (SRG) and SRK proteins, of any particular haplotype, are coordinately expressed in the papillar cells of the stigma epidermis. Takasaki *et al.* (2000) have transformed *Brassica rapa* with an *SRK28* and an *SLG28* transgene separately. They found that expression of *SRK28* alone, but not *SLG28* alone, conferred the ability to reject self (S28)-pollen on the transgenic plants. However, the ability of *SRK28* to reject S28 pollen was enhanced by *SLG28*. They concluded that SRK alone determines the specificity of S haplotype of the stigma, and that SLG acts to promote a full expression of the self-incompatibility response. A model would be that SRK can form homodimers (Giranton *et al.*, 2000) that act as a receptor for some component of the pollen grain but in combination with SLG this is more efficient. Perhaps, an analogous system occurs with the meristematic RLK CLV1; CLV2 is required for the normal accumulation of CLV1 protein and its assembly into protein complexes. This suggests that CLV2 may form a heterodimer with CLV1 to transduce extracellular signals (Jeong *et al.*, 1999).

The biology of the *Brassica* self-incompatibility system is not as simple as two-receptors (SLK and SLG) and a pollen-determined ligand. The pollen tubes themselves also express specific RLKs, apparently of the LRR type (Mu *et al.*, 1994; Lee *et al.*, 1996; Muschietti *et al.*, 1998). Cabrillac *et al.* (1999) have recently shown in *Brassica oleracea* that there are two different SLG genes, *SLGA* and *SLGB*. *SLGA* encodes both soluble and membrane-anchored forms of SLG, whereas *SLGB* encodes only soluble SLG proteins. A further regulation of self-incompatibility may occur via endogenous anti-sense RNA transcripts. It has been suggested that SRK is transcribed in both directions; naturally occurring antisense transcripts have been found and SRK was shown to be transcribed in an antisense direction. In addition, an antisense SRK transcript was shown to inhibit translation of a sense transcript *in vitro* (Cock *et al.*, 1997). Furthermore, diverse expression patterns of the *Arabidopsis* S-type RLKs suggests that close homologs of SLK/SLG function in roles more diverse than pollen self-incompability (Dwyer *et al.*, 1994).

8.4 Toward understanding receptor histidine kinase function

As already discussed, rHKs have some intriguing structural and functional features. An example of this is the detection of cytokinin, one of the phyto-hormones that direct cell division in plants. CRE1 and CKI1 are putative rHKs for cytokinin signalling (Kakimoto, 1996). CKI1 was first isolated by activation tagging, which gave *Arabidopsis* callus a cytokinin-treated phenotype (rapid cell division and shoot formation) in the absence of cytokinin. The activated gene, *CKI1*, was cloned and overexpression of the gene in *Arabidopsis* again produced a cytokinin-treated phenotype. The sequence of CKI1 showed a high level of homology to ETR1, so it was suggested that this rHK was involved in cytokinin signaling or perception (Kakimoto, 1996). However, recently another gene, *CRE1*, has been isolated by members of the same research group (Inoue *et al.*, 2001); this gene also gives *Arabidopsis* a cytokinin-treated phenotype in the absence of cytokinin. Moreover, the authors use yeast to provide convincing evidence that CRE1 is a cytokinin receptor. The yeast mutant *sln1*Δ is lethal because of its mutated endogenous osmosensing rHK, SLN1, which fails to inhibit downstream mitogen-activated protein kinase (MAPK) pathways. Expressing the *Arabidopsis CRE1* gene in these cells allows them to survive, but only in the presence of cytokinin. The sequence of CRE1 is similar to ETR1, and less so to CKI1, suggesting a complexity of cytokinin perception and signaling that rivals that of ethylene.

The phospho-relay networks that occur in response to rHK-ligand binding have been well studied. Downstream elements of cytokinin signaling are induced by changes in cytokinin levels. In addition, stress (temperature, salinity, a lack of water) can induce the expression of a number of *Arabidopsis* two-component response regulator-like proteins (Urao *et al.*, 1998). Levels of two *Arabidopsis* mRNAs, IBC7 and IBC6, are elevated by cytokinin application and contain regions similar to the receiver domain of bacterial two-component response regulators (Brandstatter and Kieber, 1998). In maize the His-to-Asp phosphorelay proteins ZmRR2 and ZmRR1 show similarity to that of HPt domains from *Arabidopsis* thaliana (AHP1–AHP3: 44 to 47% identity) and yeast (Ypd1p: 24% identity; Sakakibara *et al.*, 1999). Cytokinin or inorganic nitrogen induces the accumulation of ZmRR1 and ZmRR2 transcripts indicating that His-to-Asp phospho-transfer may be involved in the transduction of nitrogen signals mediated by cytokinin (Sakakibara *et al.*, 1999).

Because of its commercial importance, there has been much focus on ethylene perception and action during fruit ripening. The levels of ethylene receptor expression change dramatically during fruit development (Sato-Nara *et al.*, 1999). Transcript of the tomato ETR1 homologue (tETR) increases during the early stages of ripening, flower senescence and in abscission zones (Payton *et al.*, 1996). tETR is 70% identical to *Arabidopsis* ETR1 at the protein level. However tETR lacks the *C*-terminal response domain and is identical to that encoded by the tomato *Never ripe* gene. It is greatly reduced in fruit of ripening mutants deficient in ethylene synthesis or response. Melon homologs of ETR1 and ERS1, Cm-ETR1 and Cm-ERS1, have been studied during ripening. Cm-ERS1 mRNA increases in the fruit pericarp in parallel with the increase in fruit size and then decreases when the fruit is at full size. During ripening Cm-ERS1 mRNA increases only slightly, with a marked increase of Cm-ETR1 mRNA correlated with climacteric ethylene production (Sato-Nara *et al.*, 1999). It therefore appears that changes in ethylene sensitivity are mediated by modulation of receptor levels during development and ripening of fruit (Payton *et al.*, 1996).

8.5 Identification of ligands and downstream elements

As already mentioned, there is a dearth of ligands for the many predicted plant RLKs. While some of the RLK ligands are likely to be small proteins or glycoproteins (e.g. those detected by LRR and S-type RLKs), other ligands may well be oligosaccharides. Within the lectin RLK group, the lecRK-a family is most similar to glucose-mannose specific lectins, suggesting these oligosaccharides as possible ligands (Herve *et al.*, 1999). As all the receptor kinases discussed here are plasma membrane-bound, ligands can still be effective even if they cannot permeate through the membrane, though many smaller ligands may well move through this barrier for intracellular detection (e.g. ethylene). Large ligands, such as parts of the flagella of motile Eubacteria, can be detected by RLKs (Felix *et al.*, 1999; Gomez-Gomez and Boller, 2000), although unraveling or disassembly of the protein may be required. We have discovered that the walled plant cell can detect extracellular tobacco mosaic virus coat protein, and apparently even oligomers of more than one subunit (Allan *et al.*, 2001). It is attractive to speculate that RLKs could be responsible for such recognition of large non-self particles, as well as recognition of self (i.e. the neighboring cell).

Most ligands for plasma membrane (PM) located RLKs are likely to be smaller than the size exclusion limit of the cell wall, which is around 2 to 4 nm depending on cell type, age and other conditions (Titel *et al.*, 1997). One well characterized peptide ligand in plants is systemin, an 18 amino acid polypeptide, which is released from wound sites in tomato (Ryan, 2000) to activate systemically a range of defense-related genes. Systemin is processed from a 200 amino acid precursor called prosystemin. However, unprocessed prosystemin also has biological activity in its systemin domain (Dombrowski *et al.*, 1999). Systemin appears to bind a wound-inducible 160 kDa cell surface receptor with a K_d of 0.17 nM (Scheer and Ryan, 1999). This receptor binding regulates a signaling cascade including cell depolarization, an increase in intracellular Ca^{2+} (Moyen *et al.*, 1998), MAPK induction and a phospholipase A activity (Ryan, 2000). These changes appear to lead to release of linolenic acid from membranes and its subsequent conversion to jasmonic acid, an activator of defense gene transcription. Unfortunately, the systemin receptor has yet to be sequenced, so it is not yet known whether it is an RLK.

8.5.1   CLAVATA and meristem fate

Perhaps the best evidence of a proteinaceous RLK ligand for a characterized RLK is for CLV3, the protein that coordinates growth between adjacent meristems (Fletcher *et al.*, 1999). *CLV3* encodes a small, predicted extracellular protein and is expressed at the meristem surface. The receptor kinase for this apparent ligand is CLAVATA1 (CLV1). Together CLV1 and CLV3 are required to maintain the balance between cell proliferation and organ formation in the *Arabidopsis* meristems (Fletcher *et al.*, 1999). To complicate this process, a third protein, CLV2, which is a receptor-like protein without a kinase domain, is required for the normal accumulation of CLV1 protein and its assembly into protein complexes (Jeong *et al.*, 1999). Evidence has also been found that CLV3 signaling occurs exclusively through a CLV1/CLV2 receptor kinase complex (Brand *et al.*, 2000). Overexpression of CLV3 causes a certain mutant phenotype, but not in mutant *clv1* or *clv2* backgrounds (Brand *et al.*, 2000). In addition, CLV3 from plant extracts will bind CLV1 and CLV2 expressed in yeast cells, and this binding requires CLV1 kinase activity (Trotochaud *et al.*, 2000).

Earlier biochemical characterization of CLV1 has shown its presence in two protein complexes *in vivo*; one of 185 kDa and the other of 450 kDa. In each complex, CLV1 is part of a disulfide-linked multimer of approximately 185 kDa. In addition, the 450 kDa complex contains a protein phosphatase called KAPP,

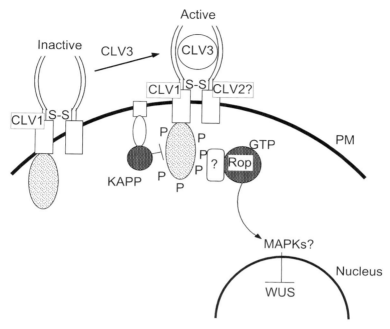

Figure 8.7 Meristem-fate signaling via the RLK CLV1 (reproduced with permission of Steve Clark, from Trotochaud *et al.*, 1999). The model depicts a heterodimer of CLV1, and possibly CLV2 (which lacks a kinase domain). This complex is activated by binding of the proposed ligand CLV3. This leads to phosphorylation of the kinase domain, and activation of KAPP and ROP, a GTPase-related protein. Signaling is then via a MAPK-like pathway to suppress the activity of the homeodomain transcription factor *WUSCHEL*, limiting the meristematic zone.

which is a negative regulator of CLV1 signaling, and ROP, a GTPase-related protein (see figure 8.7 for one possible model of CLV1–3 signaling). It was proposed that CLV1 is present as an inactive disulfide-linked heterodimer, with several additional factors (such as CLV2), and that CLV3 promotes the assembly of the 450 kDa complex, which then relays signal transduction through the GTPase (Trotochaud *et al.*, 1999). This complex then establishes a feedback loop with the homeodomain transcription factor *WUSCHEL* (Schoof *et al.*, 2000). It appears that the *CLV* genes repress *WUSCHEL* at the transcript level. In wild-type *Arabidopsis*, signaling via the CLV1–3 complex limits WUSCHEL activity to a narrow domain of cells. This is disrupted in any of the *CLV* mutants, so too many stem cells form due to inappropriate expression of *WUSCHEL*. *WUSCHEL* expression, in turn, induces the expression of CLV3, which then represses *WUSCHEL* (Schoof *et al.*, 2000).

Kinase-associated protein phosphatase(s) (KAPPs) are putative downstream components in kinase-mediated signal transduction pathways, including the CLV1–3 pathway. KAPP was first cloned as a protein that interacts with the

Arabidopsis RLK5 (or HAESA; Stone *et al.*, 1994). This phosphatase is composed of three domains: an *N*-terminal signal anchor, a kinase interaction (KI) domain, and a type 2C protein phosphatase catalytic region. RLK5 was shown to associate with the KI domain of KAPP after phosphorylation (Stone *et al.*, 1994). KAPP binds directly to autophosphorylated CLV1 *in vitro* and co-immunoprecipitates with CLV1 extracts of meristematic tissue (Stone *et al.*, 1998). CLV1 has kinase activity, in that it phosphorylates itself and KAPP. Also KAPP binds and dephosphorylates CLV1 (Williams *et al.*, 1997).

KAPPs have been found in both monocot and dicot species. The KI domain of the rice homolog of KAPP, OsKAPP, is phosphorylated *in vitro* by the kinase domain of the rice LRR-type RLK called OsTMK (van der Knaap *et al.*, 1999). In addition, OsKAPP cross-reacts with the *Arabidopsis* RLK5, but not with kinase domains of RLKs Xa21 (rice) and ZmPK1 (maize) (van der Knaap *et al.*, 1999). Observations of multiple associations of KAPP with other RLKs have been made in maize, including an RLK called KI domain interacting kinase 1 (KIK1, Braun *et al.*, 1997). Site directed mutagenesis of the KI domain of KAPP has shown that four highly conserved sites, within this domain, are essential for interaction with the RLK. Binding between the KI domain and receptor-like protein kinases is high-affinity, having a K_d of between 25–100 nM (Li *et al.*, 1999).

8.5.2   *Pollen self-incompatibility*

Another potential proteinaceous RLK ligand, which appears close to being fully characterized, is the ligand for the stigmatic receptor during pollen self-incompatibility. It has been shown that SLG and SRK are sufficient for the female side of the self-incompatible (SI) response in *Brassica*, but not the male side of the SI phenotype, and that the pollen S factor is encoded outside the S-locus (Cui *et al.*, 2000). However, another study (Schopfer *et al.*, 1999) has shown that the anther-expressed gene, SCR, is encoded at the S locus and fulfills the requirements for the male side of self-incompatible response. In addition Takayama *et al.* (2000) have identified an S locus gene, *SP11* (S locus protein 11, an SCR), of the S(9) haplotype of *Brassica campestris* and proposed that it potentially encodes the pollen ligand. The authors show that recombinant SP11 of the S(9) haplotype applied to papillar cells of S(9) stigmas, but not of S(8) stigmas, elicited an SI response. These results suggest that SP11 is the pollen S determinant in self-incompatibility (Takayama *et al.*, 2000). SP11 is a member of the pollen coat protein (PCP) family of proteins. These small PCPs have been shown to bind with high affinity to SLGs irrespective of S genotype (Doughty *et al.*, 1998), but in this studied case the selected PCPs were unlinked to the S locus. An additional 14 alleles of SP11 have been sequenced and analyzed (Watanabe *et al.*, 2000). It was found that these SP11 proteins are highly divergent, except for the presence of conserved cysteines. The phylogeny of these proteins and SRK/SRGs suggests that male and female sides of the self-incompatibility response have co-evolved (Watanabe *et al.*, 2000).

The mRNA of one pollen coat protein, PCP-A1, accumulates specifically in pollen at the late stages of development and contains a motif of eight cysteine residues, which also occur in the plant defensins (Doughty *et al.*, 1998). A putative structure for PCP-A1 has been produced. The authors used computer analysis to demonstrate that the female determinant, SLG, as a monomer or dimer, can bind PCP-A1 (Doughty *et al.*, 1998). The PCP proteins have also been analyzed for their adhesion properties, a critical step for pollen grain development. An SRG-like receptor has been shown to bind proteins, from pollen coat extracts, which are members of the PCP family (Luu *et al.*, 1999). Kinetic analysis showed that two PCPs specifically bound this stigmatic receptor with high affinity (K_d of 4–5 nM).

A protein that interacts with the kinase domain of the female SI determinant, SRK, was cloned using the yeast two-hybrid system. This protein was termed ARC1 (arm repeat containing) and was shown to interact with the phosphorylated kinase domain. *In vivo* and *in vitro* binding experiments determined that ARC1 could interact with the kinase domains from SRK-910 and SRK-A14 but not with kinase domains from other RLKs. These interactions were phosphorylation dependent. ARC1 was expressed specifically in the stigma (Gu *et al.*, 1998). Decreasing ARC1 mRNA levels reduce the SI response during interactions between pollen and stigma that should be incompatible (Stone *et al.*, 1999). This provides strong evidence that ARC1 is a positive transducer of the SI response. Surprisingly, this protein is unrelated to KAPP and contains 'arm' repeats (Pfam, http://www.sanger.ac.uk/ PF00514; armadillo/beta-catenin-like repeats) which are 40 amino acid tandem repeats forming a superhelix of helices, of the type found in *Drosophila* proteins Armadillo and Vertebrate Catenins. These *Drosophila* proteins are involved in the Wnt/Wingless signaling pathway which involves Ser–Thr phosphorylation (Gumbiner, 1995).

8.5.3   Other ligands

There is also good evidence for brassinosteroid being the ligand for BRI1. When the extracellular LRR and transmembrane domains of *Arabidopsis* BRI1 are fused to the kinase domain of Xa21 (the rice disease resistance receptor) the chimeric receptor initiates plant defense responses in rice cells when treated with brassinosteroids (He *et al.*, 2000). In addition, active brassinosteroids accumulate in plants with mutant BRI1 receptors, suggesting BRI1 is required for the homeostasis of endogenous BR levels (Noguchi *et al.*, 1999).

8.6   Comparison of similar proteins in animal cells

Receptor kinases in plants are very likely to have diverged both structurally and functionally from their mammalian counterparts. For example, although between 50–60% of protein biosynthesis genes in *Arabidopsis* are related by sequence to proteins in other eukaryotic genomes, only 8–23% of proteins involved in transcription have a characterized partner (The *Arabidopsis* Genome Initiative, 2000). In spite of this imbalance, lessons on the complexity of the signaling cascades underlying plant receptor kinases can still be learnt from mammalian systems.

Most receptor-linked kinases in animal cells are membrane proteins with an intracellular kinase domain. Unlike plant receptor kinases, the majority are protein–tyrosine kinases, i.e. they autophosphorylate on a tyrosine residue. Sequence analysis indicates that tyrosine kinases are not found in *Arabidopsis* (The *Arabidopsis* Genome Initiative, 2000). However, in the animal cell there are a small number of protein–serine–threonine kinases. These include the family called transforming growth factor-β (TGF-β). The TGF-β complex has distinct type I and type II receptors (both are protein–serine–threonine kinases). What is revealing for future work in plant science is the importance of membrane anchoring during the TGF-β pathway (well reviewed by ten Dijke and Heldin, 1999). After binding of the ligand TGF-β to the type II receptor, the type I receptor is phosphorylated, which then phosphorylates proteins of the Smad family. An anchor protein known as SARA (Smad anchor for receptor activation) acts to restrict the subcellular localization of the signaling components and also acts as a platform to bring components together. SARA interacts directly with Smad2 and Smad3 (Tsukazaki *et al.*, 1998). Phosphorylation of Smad2 by the TGF-β complex induces dissociation from SARA and formation of Smad2/Smad4 complexes in the cytoplasm. These complexes associate with other SMADs and move to the nucleus where they trigger changes in gene expression. Adding to the complexity of this single receptor kinase pathway is another class of Smads, inhibitory Smads (I-Smads), which inhibit the signals from the type I and II receptors (Miyazono *et al.*, 2000). Smads are also regulated by other signaling pathways, such as the MAPK pathway (Miyazono *et al.*, 2000). Anchor proteins like SARA, or inhibitors of proteins downstream of RLKs like I-Smads, have not yet been characterized in plants.

The receptor-linked protein-tyrosine kinases found in (non-plant) eukaryotes usually exist as monomers and, after ligand binding, form homo- or hetero-oligomers. This dimerization leads to receptor autophosphorylation on tyrosine residues. In certain receptors, the autophosphorylation regulates the catalytic activity of the kinase. Moreover, autophosphorylated tyrosine residues bind downstream proteins with SH2 or other domains. However, this 'post-ligand assembly' model (Heldin, 1995) has recently been challenged, for some receptors, by a 'pre-ligand assembly' model (Golstein, 2000). Fas, a member of the TNFR superfamily, initiates cell death by two alternative pathways, one relying on caspase-8 and the other dependent on the kinase receptor-interacting protein (RIP). The Fas ligand (FasL) forms a trimer to bind a trimer of Fas receptors. However, trimerization of the Fas receptor is required before binding of the ligand (Siegel *et al.*, 2000). For activation, other TNFR receptors appear to require the same pre-ligand dimerization via an extracellular pre-ligand-binding assembly domain (PLAD) that is distinct from the domain that contacts the ligand (Chan *et al.*, 2000). The presence of TNFR-like RLKs in plants provides the exciting possibility that such dimerization is occurring.

8.7 Conclusions

When presented with data from the *Arabidopsis* genome, several key differences from other eukaryotic cells are immediately apparent; for example, the huge number and uniqueness of plant transcription factors. It also appears that plants have evolved their own pathways of signal transduction. None of the receptor-initiated pathways that are common between humans, *C. elegans* and *Drosophila* are found in *Arabidopsis* (e.g. Wingless/Wnt, Notch/lin12, JAK, TGF-/SMADs, receptor tyrosine kinase/Ras or the nuclear steroid hormone receptors). Nor are these likely to be found in other plants.

Perhaps the reason for the number of novel signal transduction pathways (from receptor to transcription factor) found in plants, is that plants are sessile, anchored to one place. They are forced to respond to local environmental changes by changing their physiology or redirecting their growth. Plants must cope with rapid and drastic changes in light, temperature, water, nutrients, touch and gravity and, finally, attack by pathogens or herbivores. Each cell in the plant must integrate these conditions, as well as receiving signals from across the plant body, in the form of plant hormones and peptides. Some of these signals are freely permeable (e.g. weak acids such as auxin, or gaseous ethylene) while others are small impermeant molecules (e.g. brassinosteroids) or peptides (e.g. systemin or CLV3). It is now apparent that many of these molecules are perceived by receptor kinases, resulting in a signal that is transduced to the nucleus, to alter patterns of gene expression.

It is now the challenge for plant biologists to determine more ligands for these receptors and, using knock-out or overexpression techniques, assign functions to many of the predicted *Arabidopsis* receptor kinases (e.g. the 340 RLKs). Having performed this task for *Arabidopsis*, this knowledge can then be applied to the 'jungle' of other plant species.

References

Allan, A.C., Lapidot, M., Culver, J.N. and Fluhr, R. (2001) An early tobacco mosaic virus-induced oxidative burst in tobacco indicates extracellular perception of the virus coat protein. *Plant Physiol.*, **126**, 97-108.

Annen, F. and Stockhaus, J. (1999) SbRLK1, a receptor-like protein kinase of *Sorghum bicolor* (L.) Moench that is expressed in mesophyll cells. *Planta*, **208**, 420-425.

Banner, D.W., d'Arcy, A., Janes, W. *et al.* (1993) Crystal structure of the soluble human 55 kDa TNF receptor-human TNF beta complex: implication for TNF receptor activation. *Cell*, **73**, 431-445.

Bassett, C.L., Nickerson, M.L., Cohen, R.A. and Rajeevan, M.S. (2000) Alternative transcript initiation and novel post-transcriptional processing of a leucine-rich repeat receptor-like protein kinase gene that responds to short-day photoperiodic floral induction in morning glory (*Ipomoea nil*). *Plant Mol. Biol.*, **43**, 43-58.

Baumberger, N., Ringli, C. and Keller, B. (2001) The chimeric leucine-rich repeat/extensin cell wall protein LRX1 is required for root hair morphogenesis in Arabidopsis thaliana. *Genes Dev.*, **15**, 1128-1139.

Becraft, P.W. (1998) Receptor kinases in plant development. *Trends Plant Sci.*, **3**, 384-388.

Becraft, P.W., Stinard, P.S. and McCarty, D.R. (1996) CRINKLY4: A TNFR-like receptor kinase involved in maize epidermal differentiation. *Science*, **273**, 1406-1409.

Brand, U., Fletcher, J.C., Hobe, M., Meyerowitz, E.M. and Simon, R. (2000) Dependence of stem cell fate in *Arabidopsis* on a feedback loop regulated by CLV3 activity. *Science*, **289**, 617-619.

Brandstatter, I. and Kieber, J.J. (1998) Two genes with similarity to bacterial response regulators are rapidly and specifically induced by cytokinin in *Arabidopsis*. *Plant Cell*, **10**, 1009-1019.

Braun, D.M., Stone, J.M. and Walker, J.C. (1997) Interaction of the maize and *Arabidopsis* kinase interaction domains with a subset of receptor-like protein kinases: implications for transmembrane signalling in plants. *Plant J.*, **12**, 83-95.

Cabrillac, D., Delorme, V., Garin, J. *et al.* (1999) The S15 self-incompatibility haplotype in *Brassica oleracea* includes three S gene family members expressed in stigmas. *Plant Cell*, **11**, 971-986.

Chan, F.K., Chun, H.J., Zheng, L., Siegel, R.M., Bui, K.L. and Lenardo, M.J. (2000) A domain in TNF receptors that mediates ligand-independent receptor assembly and signaling. *Science*, **288**, 2351-2354.

Chang, C. and Meyerowitz, E.M. (1995) The ethylene hormone response in *Arabidopsis*: a eukaryotic two-component signaling system. *Proc. Natl Acad. Sci. USA*, **92**, 4129-4133.

Chang, C., Kwok, S.F., Bleecker, A.B. and Meyerowitz, E.M. (1993) *Arabidopsis* ethylene-response gene ETR1: similarity of product to two-component regulators. *Science*, **262**, 539-544.

Charng, Y.Y., Sun, C.-W., Yan, S.-L., Chou, S.-J., Chen, Y.-R. and Yang, S.F. (1997) cDNA sequence of a putative ethylene receptor from carnation petals (Accession No. AF016250). *Plant Physiol.*, **115**, 863.

Cho, H.S. and Pai, H.S. (2000) Cloning and characterization of ntTMK1 gene encoding a TMK1-homologous receptor-like kinase in tobacco. *Mol. Cell*, **10**, 317-324.

Cock, J.M., Swarup, R. and Dumas, C. (1997) Natural antisense transcripts of the S locus receptor kinase gene and related sequences in *Brassica oleracea*. *Mol. Gen. Genet.*, **255**, 514-524.

Cui, Y., Bi, Y.M., Brugiere, N., Arnoldo, M. and Rothstein, S.J. (2000) The S locus glycoprotein and the S receptor kinase are sufficient for self-pollen rejection in *Brassica*. *Proc. Natl Acad. Sci. USA*, **97**, 3713-3717.

Czernic, P., Visser, B., Sun, W. *et al.* (1999) Characterization of an *Arabidopsis thaliana* receptor-like protein kinase gene activated by oxidative stress and pathogen attack. *Plant J.*, **18**, 321-327.

Deeken, R. and Kaldenhoff, R. (1997) Light-repressible receptor protein kinase: a novel photo-regulated gene from *Arabidopsis thaliana*. *Planta*, **202**, 479-486.

Delorme, V., Giranton, J.L., Hatzfeld, Y. *et al.* (1995) Characterization of the S locus genes, SLG and SRK, of the *Brassica* S3 haplotype: identification of a membrane-localized protein encoded by the S locus receptor kinase gene. *Plant J.*, **7**, 429-440.

Dombrowski, J.E., Pearce, G. and Ryan, C.A. (1999) Proteinase inhibitor-inducing activity of the prohormone prosystemin resides exclusively in the *C*-terminal systemin domain. *Proc. Natl Acad. Sci. USA*, **96**, 12,947-12,950.

Doughty, J., Dixon, S., Hiscock, S.J., Willis, A.C., Parkin, I.A. and Dickinson, H.G. (1998) PCP-A1, a defensin-like *Brassica* pollen coat protein that binds the S locus glycoprotein, is the product of gametophytic gene expression. *Plant Cell*, **10**, 1333-1347.

Dubreucq, B., Berger, N., Vincent, E. *et al.* (2000) The *Arabidopsis AtEPR1* extensin-like gene is specifically expressed in endosperm during seed germination. *Plant J.*, **23**, 643-652.

Dwyer, K.G., Kandasamy, M.K., Mahosky, D.I. *et al.* (1994) A superfamily of S locus-related sequences in *Arabidopsis*: diverse structures and expression patterns. *Plant Cell*, **6**, 1829-1843.

Felix, G., Duran, J.D., Volko, S. and Boller, T. (1999) Plants have a sensitive perception system for the most conserved domain of bacterial flagellin. *Plant J.*, **18**, 265-276.

Feuillet, C., Schachermayr, G. and Keller, B. (1997) Molecular cloning of a new receptor-like kinase gene encoded at the Lr10 disease resistance locus of wheat. *Plant J.*, **11**, 45-52.

Feuillet, C., Reuzeau, C., Kjellbom, P. and Keller, B. (1998) Molecular characterization of a new type of receptor-like kinase (*wlrk*) gene family in wheat. *Plant Mol. Biol.*, **37**, 943-953.

Fletcher, J.C., Brand, U., Running, M.P., Simon, R. and Meyerowitz, E.M. (1999) Signaling of cell fate decisions by *CLAVATA3* in *Arabidopsis* shoot meristems. *Science*, **283**, 1911-1914.

Friedrichsen, D.M., Joazeiro, C.A., Li, J., Hunter, T. and Chory, J. (2000) Brassinosteroid-insensitive-1 is a ubiquitously expressed leucine-rich repeat receptor serine/threonine kinase. *Plant Physiol.*, **123**, 1247-1256.

Gamble, R.L., Coonfield, M.L. and Schaller, G.E. (1998) Histidine kinase activity of the ETR1 ethylene receptor from *Arabidopsis*. *Proc. Natl Acad. Sci. USA*, **95**, 7825-7829.

Giranton, J.L., Dumas, C., Cock, J.M. and Gaude, T. (2000) The integral membrane S-locus receptor kinase of *Brassica* has serine/threonine kinase activity in a membranous environment and spontaneously forms oligomers in planta. *Proc. Natl Acad. Sci. USA*, **97**, 3759-3764.

Golstein, P. (2000) Signal transduction. FasL binds preassembled Fas. *Science*, **288**, 2328-2329.

Gomez-Gomez, L. and Boller, T. (2000) FLS2: an LRR receptor-like kinase involved in the perception of the bacterial elicitor flagellin in *Arabidopsis*. *Mol. Cell*, **5**, 1003-1011.

Gu, T., Mazzurco, M., Sulaman, W., Matias, D.D. and Goring, D.R. (1998) Binding of an arm repeat protein to the kinase domain of the S-locus receptor kinase. *Proc. Natl Acad. Sci. USA*, **95**, 382-387.

Gumbiner, B.M. (1995) Signal transduction of beta-catenin. *Curr. Opin. Cell Biol.*, **7**, 634-640.

Hajouj, T., Michelis, R. and Gepstein, S. (2000) Cloning and characterization of a receptor-like protein kinase gene associated with senescence. *Plant Physiol.*, **124**, 1305-1314.

Hall, A.E., Chen, Q.G., Findell, J.L., Schaller, G.E. and Bleecker, A.B. (1999) The relationship between ethylene binding and dominant insensitivity conferred by mutant forms of the ETR1 ethylene receptor. *Plant Physiol.*, **121**, 291-300.

Hardie, D.G. (1999) Plant protein serine/threonine kinases: classification and functions. *Annu. Rev. Plant Physiol. Plant Mol. Biol.*, **50**, 97-131.

He, Z.H., Fujiki, M. and Kohorn, B.D. (1996) A cell wall-associated, receptor-like protein kinase. *J. Biol. Chem.*, **271**, 19,789-19,793.

He, Z.H., He, D. and Kohorn, B.D. (1998) Requirement for the induced expression of a cell wall associated receptor kinase for survival during the pathogen response. *Plant J.*, **14**, 55-63.

He, Z.H., Cheeseman, I., He, D. and Kohorn, B.D. (1999) A cluster of five cell wall-associated receptor kinase genes, *Wak1-5*, are expressed in specific organs of *Arabidopsis*. *Plant Mol. Biol.*, **39**, 1189-1196.

He, Z., Wang, Z.Y., Li, J., Zhu, Q., Lamb, C., Ronald, P. and Chory, J. (2000) Perception of brassinosteroids by the extracellular domain of the receptor kinase BRI1. *Science*, **288**, 2360-2363.

Heldin, C.H. (1995) Dimerization of cell surface receptors in signal transduction. *Cell*, **80**, 213-223.

Herve, C., Dabos, P., Galaud, J.P., Rouge, P. and Lescure, B. (1996) Characterization of an *Arabidopsis thaliana* gene that defines a new class of putative plant receptor kinases with an extracellular lectin-like domain. *J. Mol. Biol.*, **258**, 778-788.

Herve, C., Serres, J., Dabos, P. *et al.* (1999) Characterization of the *Arabidopsis lecRK-alpha* genes: members of a superfamily encoding putative receptors with an extracellular domain homologous to legume lectins. *Plant Mol. Biol.*, **39**, 671-682.

Hong, S.W., Jon, J.H., Kwak, J.M. and Nam, H.G. (1997) Identification of a receptor-like protein kinase gene rapidly induced by abscisic acid, dehydration, high salt, and cold treatments in *Arabidopsis thaliana*. *Plant Physiol.*, **113**, 1203-1212.

Hua, J., Chang, C., Sun, Q. and Meyerowitz, E.M. (1995) Ethylene insensitivity conferred by *Arabidopsis ERS* gene. *Science*, **269**, 1712-1714.

Hua, J., Sakai, H., Nourizadeh, S. *et al.* (1998) *EIN4* and *ERS2* are members of the putative ethylene receptor gene family in *Arabidopsis*. *Plant Cell*, **10**, 1321-1332.

Imamura, A., Hanaki, N., Umeda, H. *et al.* (1998) Response regulators implicated in His-to-Asp phosphotransfer signaling in *Arabidopsis*. *Proc. Natl Acad. Sci. USA*, **95**, 2691-2696.

Imamura, A., Hanaki, N., Nakamura, A. *et al.* (1999) Compilation and characterization of *Arabiopsis thaliana* response regulators implicated in His-Asp phosphorelay signal transduction. *Plant Cell Physiol.*, **40**, 733-742.

Inoue, T., Higuchi, M., Hashimoto, Y. *et al.* (2001) Identification of CRE1 as a cytokinin receptor from *Arabidopsis*. *Nature*, **409**, 1060-1063.

Jeong, S., Trotochaud, A.E. and Clark, S.E. (1999) The *Arabidopsis CLAVATA2* gene encodes a receptor-like protein required for the stability of the CLAVATA1 receptor-like kinase. *Plant Cell*, **11**, 1925-1934.

Jin, P., Guo, T. and Becraft, P.W. (2000) The maize CR4 receptor-like kinase mediates a growth factor-like differentiation response. *Genesis*, **27**, 104-116.

Jinn, T.L., Stone, J.M. and Walker, J.C. (2000) HAESA, an *Arabidopsis* leucine-rich repeat receptor kinase, controls floral organ abscission. *Genes Dev.*, **14**, 108-117.

John, M., Rohrig, H., Schmidt, J., Walden, R. and Schell, J. (1997) Cell signalling by oligosaccharides. *Trends Plant Sci.*, **2**, 111-115.

Kakimoto, T. (1996) CKI1, a histidine kinase homolog implicated in cytokinin signal transduction. *Science*, **274**, 982-985.

Kennedy, J.F., Palva, P.M.G., Corella, M.T.S., Cavalcanti, M.S.M. and Coelho, L.C.B.B. (1995) Lectins, versatile proteins of recognition: a review. *Carbohydrate Polymers*, **26**, 219-230.

Kieber, J.J., Rothenberg, M., Roman, G., Feldmann, K.A. and Ecker, J.R. (1993) CTR1, a negative regulator of the ethylene response pathway in *Arabidopsis*, encodes a member of the raf family of protein kinases. *Cell*, **72**, 427-441.

Kim, Y.S., Lee, J.H., Yoon, G.M. *et al.* (2000) CHRK1, a chitinase-related receptor-like kinase in tobacco. *Plant Physiol.*, **123**, 905-915.

Klarzynski, O., Plesse, B., Joubert, J.-M. *et al.* (2000) Linear B-1,3 glucans are elicitors of defense responses in tobacco. *Plant Physiol.*, **124**, 1027-1037.

van der Knaap, E., Song, W.Y., Ruan, D.L., Sauter, M., Ronald, P.C. and Kende, H. (1999) Expression of a gibberellin-induced leucine-rich repeat receptor-like protein kinase in deepwater rice and its interaction with kinase-associated protein phosphatase. *Plant Physiol.*, **120**, 559-570.

Kobe, B. and Deisenhofer, J. (1993) Crystal structure of porcine ribonuclease inhibitor, a protein with leucine-rich repeats. *Nature*, **366**, 751-756.

Kobe, B. and Deisenhofer, J. (1995) A structural basis of the interactions between leucine-rich repeats and protein ligands. *Nature*, **374**, 183-186.

Kusaba, M. and Nishio, T. (1999) Comparative analysis of S haplotypes with very similar SLG alleles in *Brassica rapa* and *Brassica oleracea*. *Plant J.*, **17**, 83-91.

Lange, J., Xie, Z.P., Broughton, W.J., Vogeli Lange, R. and Boller, T. (1999) A gene encoding a receptor-like protein kinase in the roots of common bean is differentially regulated in response to pathogens, symbionts and nodulation factors. *Plant Sci.*, **142**, 133-145.

Lashbrook, C.C., Tieman, D.M. and Klee, H.J. (1998) Differential regulation of the tomato *ETR* gene family throughout plant development. *Plant J.*, **15**, 243-252.

Lee, H.S., Karunanandaa, B., McCubbin, A., Gilroy, S. and Kao, T.H. (1996) PRK1, a receptor-like kinase of *Petunia inflata*, is essential for postmeiotic development of pollen. *Plant J.*, **9**, 613-624.

Li, J. and Chory, J. (1997) A putative leucine-rich repeat receptor kinase involved in brassinosteroid signal transduction. *Cell*, **90**, 929-938.

Li, H.Y. and Gray, J.E. (1997) Pollination-enhanced expression of a receptor-like protein kinase related gene in tobacco styles. *Plant Mol. Biol.*, **33**, 653-665.

Li, J., Smith, G.P. and Walker, J.C. (1999) Kinase interaction domain of kinase-associated protein phosphatase, a phosphoprotein-binding domain. *Proc. Natl Acad. Sci. USA*, **96**, 7821-7826.

Luu, D.T., Marty-Mazars, D., Trick, M., Dumas, C. and Heizmann, P. (1999) Pollen-stigma adhesion in *Brassica* spp involves SLG and SLR1 glycoproteins. *Plant Cell*, **11**, 251-262.

Melchers, L.S., Apotheker de Groot, M., van der Knaap, J.A. *et al.* (1994) A new class of tobacco chitinases homologous to bacterial exo-chitinases displays antifungal activity. *Plant J.*, **5**, 469-480.

Miyata, S., Urao, T., Yamaguchi-Shinozaki, K. and Shinozaki, K. (1998) Characterization of genes for two-component phosphorelay mediators with a single HPt domain in *Arabidopsis thaliana*. *FEBS Lett.*, **437**, 11-14.

Miyazono, K., ten Dijke, P. and Heldin, C.H. (2000) TGF-beta signaling by Smad proteins. *Adv. Immunol.*, **75**, 115-157.

Moyen, C., Hammond Kosack, K.E., Jones, J., Knight, M.R. and Johannes, E. (1998) Systemin triggers an increase of cytoplasmic calcium in tomato mesophyll cells: Ca^{2+} mobilization from intro- and extracellular compartments. *Plant Cell Environ.*, **21**, 1101-1111.

Mu, J.H., Lee, H.S. and Kao, T. (1994) Characterization of a pollen-expressed receptor-like kinase gene of *Petunia inflata* and the activity of its encoded kinase. *Plant Cell*, **6**, 709-721.

Muschietti, J., Eyal, Y. and McCormick, S. (1998) Pollen tube localization implies a role in pollen-pistil interactions for the tomato receptor-like protein kinases LePRK1 and LePRK2. *Plant Cell*, **10**, 319-330.

Noguchi, T., Fujioka, S., Choe, S. *et al* (1999) Brassinosteroid-insensitive dwarf mutants of *Arabidopsis* accumulate brassinosteroids. *Plant Physiol.*, **121**, 743-752.

Ohtake, Y., Takahashi, T. and Komeda, Y. (2000) Salicylic acid induces the expression of a number of receptor-like kinase genes in *Arabidopsis thaliana*. *Plant Cell Physiol.*, **41**, 1038-1044.

Pastuglia, M., Roby, D., Dumas, C. and Cock, J.M. (1997) Rapid induction by wounding and bacterial infection of an S gene family receptor-like kinase gene in *Brassica oleracea*. *Plant Cell*, **9**, 49-60.

Payton, S., Fray, R.G., Brown, S. and Grierson, D. (1996) Ethylene receptor expression is regulated during fruit ripening, flower senescence and abscission. *Plant Mol. Biol.*, **31**, 1227-1231.

Rao, Z., Handford, P., Mayhew, M., Knott, V., Brownlee, G.G. and Stuart, D. (1995) The structure of a Ca^{2+}-binding epidermal growth factor-like domain: its role in protein–protein interactions. *Cell*, **82**, 131-141.

Ronald, P.C., Song, W.Y., Pi, L.Y., Wang, G.L. and Ruan, D.L. (1996) The cloned gene, *Xa21*, encodes a receptor like kinase and confers resistance to multiple *Xanthomonas oryzae* pv. *oryzae* isolates in transgenic rice plants. *Curr. Top. Plant Biochem. Physiol.*, **15**, 73-74.

Ryan, C.A. (2000) The systemin signaling pathway: differential activation of plant defensive genes. *Biochim. Biophys. Acta*, **7**, 1-2.

Sakai, H., Aoyama, T., Bono, H. and Oka, A. (1998a) Two-component response regulators from *Arabidopsis thaliana* contain a putative DNA-binding motif. *Plant Cell Physiol.*, **39**, 1232-1239.

Sakai, H., Hua, J., Chen, Q.G. *et al.* (1998b) ETR2 is an ETR1-like gene involved in ethylene signaling in *Arabidopsis*. *Proc. Natl Acad. Sci. USA*, **95**, 5812-5817.

Sakai, T., Wada, T., Ishiguro, S. and Okada, K. (2000) RPT2: a signal transducer of the phototropic response in *Arabidopsis*. *Plant Cell*, **12**, 225-236.

Sakakibara, H., Suzuki, M., Takei, K., Deji, A., Taniguchi, M. and Sugiyama, T. (1998) A response-regulator homologue possibly involved in nitrogen signal transduction mediated by cytokinin in maize. *Plant J.*, **14**, 337-344.

Sakakibara, H., Hayakawa, A., Deji, A., Gawronski, S.W. and Sugiyama, T. (1999) His-Asp phosphotransfer possibly involved in the nitrogen signal transduction mediated by cytokinin in maize: molecular cloning of cDNAs for two-component regulatory factors and demonstration of phosphotransfer activity *in vitro*. *Plant Mol. Biol.*, **41**, 563-573.

Salomon, M., Christie, J.M., Knieb, E., Lempert, U. and Briggs, W.R. (2000) Photochemical and mutational analysis of the FMN-binding domains of the plant blue light receptor, phototropin. *Biochemistry*, **39**, 9401-9410.

Sato-Nara, K., Yuhashi, K.I., Higashi, K., Hosoya, K., Kubota, M. and Ezura, H. (1999) Stage- and tissue-specific expression of ethylene receptor homolog genes during fruit development in muskmelon. *Plant Physiol.*, **120**, 321-330.

Savage, C.R.J., Inagaric, T. and Cohen, S.J. (1972) The primary structure of epidermal growth factor. *J. Biol. Chem.*, **247**, 7612-7621.

Schaller, G.E. and Bleecker, A.B. (1995) Ethylene-binding sites generated in yeast expressing the *Arabidopsis ETR1* gene. *Science*, **270**, 1809-1811.

Scheer, J.M. and Ryan, C.A. (1999) A 160-kD systemin receptor on the surface of *Lycopersicon peruvianum* suspension-cultured cells. *Plant Cell*, **11**, 1525-1536.

Schmidt, E.D., Guzzo, F., Toonen, M.A. and de Vries, S.C. (1997) A leucine-rich repeat containing receptor-like kinase marks somatic plant cells competent to form embryos. *Development*, **124**, 2049-2062.

Schoof, H., Lenhard, M., Haecker, A., Mayer, K.F., Jurgens, G. and Laux, T. (2000) The stem cell population of *Arabidopsis* shoot meristems is maintained by a regulatory loop between the *CLAVATA* and *WUSCHEL* genes. *Cell*, **100**, 635-644.

Schopfer, C.R., Nasrallah, M.E. and Nasrallah, J.B. (1999) The male determinant of self-incompatibility in *Brassica*. *Science*, **286**, 1697-1700.

Shibuya, K., Satoh, S. and Yoshioka, T. (1998) A cDNA encoding a putative ethylene receptor related to petal senescence in carnation (*Dianthus caryophyllus* L.) flowers (Accession No. AF034770). *Plant Physiol.*, **116**, 867.

Siegel, R.M., Frederiksen, J.K., Zacharias, D.A. *et al.* (2000) Fas preassociation required for apoptosis signaling and dominant inhibition by pathogenic mutations. *Science*, **288**, 2354-2357.

Song, W.Y., Wang, G.L., Chen, L.L. *et al.* (1995) A receptor kinase-like protein encoded by the rice disease resistance gene, *Xa21*. *Science*, **270**, 1804-1806.

Stone, J.M., Collinge, M.A., Smith, R.D., Horn, M.A. and Walker, J.C. (1994) Interaction of a protein phosphatase with an *Arabidopsis* serine–threonine receptor kinase. *Science*, **266**, 793-795.

Stone, J.M., Trotochaud, A.E., Walker, J.C. and Clark, S.E. (1998) Control of meristem development by CLAVATA1 receptor kinase and kinase-associated protein phosphatase interactions. *Plant Physiol.*, **117**, 1217-1225.

Stone, S.L., Arnoldo, M. and Goring, D.R. (1999) A breakdown of *Brassica* self-incompatibility in ARC1 antisense transgenic plants. *Science*, **286**, 1729-1731.

Suzuki, T., Imamura, A., Ueguchi, C. and Mizuno, T. (1998) Histidine-containing phosphotransfer (HPt) signal transducers implicated in His-to-Asp phosphorelay in *Arabidopsis*. *Plant Cell Physiol.*, **39**, 1258-1268.

Takasaki, T., Hatakeyama, K., Suzuki, G., Watanabe, M., Isogal, A. and Hinata, K. (2000) The S receptor kinase determines self-incompatibility in *Brassica* stigma. *Nature*, **403**, 913-916.

Takayama, S., Shiba, H., Iwano, M., *et al.* (2000) The pollen determinant of self-incompatibility in *Brassica campestris*. *Proc. Natl Acad. Sci. USA*, **97**, 1920-1925.

ten Dijke, P. and Heldin, C.H. (1999) Signal transduction. An anchor for activation. *Nature*, **397**, 109-111.

The *Arabidopsis* Genome Initiative (2000) Analysis of the genome sequence of the flowering plant *Arabidopsis thaliana*. *Nature*, **408**, 796-815.

Tieman, D.M. and Klee, H.J. (1999) Differential expression of two novel members of the tomato ethylene-receptor family. *Plant Physiol.*, **120**, 165-172.

Titel, C., Woehlecke, H., Afifi, I. and Ehwald, R. (1997) Dynamics of limiting cell wall porosity in plant suspension cultures. *Planta*, **203**, 320-326.

Tobias, C.M., Howlett, B. and Nasrallah, J.B. (1992) An *Arabidopsis thaliana* gene with sequence similarity to the S-locus receptor kinase of *Brassica oleracea*—sequence and expression. *Plant Physiol.*, **99**, 284-290.

Torii, K.U., Mitsukawa, N., Oosumi, T. *et al.* (1996) The *Arabidopsis ERECTA* gene encodes a putative receptor protein kinase with extracellular leucine-rich repeats. *Plant Cell*, **8**, 735-746.

Trotochaud, A.E., Hao, T., Wu, G., Yang, Z. and Clark, S.E. (1999) The CLAVATA1 receptor-like kinase requires CLAVATA3 for its assembly into a signaling complex that includes KAPP and a Rho-related protein. *Plant Cell*, **11**, 393-406.

Trotochaud, A.E., Jeong, S. and Clark, S.E. (2000) CLAVATA3, a multimeric ligand for the CLAVATA1 receptor-kinase. *Science*, **289**, 613-617.

Tsukazaki, T., Chiang, T.A., Davison, A.F., Attisano, L. and Wrana, J.L. (1998) SARA, a FYVE domain protein that recruits Smad2 to the TGFbeta receptor. *Cell*, **95**, 779-791.

Urao, T., Yakubov, B., Yamaguchi-Shinozaki, K. and Shinozaki, K. (1998) Stress-responsive expression of genes for two-component response regulator-like proteins in *Arabidopsis thaliana. FEBS Lett.*, **427**, 175-178.

Urao, T., Yakubov, B., Satoh, R. *et al.* (1999) A transmembrane hybrid-type histidine kinase in *Arabidopsis* functions as an osmosensor. *Plant Cell*, **11**, 1743-1754.

Urao, T., Yamaguchi Shinozaki, K. and Shinozaki, K. (2000) Two-component systems in plant signal transduction. *Trends Plant Sci.*, **5**, 67-74.

Vriezen, W.H., Rijn, C.P.E., Voesenek, L.A.C.J. and Mariani, C. (1997) A homolog of the *Arabidopsis thaliana ERS* gene is actively regulated in *Rumex palustris* upon flooding. *Plant J.*, **11**, 1265-1271.

Wagner, T.A. and Kohorn, B.D. (2001) Wall-associated kinases are expressed throughout plant development and are required for cell expansion. *Plant Physiol.*, **13**, 303-318.

Walker, J.C. (1994) Structure and function of the receptor-like protein kinases of higher plants. *Plant Mol. Biol.*, **26**, 1599-1609.

Wang, X., Zafian, P., Choudhary, M. and Lawton, M. (1996) The PR5K receptor protein kinase from *Arabidopsis thaliana* is structurally related to a family of plant defense proteins. *Proc. Natl Acad. Sci. USA*, **93**, 2598-2602.

Watanabe, M., Ito, A., Takada, Y. *et al.* (2000) Highly divergent sequences of the pollen self-incompatibility (S) gene in class-I S haplotypes of *Brassica campestris* (syn. *rapa*) L. *FEBS Lett.*, **473**, 139-144.

Wilkinson, J.Q., Lanahan, M.B., Yen, H.C., Giovannoni, J.J. and Klee, H.J. (1995) An ethylene-inducible component of signal transduction encoded by *Never-ripe. Science*, **270**, 1807-1809.

Williams, R.W., Wilson, J.M. and Meyerowitz, E.M. (1997) A possible role for kinase-associated protein phosphatase in the *Arabidopsis* CLAVATA1 signaling pathway. *Proc. Natl Acad. Sci. USA*, **94**, 10,467-10,472.

Yeh, K.C. and Lagarias, J.C. (1998) Eukaryotic phytochromes: light-regulated serine/threonine protein kinases with histidine kinase ancestry. *Proc. Natl Acad. Sci. USA*, **95**, 13,976-13,981.

Zhou, D., Kalaitzis, P., Mattoo, A.K. and Tucker, M.L. (1996) The mRNA for an *ETR1* homologue in tomato is constitutively expressed in vegetative and reproductive tissues. *Plant Mol. Biol.*, **30**, 1331-1338.

9 Cytoplasmatic protein kinases in signal transduction

Claudia Jonak and Heribert Hirt

9.1 Introduction

Almost 50 years ago, protein phosphorylation was detected as a regulatory mechanism of mammalian glycogen metabolism. Soon after, it was realized to be implicated broadly in information processing in all organisms. Proteins can become phosphorylated on serine, threonine, tyrosine, histidine and aspartate residues. Whereas the first three phosphorylated amino acids represent the predominant modifications in animals, the phosphohistidine and phosphoaspartate residues are not found in animals but are typical for bacteria. Plants do not obey these 'rules' and all five types of phosphorylated amino acids are found in this kingdom.

Protein phosphorylation does not occur spontaneously but is a covalent modification that is carried out by specific enzymes, the protein kinases. The protein kinases are classified into three groups: serine–threonine, tyrosine and histidine kinases. There are no aspartate kinases because aspartate phosphorylation is just an intermediate step of the activity of histidine kinases. This chapter will only deal with the cytoplasmic protein kinases of plants. For this reason, the plasma membrane-integrated plant histidine kinases will not be discussed here. The very large majority of protein kinases in plants belongs to the class of serine–threonine protein kinases and will therefore be the subject of this chapter.

In the discussion of protein phosphorylation, protein kinases were once thought to be the major players and the hydrolysis of the phosphorylated amino acids was considered to be a continuous unregulated process. This picture has changed, however, and it is now clear that protein phosphatases are as equally important regulators as protein kinases and that the phosphorylation state of a protein is often regulated through a number of different kinases and phosphatases. With the availability of sequences of entire genomes, it is clear that all organisms, including plants, contain hundreds of protein kinase and phosphatase genes, making up more than 5% of the total coding capacity. These numbers clarify why protein phosphorylation is involved in almost every single enzyme reaction and regulatory process that has been investigated in biology, but they also show that it cannot be within the scope of this chapter to give a comprehensive review of all protein kinases. However, a discussion of two protagonists of protein kinases, the mitogen-activated protein kinases (MAPKs) and the cyclin-dependent protein kinases (CDKs), should nonetheless give an appreciation of the basic rules and also show the potential complexity of these enzymes.

9.2 Mitogen-activated protein kinases

Mitogen-activated protein kinases (MAPKs) are evolutionary conserved serine–threonine protein kinases. They are an integral part of diverse signal transduction pathways involved in mediating different extracellular signals to various cellular targets. In mammals, MAPKs regulate diverse cellular processes such as cell growth, differentiation and stress responses. In *Drosophila melanogaster* and *Caenorhabditis elegans* MAPKs are required for development. In *Saccharomyces cerevisiae*, MAPK pathways control mating, sporulation, invasive growth, pseudohyphal growth and cell wall integrity, as well as osmoregulation. In plants, MAPKs have been implicated in biotic and abiotic stress response, hormone signaling and cell division.

9.2.1 MAPK modules

The activity of MAPKs is tightly controlled by dual phosphorylation, which induces a conformational change rendering the enzyme active. Upon stimulation of cells, MAPKs are phosphorylated on a threonine and a tyrosine residue within the activation loop at the signature sequence TXY (single letter code) by a dual specificity MAPK kinase (MAPKK or MEK). These MAPKKs are in turn phosphorylated on serine and/or threonine residues (S-X_{3-5}-S/T) being activated by the diverse family of MAPKK kinases (MAPKKK or MEKK) (figure 9.1). Thus the signal is transduced in the form of a phosphorylation cascade from upstream kinases to the downstream MAPK. Within this MAPK module (MAPKKK–MAPKK–MAPK) direct enzyme substrate interactions play a critical role: a MAPKKK directly binds a MAPKK, and a MAPKK binds to a MAPK, which itself interacts with distinct protein substrates.

9.2.2 Diversity of MAPK pathways

In a single cell several different MAPK pathways exist in parallel. There are multiple members of each component of the MAPK cascade which are organized in distinct MAPK modules. To maintain specificity of the various MAPK modules protein–protein interactions play a crucial role. MAPK signaling is best understood in the budding yeast *S. cerevisiae* and in mammals, and so examples of these organisms will be discussed first. However, analogous mechanisms do appear to operate in plants.

Presently, five MAPK pathways have been characterized in *S. cerevisiae* (for an overview see figure 9.2). A specific extracellular stimulus activates a

Activator

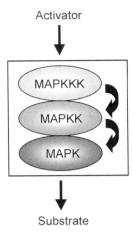

Substrate

Figure 9.1 The MAPK module. Extra- and intracellular signals can activate MAPKKKs, inducing a set of phosphorylation events. First, the MAPKKK phosphorylates and thus activates the corresponding MAPKK, which in turn phosphorylates and activates MAPK. At the other end of the cascade, the active MAPK can activate a number of substrates, giving rise to the cellular response to a signal.

particular MAPK module which signals independently to initiate a unique physiological response (for reviews see, for example, Herskowitz, 1995; Widmann *et al.*, 1999). However, some components of these modules can participate in more than one signal transduction pathway. For example, stimulation of haploid cells with pheromone leads to activation of the STE11(MAPKKK)–STE7(MAPKK)–FUS3(MAPK) pathway resulting in cell cycle arrest and transcriptional activation of mating-specific genes. Two components of the mating MAPK module, STE11 and STE7, are also shared with the filamentation pathway activating KSS1 (MAPK). Moreover, STE11 is activated by high osmolarity conditions in the PBS2(MAPKK)–HOG1(MAPK) cascade. However, activation of STE11 by pheromone induces mating-specific genes, but not those of the filamentation pathway or the high-osmolarity response, and *vice versa*. Thus, pathway specificity is maintained even though individual components participate in more than one signaling pathway.

In multicellular organisms like mammals and plants, MAPK signaling is more complex. A specific signal can activate several different MAPK pathways. On the other hand, a particular MAPK module can be activated by a variety of stimuli and cross-talk between different pathways is common (figure 9.3). Additionally, the biological context of a signal plays a determinative role on the final cellular response.

The number of mammalian MAPK modules is substantial (for a review see, for example, Widmann *et al.*, 1999). Only a few will be introduced briefly to illustrate some basic features of the highly complex mammalian MAPK signaling

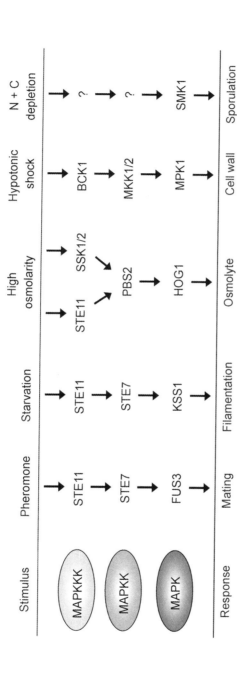

Figure 9.2 Schematic overview of yeast MAPK modules. In *Saccharomyces cerevisiae* MAPK cascades regulate mating, filamentation, high-osmolarity response, cell-wall remodelling and sporulation.

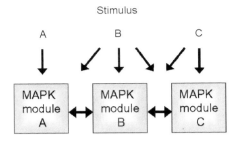

Figure 9.3 Schematic representation of signal divergence, convergence and cross-talk. A specific stimulus can activate a single or several MAPK pathways. A particular MAPK cascade can be activated by different stimuli. Communication between MAPK pathways is common and can occur at different levels.

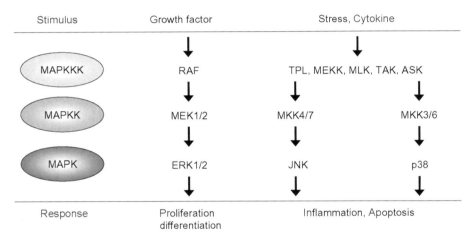

Figure 9.4 Schematic overview of mammalian MAPK cascades. MAPK pathways mediate signaling of diverse extracellular stimuli and regulate cell growth, differentiation and stress responses.

without discussing cross-talk and multisignal activation (figure 9.4). Extracellular signal-regulated kinases (ERK1 and ERK2) are involved in cell proliferation, differentiation, cell cycle control and cell survival. ERK1 and ERK2 are activated by MEK1 and MEK2 (MAPKK) which in turn are predominantly activated by Raf (MAPKKK). However, several other MAPKKKs can also be part of the ERK MAPK module. In response to diverse stresses a MAPK cascade, comprising different MAPKKKs, MKK4 and MKK7 (MAPKK) and JNK (c-jun *N*-terminal kinase; MAPK), is activated. Stress and cytokines also lead to activation of a pathway composed of p38 (MAPK)—MKK3/6 (MAPKK)—and different MAPKKKs.

Conceptually, MAPKKs display a high specificity for a distinct MAPK subtype, whereas MAPKKKs regulate various MAPKKs. MAPKKKs are structurally highly divergent and contain different regulatory motifs. Thus MAPKKKs can be regulated by a wide range of different stimuli and are able to integrate various signals to MAPK modules. Additionally, MAPKKs may regulate pathways not involved in MAPK signaling.

9.2.3   Protein–protein interactions in MAPK modules

Recognition between MAPKs and the interacting molecules involves transient enzyme–substrate interaction at the active centre as well as docking interactions that help to regulate the efficiency and specificity of the enzymatic reactions. Analysis of chimeric molecules between growth factor-stimulated ERK and stress-activated p38 have revealed that distinct domains confer specificity to upstream activation and to particular substrates. A chimera consisting of the *N*-terminal region of p38 and the *C*-terminal region of ERK can be activated, like p38, by stress and can phosphorylate an ERK-specific substrate (Brunet and Pouyssegur, 1996).

MAPKs phosphorylate an array of targets both in the cytoplasm as well as in the nucleus leading to the cellular response to the input signal. Cytoplasmic ERK can translocate to the nucleus after activation. In the inactive state, ERK associates with MEKs which anchor the complex in the cytoplasm (Fukuda *et al.*, 1997). Upon phosphorylation of ERK, ERK dissociates from MEK and dimerizes (Khokhlatchev *et al.*, 1998; Adachi *et al.*, 1999). ERK dimers are subsequently translocated to the nucleus and phosphorylate nuclear substrates, most importantly transcription factors.

In the cellular context, the magnitude as well as the duration of MAPK activity determines the physiological outcome of the pathway. The kinetics of MAPK activity is regulated by a balance of upstream kinases, which phosphorylate and activate MAPKs and phosphatases, dephosphorylating and thus inactivating MAPKs. Since phosphorylation of both threonine and tyrosine is necessary for MAPK activity, dephosphorylation of either residue is sufficient for inactivation (Keyse, 2000). Inactivation is achieved by induction of feedback inhibition mechanisms including the action of tyrosine phosphatases, serine–threonine phosphatases as well as dual-specificity phosphatases. Similar to the kinase-substrate binding, specific protein–protein interactions determine the phosphatase-substrate specificity.

9.2.4   MAPK complexes and scaffolds

Complex formation is an important mechanism to regulate MAPK signal transduction pathways. Signaling molecules can directly bind to each other or they are tethered by scaffold proteins. Scaffold proteins can bind several signaling molecules bringing them into close vicinity, and thus facilitate optimal information flow.

In *S. cerevisiae*, the STE5 scaffold protein binds components of the mating-response MAPK module (figure 9.5). Distinct regions of STE5 interact with STE11, STE7 and FUS3 and link the kinase cascade to upstream activators (Choi *et al.*, 1994; Printen and Sprague, 1994; Marcus *et al.*, 1994). Similar to STE5, PBS2 functions to organize a multicomponent signaling complex. PBS2 coordinates the osmoregulatory MAPK module binding STE11, HOG1 and the upstream osmosensor SHO1 (Posas and Saito, 1997). In contrast to STE5, which is an accessory protein, PBS2 is both a component of the MAPK cascade, transducing the high-osmolarity signal, and a scaffold protein binding the different components of the MAPK module (figure 9.5). The co-localization of successive members of a MAPK cascade by scaffold proteins does not only enhance the efficiency of enzyme substrate interactions by increasing the local concentration, but also prevents illegitimate cross-talk with components of other MAPK pathways. The latter is clearly demonstrated by STE11 which is shared by distinct MAPK pathways (figure 9.2). In response to pheromone STE5 restricts STE11 to phosphorylate STE7 leading to FUS3 activation and subsequently to mating. However, in response to high extracellular osmolarity, PBS2 binds STE11 and HOG1 resulting in osmolyte synthesis.

Scaffold proteins not only assemble distinct components but also enhance the efficiency of signal propagation of specific MAPK modules (figure 9.5). JNK-interacting protein 1 (JIP1) selectively binds JNK, MKK7 and MLK3 (MAPKKK) as well as an upstream activator and facilitates signaling by the bound kinases (Whitmarsh *et al.*, 1998). In contrast, MEKK1 (MAPKKK) tethers JNK and MKK4 in a PBS2-like manner (Xu and Cobb, 1997). MP1 is distinct from JIP1 and MEKK1. Although MP1 only associates with ERK1 and MEK1, but not with the upstream MAPKKK, it still enhances the activation of ERK1 (Schaeffer *et al.*, 1998).

Scaffold proteins do not have any obvious similarities in their primary structures and they can be enzymatic components or accessory proteins of MAPK modules. Their most important property might be routing signals through the 'jungle' of signal transduction.

9.2.5 Plant MAPK pathways

Different MAPKs, MAPKKs and MAPKKKs have been identified that constitute various MAPK cascades in plants (Ligterink and Hirt, 2001). Similar to mammals, MAPK signaling is highly complex, involving cross-talk between pathways and feedback mechanisms. A particular MAPK can be activated by various stimuli and a specific stimulus can activate different MAPKs. MAPK pathways are activated in response to environmental challenges as well as by

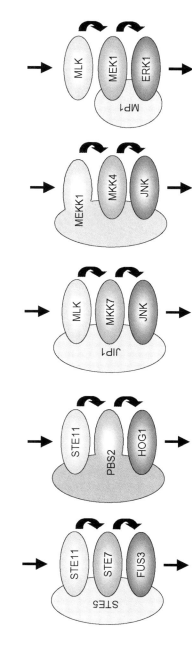

Figure 9.5 MAPK scaffold complexes. In *Saccharomyces cerevisiae* STE11 can function in both the mating and the high osmolarity pathway. For the mating pathway, STE5 tethers FUS3, STE7 and STE11 into a complex while, for the high osmolarity pathway, PBS2 binds STE11 and HOG1. In mammals JIP1 tethers JNK, MKK7 and MLK into a complex. MEKK1 binds MKK4 and JNK, and MP1 associates with MEK1 and ERK1.

hormones and during cell division, emphasizing their importance in plant signal transduction (Jonak *et al.*, 1999; Meskiene and Hirt, 2000).

From previous studies, stress-induced MAPK pathways are best understood and will be discussed here as an example. Data of different plant species show that at least two MAPK pathways are activated by various types of stress. The MAPKs SIMK and SAMK from *Medicago sativa* and their orthologs from tobacco and *Arabidopsis* are activated in response to pathogen-associated stimuli as well as by abiotic stresses, including mechanical stimulation, wounding, drought and cold (Jonak *et al.*, 1999; Mizoguchi *et al.*, 2000; Hirt and Scheel, 2000; Seo and Ohashi, 2000; Zhang and Klessig, 2000). SIMK and SAMK are both involved in MAPK cascades mediating these diverse stresses whereas only SIMK activity is induced by hyperosmotic stress (Munnik *et al.*, 1999) (figure 9.6). Plants use sophisticated mechanisms for pathogen recognition and induction of defence responses. Thus it is not surprising that two additional alfalfa MAPKs, MMK2 and MMK3, are insensitive to abiotic stimuli but respond to pathogen-derived elicitors (Cardinale *et al.*, 2000). The involvement of several MAPK pathways in pathogen response might contribute to the flexibility of plants towards the multitude of pathogens to induce the appropriate response by regulation of gene expression and cytoplasmic targets.

Interestingly, MMK3 not only plays a role in pathogen response but also in cytokinesis. In dividing cells, MMK3 and the tobacco ortholog Ntf6 are active in anaphase and telophase and localize to the cell plate (Calderini *et al.*, 1998; Bögre *et al.*, 1999). NPK1, a tobacco MAPKKK, co-localizes with MMK3/Ntf6 and is essential for cell plate formation in dividing cells (Nishihama *et al.*, 2001).

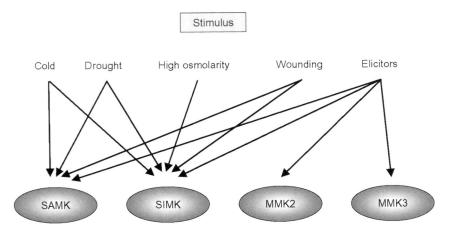

Figure 9.6 Stress-responsive MAPKs in *Medicago sativa*. Cold, drought, hyperosmotic stress, wounding and elicitors differentially activate the MAPKs SAMK, SIMK, MMK2 and MMK3.

Thus NPK1 may act upstream of Ntf6 in the same MAPK module. Similar to mammalian MAPKKKs, NPK1 seems to participate in distinct MAPK pathways. Overexpression of a constitutively active version of ANP (*Arabidopsis* ortholog of NPK1) activates two stress-related MAPKs of the SIMK and SAMK family and induces stress-responsive genes but inhibits auxin-induced gene expression (Kovtun *et al.*, 2000).

Direct protein–protein interactions are also important for constituting plant MAPK modules. For example, SIMKK, a dual-specific MAPKK that phosphorylates SIMK on the regulatory threonine and tyrosine residues and mediates salt-induced activation of SIMK, binds directly to SIMK. A MAPK docking site at the *N*-terminus of SIMK promotes but is not sufficient for MAPKK interaction (Kiegerl *et al.*, 2000).

Signal transduction by MAPK cascades is a universal tool to transmit an enormous variety of signals. MAPK pathways are highly complex using specific components for transmission of different stimuli. Combinatorial activation of several MAPK pathways by the same stimulus may determine the physiological response. Even though different organisms have adapted MAPK modules for specific needs the principal mechanisms appear to be evolutionarily conserved. MAPKs are organized into higher order complexes including protein kinases, phosphatases and non-catalytic proteins, which are critical for proper information transmission by reversible protein phosphorylation.

9.3 Cyclin-dependent protein kinases: the central regulators of cell division

Cell division is a highly complex process, and so our understanding was originally limited to simple unicellular eukaryotes such as budding and fission yeast. However, once it was clear that the basic molecular players of cell division were highly conserved in all eukaryotes, attention was soon focused on both the animal and plant kingdoms. However, due to differences in the life cycles of various organisms, specific adaptations and modifications had to be integrated into the cell division machinery (Hirt and Heberle-Bors, 1994). Compared with animals, for example, the body plan of a plant is composed of cells that are layered, and is not due to migration of certain populations of cells with respect to each other, which is a common feature of animal development.

Another difference between animals and plants can be found in mitosis. Before spindle formation, a specific mirotubular structure called the pre-prophase band is laid down and sets the plane of subsequent cell division. Later in mitosis, when cytokinesis occurs, another plant-specific microtubule array, the phragmoplast, appears and forms the cell wall between the newly generated daughter cells. Mutants in either the pre-prophase band or the phragmoplast have been identified and show strong developmental defects.

During normal development, most cell division is restricted to certain regions in the plant, the meristems. In contrast to most animal cells, however, many plant cells retain a certain degree of totipotency, allowing regeneration of entire organs or even complete embryogenesis to occur. At the heart of the decision of a cell to divide are the CDKs. Plants contain several types of CDKs and associated proteins, allowing cells to proceed though the cell cycle in a manner that can be stopped whenever necessary (Jacobs, 1995; Doonan and Fobert, 1997; Mironov *et al.*, 1999). The cell cycle consists of four phases: G1, S, G2 and M (figure 9.7). Whereas S and M represent DNA synthesis and mitosis, respectively, G1 and G2 are gap phases for growth and repair. A cell progresses through the cell cycle in a stepwise manner. Checkpoints ensure that one phase is completed before entry into the next one is allowed (Murray, 1992). Entry into the S and the M phases corresponds with activity peaks of certain CDK complexes (figure 9.7, Nurse, 1997). The CDK complexes minimally consist of a catalytic kinase subunit (the CDK) and a regulatory cyclin subunit.

In all higher eukaryotes, including plants, several types of CDKs are present and function at different points in the cell cycle. All CDKs have eleven conserved domains, labeled with Roman letters I–XI, found in all other serine–threonine protein kinases (Hanks and Hunter, 1995). However, several structural motifs allow classification of a protein kinase to be a member of the CDK family. The CDK signatures include the GEGTYG motif in subdomain I which is involved in ATP binding. In subdomain III, the so-called PSTAIRE motif (EGVPSTAIREISLLKE) is found. Moreover, subdomain VI contains the HRDLKPQN and subdomian VIII the WYRAPE motifs.

CDKs are enzymes that catalyze the transfer of a phosphate from ATP to specific serine or threonine residues of protein substrates. The consensus

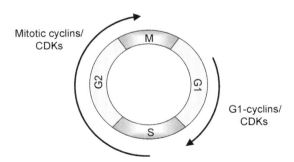

Figure 9.7 The cell cycle. The cell cycle consists of S (DNA synthesis) and M (mitosis) separated by the gap phases G1 and G2, during which growth and repair occurs. Under unfavorable conditions, cells become arrested in G1 or G2. Besides other regulation, entry into S phase and M phase requires activation of CDKs by either G1 or mitotic cyclins, respectively.

phosphorylation motif for all CDKs is S/T-P-X-Z, where X is a polar and Z a basic amino acid. Phosphorylation can occur at this motif on either the serine or the threonine. Phosphorylation changes the structure and thereby the activity of the respectively modified proteins and drives cells through the cell cycle checkpoints into the next phase.

The prototype CDK has been studied in fission yeast where it is called Cdc2. Most of the knowledge gained on the function of Cdc2 also holds true for higher eukaryote CDKs (Nurse, 1997). Mutants in Cdc2 stop the cell cycle at either the G1–S or the G2–M phase transition, indicating that Cdc2 functions at the onset of DNA synthesis as well as mitosis (Nurse, 1991). Budding and fission yeast only possess one type of CDK that is responsible for all cell-cycle transitions (Nurse, 1997). In higher plants, several types of specialized CDKs are present (Doonan and Fobert, 1997; Mironov *et al.*, 1999). The plant CDC2A kinases are most similar to the yeast CDKs and also contain a PSTAIRE motif. However, other CDK variants exist that have either a PPTALRE or a PPTTLRE motif and appear to perform distinct cell-cycle phase-specific functions (see below). The importance of the PSTAIRE domain lies in the fact that these motifs are necessary for interaction of CDKs with the regulatory proteins of the cyclin class which are major determinants of CDK substrate specificity.

The plant CDC2A kinases show maximal kinase activity at the G1–S and G2–M transitions (Doonan and Fobert, 1997; Mironov *et al.*, 1999). In contrast, the CDC2B kinases are less well understood. CDC2B has several subtypes and shows regulated expression in S, G2 and M phases (Fobert *et al.*, 1996; Magyar *et al.*, 1997; Segers *et al.*, 1996). Yet another class of CDKs exists and contains the CDK-activating kinase (CAK) (Sauter, 1997).

9.3.1   Cyclins: regulatory subunits of CDKs

In fission and budding yeast, one single CDK is able to perform two very different functions. At one developmental stage, the CDK initiates DNA synthesis, whereas at another stage of development, the same CDK initiates mitosis. This can be reconciled by the fact that CDKs do not function alone but as protein kinase complexes in association with regulatory subunits, called cyclins. Because the CDK catalytic subunits are almost completely inactive as protein kinases, no non-target proteins can become phosphorylated in the absence of the appropriate cyclin. It is clear that cyclin expression, abundance and localization are key regulatory events in the lifetime of a cell and so cyclin expression and abundance are under tight cell-cycle phase-specific control (Nurse, 1997). Moreover, the localization of cyclin proteins is regulated by yet other control elements, coupling cell division to a highly complex network of signal transduction events (Yang and Kornbluth, 1999).

In yeast, several genes have been uncovered that either function in G1, therefore called 'G1 cyclins', or in S and M, loosely called 'mitotic cyclins'.

Yeast and higher eukaryotes share mitotic cyclins, also called A and B type cyclins, but show distinct families of G1 cyclins. In plants, the G1 cyclins that have been identified so far all belong to one family of proteins that is related to the mammalian G1 cyclins of the D type (Mironov *et al.*, 1999).

As mentioned above, the structure of mitotic cyclins is considerably conserved and contains two key features, a cyclin box that is involved in binding to the respective CDK partner, and an N-terminally located motif called the destruction box, ensuring the degradation of the mitotic cyclins by ubiquitin-mediated proteolysis and thereby exit from mitosis into the next G1 phase. The molecular mechanism that regulates the degradation of mitotic cyclins has not yet been identified in plants. In yeast, however, the responsible factor is called APC, for anaphase promoting complex, and has been characterized by both biochemical and genetic means (Nasmyth, 1999).

The structures of G1 cyclins are much less well-defined. In addition to cyclin boxes, G1 cyclins contain an element that regulates the degradation at the G1–S transition. These motifs are called PEST motifs for their richness in proline, glutamic acid, serine and threonine residues, and are known to confer instability to a number of other proteins. In animals, a major target of G1 CDKs is RB (retinoblastoma protein) which usually binds the transcription factor E2F and thereby blocks transcription of genes that require E2F activity (figure 9.8). G1 cyclins have an N-terminal L-X-C-X-E motif that mediates binding to RB. CDK phosphorylation of RB leads to dissociation of E2F from RB coupled with subsequent gene activation and entry into S phase. In G1 CDK activity mainly depends on the abundance of cyclin D (figure 9.8). Cyclin D is under transcriptional control and directly dependent on the presence of cytokinins and glucose (Riou Khamlichi *et al.*, 1999), ensuring that cell division only proceeds under sufficient nutrient conditions and growth factor control by the plant.

The localization of cyclins has been extensively studied for mitotic cyclins in mammals and is thought to occur in a similar manner in plants and other eukaryotes (Yang and Kornbluth, 1999). Cyclin A is predominantly nuclear and can interact with at least two CDKs. At the onset of S phase, it interacts with mammalian CDK2 and initiates DNA synthesis. Shortly before entry into mitosis, cyclin A associates with CDK1 and regulates nuclear envelope breakdown, which is a prerequisite for access of cytoplasmically localized cyclin B, which then complexes with CDK1 and drives cells through mitosis.

9.3.2 *Regulation of CDK activity by phosphorylation*

Post-translational modification is one of the major mechanisms of cells to regulate the activity of protein kinases. In contrast to modifying the transcription rate or the stablity of a protein, phosphorylation has the advantage that it can act immediately on a protein that is already present in a cell. Moreover, the

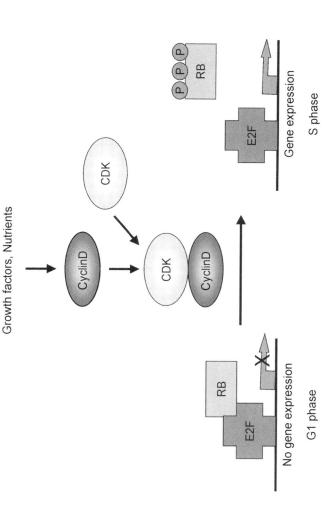

Figure 9.8 CDK regulation of G1–S phase transition. In G1 phase, expression of S phase genes by the transcription factor E2F is blocked by retinoblastoma protein (RB). Growth factors and nutrients induce expression of D-type cyclins. Upon association with cyclin D, CDK becomes activated and phosphorylates RB, resulting in dissociation of RB from E2F and thereby activation of S-phase genes.

decision to activate a protein kinase by phosphorylation is readily reversible through dephosphorylation by phosphatases. It is not surprising that important regulators such as the CDKs are key targets of multiple protein kinases and phosphatases. As before, the fission yeast Cdc2 can be taken as a paradigm for all eukaryote CDKs (Nurse, 1990). Cdc2 is activated by phosphorylation at T_{167}, but inhibited by phosphorylation at T_{14} and Y_{15} (figure 9.9). The enzyme that is responsible for T_{167} phosphorylation is a serine–threonine protein kinase called CAK and is structurally related to the family of CDKs. In fission yeast, T_{167} phosphorylation occurs at the G2–M transition and is sufficient to activate Cdc2 when it is complexed with a cyclin (figure 9.8). However, usually Cdc2 is also phosphorylated at T_{14} and Y_{15} at this cell-cycle phase. The protein kinase that phosphorylates T_{14} and Y_{15} of Cdc2 is called Wee1 and is a dual-specificity protein kinase because it can act on both threonine and tyrosine residues on the same substrate (figure 9.9). In contrast to T_{167} phosphorylation, T_{14} and Y_{15} phosphorylation inhibits the kinase activity of the Cdc2-cyclin complex. The biological rationale behind this complex regulatory mechanism appears to be that Wee1 is the checkpoint controlling whether DNA replication has finished correctly, or if any damage to the DNA has occurred. If DNA damage has occurred, Wee1 becomes activated and phosphorylates the Cdc2–cyclin complex, effectively blocking entry into mitosis until the underlying problems have been solved. However, rendering Wee1 inactive is insufficient for activating an already T_{14}- and Y_{15}-phosphorylated Cdc2–cyclin complex. This job requires the functioning of the dual-specificity phosphatase Cdc25 that dephosphorylates T_{14} and Y_{15} and thereby enables cells to enter mitosis (figure 9.9).

Whereas homologs of CAK, Wee1 and Cdc25 have been identified in animals (Nurse, 1997), only proteins with similarity to CAK and Wee1 have been detected so far in plants (Mironov *et al.*, 1999), suggesting that dephosphorylation and thereby activation of plant CDK–cyclin complexes might be mediated by phosphatases other than a Cdc25.

9.3.3   CDK inhibitors

Proteins of 16–40 kDa have been identified in yeast and animals that associate with CDKs, cyclins or the CDK–cyclin complexes. The outcome of such an association is not a post-translational modification but a block of the protein kinase activity of the CDK. This class of proteins have been named the CDK inhibitors and can be present constantly or show a cell-cycle phase-specific expression pattern. Several types of CDK inhibitors have also been identified recently in plants (Mironov *et al.*, 1999), but only *Arabidopsis* ICK1 has been analyzed in detail, revealing inhibition of CDK activity through its binding to a D-type cyclin (Wang *et al.*, 1998).

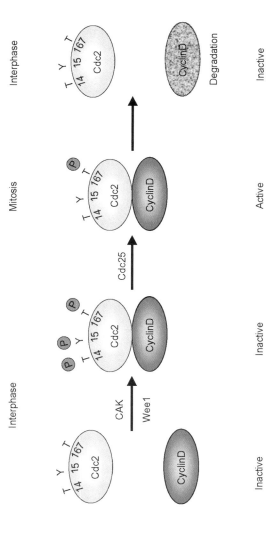

Figure 9.9 CDK regulation of G2–M phase transition. In interphase, Cdc2 is present either as a monomer or associated with cyclin B. CAK phosphorylates Cdc2 on T_{167}, whereas Wee1 phosphorylates Cdc2 on T_{14} and Tyr_{15}. Upon Cdc2 dephosphorylation at T_{14} and Y_{15} by Cdc25, the Cdc2–cyclin complex is activated and cells enter mitosis. Exit of mitosis requires inactivation of Cdc2. This is brought about by degradation of cyclin D and dephosphorylation of Cdc2 at T_{167}.

9.4 Conclusions

Cells as a whole constantly have to measure and respond to a large variety of stimuli. These stimuli can be external factors, including developmental cues, nutrient availability and growth factors, but may also report on the internal cellular state such as the presence of damaged DNA or the chromosomal replication or division status. All this information is transduced by signal transduction cascades that are finally integrated at the level of key regulatory protein kinases. Because almost any condition that affects plant growth also affects decisions regarding adaptation, cell division and development, two of the regulators that are commonly targeted under a variety of these conditions are MAPKs and CDKs. A full understanding of the coordination of cell division, metabolism and development is the challenge of this century and requires new approaches, including cell- or even organism-wide genomic and proteomic data aquisition techniques coupled to new bioinformatic algorithms.

Acknowledgments

This work was supported by grants P14631-GEN, P14114-GEN, and P13535-GEN of the Austrian Science Foundation. C. J. is a Hertha Firnberg-Fellow supported by T-93 of the Austrian Science Foundation.

References

Adachi, M., Fukuda, M. and Nishida, E. (1999) Two co-existing mechanisms for nuclear import of MAP kinase: passive diffusion of a monomer and active transport of a dimer. *EMBO J.*, **18**, 5347-5358.

Bögre, L., Calderini, O., Binarova, P. *et al.* (1999) A MAP kinase is activated late in plant mitosis and becomes localized to the plane of cell division. *Plant Cell*, **11**, 101-113.

Brunet, A. and Pouyssegur, J. (1996) Identification of MAP kinase domains by redirecting stress signals into growth factor responses. *Science*, **272**, 1652-1655.

Calderini, O., Bögre, L., Vicente, O., Binarova, P., Heberle-Bors, E. and Wilson, C. (1998) A cell cycle regulated MAP kinase with a possible role in cytokinesis in tobacco cells. *J. Cell Sci.*, **111**, 3091-3100.

Cardinale, F., Jonak, C., Ligterink, W., Niehaus, K., Boller, T. and Hirt, H. (2000) Differential activation of four specific MAPK pathways by distinct elicitors. *J. Biol. Chem.*, **275**, 36,734-36,740.

Choi, K.Y., Satterberg, B., Lyons, D.M. and Elion, E.A. (1994) Ste5 tethers multiple protein kinases in the MAP kinase cascade required for mating in *S. cerevisiae*. *Cell*, **78**, 499-512.

Doonan, J. and Fobert, P. (1997) Conserved and novel regulators of the plant cell cycle. *Curr. Opin. Cell Biol.*, **9**, 824-830.

Fobert, P.R., Gaudin, V., Lunness, P., Coen, E.S. and Doonan, J.H. (1996) Distinct classes of *cdc2*-related genes are differentially expressed during the cell division cycle in plants. *Plant Cell*, **8**, 1465-1476.

Fukuda, M., Gotoh, Y. and Nishida, E. (1997) Interaction of MAP kinase with MAP kinase kinase: its possible role in the control of nucleocytoplasmic transport of MAP kinase. *EMBO J.*, **16**, 1901-1908.

Hanks, S.K. and Hunter, T. (1995) Protein kinases 6. The eukaryotic protein kinase superfamily: kinase (catalytic) domain structure and classification. *FASEB J.*, **9**, 576-596.

Herskowitz, I. (1995) MAP kinase pathways in yeast: for mating and more. *Cell*, **80**, 187-197.

Hirt, H. and Heberle-Bors, E. (1994) Cell cycle regulation in higher plants. *Sem. Dev. Biol.*, **5**, 147-154.

Hirt, H. and Scheel, D. (2000) Receptor-mediated MAP kinase activation in plant defense. *Results Probl. Cell Differ.*, **27**, 85-93.

Jacobs, T.W. (1995) Cell cycle control. *Annu. Rev. Plant Physiol. Plant Mol. Biol.*, **46**, 317-339.

Jonak, C., Ligterink, W. and Hirt, H. (1999) MAP kinases in plant signal transduction. *Cell Mol. Life Sci.*, **55**, 204-213.

Keyse, S.M. (2000) Protein phosphatases and the regulation of mitogen-activated protein kinase signalling. *Curr. Opin. Cell Biol.*, **12**, 186-192.

Khokhlatchev, A.V., Canagarajah, B., Wilsbacher, J. *et al.* (1998) Phosphorylation of the MAP kinase ERK2 promotes its homodimerization and nuclear translocation. *Cell*, **93**, 605-615.

Kiegerl, S., Cardinale, F., Siligan, C. *et al.* (2000) SIMKK, a mitogen-activated protein kinase (MAPK) kinase, is a specific activator of the salt stress-induced MAPK, SIMK. *Plant Cell*, **12**, 2247-2258.

Kovtun, Y., Chiu, W.L., Tena, G. and Sheen, J. (2000) Functional analysis of oxidative stress-activated mitogen-activated protein kinase cascade in plants. *Proc. Natl Acad. Sci. USA*, **97**, 2940-2945.

Ligterink, W. and Hirt, H. (2001) Mitogen-activated protein (MAP) kinase pathways in plants: versatile signaling tools. *Int. Rev. Cytol.*, **201**, 209-275.

Magyar, Z., Meszaros, T., Miskolczi, P. *et al.* (1997) Cell cycle phase specificity of putative cyclin-dependent kinase variants in synchronized alfalfa cells. *Plant Cell*, **9**, 223-235.

Marcus, S., Polverino, A., Barr, M. and Wigler, M. (1994) Complexes between STE5 and components of the pheromone-responsive mitogen-activated protein kinase module. *Proc. Natl Acad. Sci. USA*, **91**, 7762-7766.

Meskiene, I. and Hirt, H. (2000) MAP kinase pathways: molecular plug-and-play chips for the cell. *Plant Mol. Biol.*, **42**, 791-806.

Mironov, V., De Veylder, L., Van Montagu, M. and Inze, D. (1999) Cyclin-dependent kinases and cell division in plants—the nexus. *Plant Cell*, **11**, 509-521.

Mizoguchi, T., Ichimura, K., Yoshida, R. and Shinozaki, K. (2000) MAP kinase cascades in *Arabidopsis*: their roles in stress and hormone responses. *Results Probl. Cell Differ.*, **27**, 29-38.

Munnik, T., Ligterink, W., Meskiene, I. *et al.* (1999) Distinct osmo-sensing protein kinase pathways are involved in signalling moderate and severe hyper-osmotic stress. *Plant J.*, **20**, 381-388.

Murray, A.W. (1992) Creative blocks: cell-cycle checkpoints and feedback controls. *Nature*, **359**, 599-604.

Nasmyth, K. (1999) Separating sister chromatids. *Trends Biochem. Sci.*, **24**, 98-104.

Nishihama, R., Ishikawa, M., Araki, S., Soyano, T., Asada, T. and Machida, Y. (2001) The NPK1 mitogen-activated protein kinase kinase kinase is a regulator of cell-plate formation in plant cytokinesis. *Genes Dev.*, **15**, 352-363.

Nurse, P. (1990) Universal control mechanism regulating onset of M-phase. *Nature*, **344**, 503-508.

Nurse, P. (1991) Cell cycle. Checkpoints and spindles. *Nature*, **354**, 356-358.

Nurse, P. (1997) Regulation of the eukaryotic cell cycle. *Eur. J. Cancer*, **33**, 1002-1004.

Posas, F. and Saito, H. (1997) Osmotic activation of the HOG MAPK pathway via Ste11p MAPKKK: scaffold role of Pbs2p MAPKK. *Science*, **276**, 1702-1705.

Printen, J.A. and Sprague, G.F. (1994) Protein–protein interactions in the yeast pheromone response pathway: Ste5p interacts with all members of the MAP kinase cascade. *Genetics*, **138**, 609-619.

Riou Khamlichi, C., Huntley, R., Jacqmard, A. and Murray, J.A.H. (1999) Cytokinin activation of *Arabidopsis* cell division through a D-type cyclin. *Science*, **283**, 1541-1544.

Sauter, M. (1997) Differential expression of a CAK (cdc2-activating kinase)-like protein kinase, cyclins and *cdc2* genes from rice during the cell cycle and in response to gibberellin. *Plant J.*, **11**, 181-190.

Schaeffer, H.J., Catling, A.D., Eblen, S.T., Collier, L.S., Krauss, A. and Weber, M.J. (1998) MP1: a MEK binding partner that enhances enzymatic activation of the MAP kinase cascade. *Science*, **281**, 1668-1671.

Segers, G., Gadisseur, I., Bergounioux, C. *et al.* (1996) The *Arabidopsis* cyclin-dependent kinase gene *cdc2bAt* is preferentially expressed during S and G2 phases of the cell cycle. *Plant J.*, **10**, 601-612.

Seo, S. and Ohashi, Y. (2000) Mitogen-activated protein kinases and wound stress. *Results Probl. Cell Differ.*, **27**, 53-63.

Wang, H., Qi, Q., Schorr, P., Cutler, A.J., Crosby, W.L. and Fowke, L.C. (1998) ICK1, a cyclin-dependent protein kinase inhibitor from *Arabidopsis thaliana* interacts with both Cdc2a and CycD3, and its expression is induced by abscisic acid. *Plant J.*, **15**, 501-510.

Whitmarsh, A.J., Cavanagh, J., Tournier, C., Yasuda, J. and Davis, R.J. (1998) A mammalian scaffold complex that selectively mediates MAP kinase activation. *Science*, **281**, 1671-1674.

Widmann, C., Gibson, S., Jarpe, M.B. and Johnson, G.L. (1999) Mitogen-activated protein kinase: conservation of a three-kinase module from yeast to human. *Physiol Rev.*, **79**, 143-180.

Yang, J. and Kornbluth, S. (1999) All aboard the cyclin train: subcellular trafficking of cyclins and their CDK partners. *Trends Cell Biol.*, **9**, 207-210.

Xu, S. and Cobb, M.H. (1997) MEKK1 binds directly to the c-Jun N-terminal kinases/stress-activated protein kinases. *J. Biol. Chem.*, **272**, 32,056-32,060.

Zhang, S. and Klessig, D.F. (2000) Pathogen-induced MAP kinases in tobacco. *Results Probl. Cell Differ.*, **27**, 65-84.

10 Protein–protein interactions in the regulation of plant gene transcription

Dao-Xiu Zhou

10.1 Introduction

Gene expression is regulated largely at the transcription level in plants. Transcriptional regulation is one of the most complex cellular activities, because it involves the integration of various cellular signals by a large number of transcription factors and other regulators. In eukaryotic cells, genomic DNA is complexed with nuclear proteins to form chromatin that allows the packaging of the DNA in the nucleus. The chromatin structure rends the DNA inaccessible to the transcriptional apparatus, leading to general repression of genomic DNA expression. For gene transcription to occur, the chromatin in the vicinity of the gene must be remodeled to provide access for the transcription initiation complex of RNA polymerases and transcription factors play an important role in this process. Transcription factors include a heterogeneous group of proteins that can be functionally divided into several classes (Zawel and Reinberg, 1995). A major class includes those binding specifically to gene promoter and enhancer sequences. These factors can behave as activators or repressors and are largely responsible for specific gene expression. The activities of DNA-specific binding transcription factors are largely regulated by diverse mechanisms so as to integrate cellular signals to promoters and confer specific gene expression profiles. Covalent modifications such as phosphorylation play an important role in regulating transcription factor activities such as dimerization, subcellular localization and transactivation potential.

A second class of transcription factors includes general transcription factors that bind to the TATA box or initiator sequences of the promoters through intensive protein–protein interactions to form the RNA polymerase II pre-initiation complex. Transcriptional activation involves recruitment of the pre-initiation complex directly or indirectly by promoter-specific activators through complex protein–protein interactions. A third class consists of co-activators or co-repressors. These proteins modulate gene specific transcriptional activation or repression by interacting physically with promoter-specific binding factors and/or with components of the initiation complex. In many cases, these factors modulate transcription by remodeling chromatin structure (Brown *et al.*, 2000; Ng and Bird, 2000).

10.2   Plant DNA-binding transcription factors

The analysis of *Arabidopsis* genomic sequence data has revealed that there are more than 1500 transcription factor genes, corresponding to more than 5% of the total genes of this organism. Both the number and the percentage of transcription factor genes in *Arabidopsis* genome are larger than in *Drosophila* and *Caenorhabditis elegans* genomes (Riechmann *et al.*, 2000). Therefore, the regulation of transcription in *Arabidopsis* is at least as complex as in those animal systems. Sequence-specific transcription factors are classified according to their DNA-binding motifs (Luscombe *et al.*, 2000). There are several major classes of DNA-binding motifs including the basic-leucine zipper (bZIP), zinc finger (ZnF), basic-helix-loop-helix (bHLH), MYB (see below), MADS (see section 10.3.4) and helix-turn-helix (HTH or homeodomain), all of which are conserved in different eukaryotic organisms (table 10.1). Plant genomes usually contain large families of transcription factors that have been produced as a result of extensive gene duplication (Riechmann *et al.*, 2000) (table 10.1).

In many cases, duplicated transcription families members are capable of binding to DNA as homo- and heterodimers. Duplicated transcription factor genes are scattered on different chromosomes or distally on the same one, but transcription factor gene clusters are rare in the genome. An initial state of functional redundancy changes as individual member genes acquire altered expression patterns and new functions. The combinatorial interactions of transcription

Table 10.1 Major classes of plant transcription factors

Family	Estimated members in *Arabidopsis*[a]	Known protein dimerization motif	Major physiological functions[c]
Zinc finger (ZnF)			
C_2H_2	105		Broad
C_2C_2	104		Development
WRKY	72[b]	zip domain?	Pathogen resistance
C_3H_1	33		Unknown
MYB	190		Broad
AP2/EREBP	144		Stress response
bHLH	139[b]	helix-loop-helix domain	Broad
NAC	109[b]		Development
Homeodomain	89		Development
MADS	82	K-box	Development
bZIP	81	zip domain	Broad
B3/ARF/ABI3	37[b]	motifs III and IV	Auxin, ABA
Trihelix	28[b]		Broad
EIN3/EIL	6[b]		Ethylene response

[a] According to Riechmann *et al.* (2000).
[b] Plant-specific transcription factors.
[c] Found in many members of the family.

factors in overlapping expression regions define unique regulatory domains within the plant body.

The largest transcription factor family is the MYB group (so-called from its first characterization as a proto-oncogene from *myelo*blastosis virus), with at least 190 members in *Arabidopsis*. Other large families in *Arabidopsis* comprise the bHLH, NAC, C2H2 type of ZnF and APETALA2 (AP2)/EREPB protein families with more than 100 members each. The MADS, homeodomain and bZIP families each have more than 80 members. The amplification of these transcription factors in plants implies their functional redundancy and/or regulatory diversity obtained through differential expression and combinatory interaction. Functional redundancy makes the analysis of transcriptional factors complicated. DNA binding motifs unique to plants, such as AP2, NAC, B3, Trihelix and subfamilies of ZnF such as Dof, WRKY and YABBY have been identified (McCarty and Chory, 2000). Animal genomes similarly encode a number of lineage-specific DNA-binding proteins (Riechmann *et al.*, 2000; McCarty and Chory, 2000). Plant-specific factors were initially discovered almost exclusively in developmental pathways that are unique to plants. For instance, the B3 domain proteins are implicated in pathways regulated by plant hormones such as ABA and auxin (McCarty *et al.*, 1991; Ulmasov *et al.*, 1997a). The AP2 domain factors are involved in plant stress responses as well as in plant floral development (Weigel, 1995). On the other hand, some transcription factors or transcription factor families found in animals and in yeast are absent in plants (Riechmann *et al.*, 2000). This suggests that innovation of transcriptional regulators may have played a role in the evolution of these lineages. Shuffling of eukaryotic DNA-binding domains has also contributed to create novel plant-specific transcription factors during evolution. For instance, in about 50% of *Arabidopsis* homeodomain family members, the homeodomain is followed by a leucine zipper. This would have increased the complexity of protein–protein interactions among plant transcription factors. About 5% of the *Arabidopsis* transcription factors have been functionally or genetically characterized, compared with more than 25% in *Drosophila* and in *C. elegans*. Fewer plant transcription factors have been tested for their biochemical properties and transcriptional function, due to lack of efficient *in vitro* plant transcription systems.

The major DNA-binding transcription factor families identified in plants are now introduced, members of which have been shown to be involved in protein interactions to regulate plant gene transcription.

10.2.1   bZIP proteins

In bZIP transcription factors, protein dimerization and DNA binding are mediated by the bZIP motif, which consists of a region rich in basic amino acids and an adjacent leucine zipper (ZIP) domain that consists of a 4-heptad repeat of hydrophobic and non-polar residues (Hurst, 1995). The ZIP domain is required for the protein dimerization prior to binding to DNA, whereas the basic region binds to the DNA sequence (Ellenberger *et al.*, 1992). Dimerization between different bZIP proteins can generate functional diversity, such as binding specificity and affinity, and potential for transactivation.

In plants, the bZIP proteins are involved in the control of gene expression of diverse physiological processes, such as morphogenesis, organ establishment, seed development and adaptation to environmental conditions (Schmidt *et al.*, 1990; Oyama *et al.*, 1997; Hobo *et al.*, 1999; Després *et al.*, 2000). Individual plants can express a large number of bZIP proteins; for example, in *Arabidopsis*, at least 80 bZIP protein genes have been identified (Riechmann *et al.*, 2000). Plant bZIP proteins have related DNA-binding activities, binding to the nearly universal recognition sequence of an ACGT core. However, the bases flanking the core sequence largely determine the binding affinity and/or the specificity of individual bZIP proteins. Among plant bZIP proteins, three major groups can be distinguished based on DNA binding specificity. The first group prefers the G-box (C-C-A-C-G-T-**G**-G), the second the C-box (T-G-A-C-G-T-**C**-A) and the third binds the G-box and C-box with about the same affinity (Niu *et al.*, 1999). For example, EmBP1, a bZIP protein isolated from wheat by interaction with an abscisic acid responsive element (Guiltinan *et al.*, 1990), is the strongest and most specific G-box-binding protein among the tested bZIP proteins, whereas TGA1a is the most specific C-box-binding protein. TGA1a was cloned from tobacco and originally shown to bind the *as-1* element of the 35S promoter of the cauliflower mosaic virus (Katagiri *et al.*, 1989). The *as-1* element is also involved in auxin and salicylic acid-inducible transcription (Qin *et al.*, 1994; Zhang and Singh, 1994). Most of the G-box binding proteins isolated from *Arabidopsis* (GBFs) and parsley (CPRFs) can form homodimers (Menkens *et al.*, 1995). However, GBF4 from *Arabidopsis* is unique in that its leucine zipper is apparently incapable of forming stable homodimers (Menkens and Cashmore, 1994), in a similar manner to that observed for Fos oncoproteins in mammalian cells (Ryseck and Bravo, 1991). In addition to homodimerization, intra-family heterodimer formation within the GBF family of *Arabidopsis* and the CPRF family of parsley has also been examined. In *Arabidopsis*, the GBF1, GBF2 and GBF3 proteins heterodimerize promiscuously and GBF4 can form heterodimers with GBF2 and GBF3 (Schindler *et al.*, 1992; Menkens and Cashmore, 1994). In the case of the CPRF family, only certain combinations of CPRF1, CPRF2 and CPRF3 can form heterodimers (Armstrong *et al.*, 1992). Therefore, it appears that heterodimer formation within the bZIP family can create new combinations within the plant cell, although the functional significance is largely unknown at this stage. Interaction between bZIP proteins and other classes of transcription factors or protein in regulating specific gene transcription has also been reported and examples will be discussed as appropriate.

10.2.2   The MYB protein family

The MYB DNA binding domain is a region of about 52 amino acids that adopts a helix-helix-turn-helix conformation (Rosinski and Atchley, 1998). In plants, the predominant family of MYB proteins has two repeats (R2, R3) in comparison to the three repeats R1, R2 and R3 in animal c-MYB proteins (Romero *et al.*, 1998). Single MYB domain proteins have also been characterized in plants (Wada *et al.*, 1997). Structural studies of c-MYB have shown that both R2 and R3 are required for sequence-specific binding, with the *C*-terminal helix of each repeat being the recognition helix for DNA binding. It is suggested that the recognition helix of R3 specifically interacts with the core of the recognition sequence, whereas the recognition helix of R2 is involved in less specific interaction with flanking nucleotides to the core sequences (Ogata *et al.*, 1995; Jin and Martin, 1999). There are major differences in the way that MYB proteins bind to DNA with distinct target recognition sequences for different groups of MYB proteins, even within the same groups. Some plant two-repeat MYB proteins can recognize the so-called MBSI site (T-C)A-A-C(G-T)G(A-C-T)(A-C-T), which is bound by mammalian three repeat proteins, while the majority of plant two-repeat proteins bind to the MBSII sequence T-A-A-C-T-A-A-C (Romero *et al.*, 1998). In addition to their well-established roles in DNA binding, MYB domains are also emerging as important protein–protein interaction motifs. In terms of function, MYB proteins belong to different structural groups or subgroups and are unlikely to have similar functions. As mentioned previously, there are a large number of MYB protein genes in plants. Most MYB proteins are supposed to be transcriptional activators with activation domains located in the region *C*-terminal to the DNA-binding domain, as deduced from c-MYB proteins, which contain an acidic activation domains (Weston, *et al.*, 1998). A survey of the predicted *C*-terminal sequences of the *Arabidopsis* R2R3 MYB family revealed over 20 different subgroups (Kranz *et al.*, 1998). These conserved motifs might represent an activation domains, an idea supported by the fact that some are relatively acidic and others are rich in amino acids frequently associated with such domains (proline, glutamine) (Weston, 1999). Alternatively, these regions might represent an interaction domain with other proteins. Interaction between plant MYB proteins has not been reported, but there are a few cases where a MYB protein interacts with different transcription factors and these are discussed below.

10.2.3   bHLH proteins

The basic helix-loop-helix (bHLH) motif was first discovered in regulatory genes of myogenesis (Murre and Baltimore, 1989). The basic region of the bHLH motif is responsible for binding to DNA, while the HLH (helix-loop-helix) region is involved in homo- or heterodimerization with other HLH containing proteins. Differential dimerization of HLH proteins plays an important role in muscle cell differentiation and development in vertebrates and neurogenesis in *Drosophila* (Murre and Baltimore, D., 1989). Plant genomes also encode a large number of bHLH (myc) proteins, with more than 130 in *Arabidopsis* (Riechmann *et al.*, 2000). However, only a few plant bHLH protein genes have been studied functionally. Genetically identified bHLH proteins have been shown to be involved in the regulation of anthocyanin biosynthesis, trichome development and light-inducible gene expression through direct protein–protein interactions with other transcription factors or signaling proteins (Mol *et al.*, 1998; Szymanski *et al.*, 1998; Ni *et al.*, 1998) (see below).

10.2.4   Auxin response factors

Auxin reponse factors (ARFs) are transcription factors that bind with specificity to T-G-T-C-T-C auxin response elements (AuxREs) found in promoters of early auxin response genes (Ulmasov *et al.*, 1997a). ARFs are encoded by a multigene family, consisting of more than 20 members in *Arabidopsis* (Ulmasov *et al.*, 1999b; Riechmann *et al.*, 2000). ARFs possess an *N*-terminal DNA binding domain which has sequence similarity to a *C*-terminal B3 domain found in the maize transcription factor VP1 and its relatives in *Arabidopsis* (Ulmasov *et al.*, 1997a; McCarty *et al.*, 1991; Giraudat *et al.*, 1992). The *C*-terminal domain of ARFs is related to motifs III and IV found in Aux/IAA proteins (Ulmasov *et al.*, 1997a; Kim *et al.*, 1997). The *Aux/IAA* genes are rapidly and specifically induced by auxin (Abel and Theologis, 1996). There are more than 20 members of this gene family in *Arabidopsis*, and many members display distinct basal and induced expression levels and profiles. The Aux/IAA proteins are 20–35 kDa and have four conserved motifs. These proteins contain a domain related to the Arc and MetJ prokaryotic transcriptional repressors, suggesting that they might also function as transcriptional regulators. Motifs III and IV facilitate protein–protein interaction among the members of both ARF and Aux/IAA protein families. ARFs must form dimers on palindromic T-G-T-C-T-C sequence to bind stably, and this dimerization is facilitated by the motif III and IV located in the *C*-terminal domain. It has been suggested that Motif III and IV in ARFs may promote specific interactions among these transcription factors (i.e. homodimers or specific heterodimers) that determine which ARFs bind to an AuxRE target site. Ulmasov *et al.* (1999b) have demonstrated by co-transfection of carrot protoplasts that some members of the ARF family activate transcription of auxin-inducible reporters, whereas others repress expression of these reporter genes. Furthermore, overexpression of some Aux/IAA proteins that are not supposed to bind to DNA represses the transcription from an AuxRE controlled promoter (Ulmasov *et al.*, 1997b). Therefore, the negative function of those Aux/IAA proteins could be achieved through protein–protein interactions with DNA-binding ARFs. Numerous combinations of different Aux/IAA and ARF proteins could account for the modulation of auxin-regulated gene expression in response to changes in auxin concentration and other developmental cues. Genetic evidence supports the importance of Aux/IAA proteins in auxin-induced gene expression. For instance, mutations in the *AXR3* gene that encodes a member of the Aux/IAA family leads to elevated auxin responses, including ectopic auxin-inducible gene expression (Rouse *et al.*, 1998). Recent results have shown that protein degradation by the ubiquitin pathway has a central role in auxin response (reviewed in Gray and Estelle, 2000). Aux/IAA proteins have short half-lives, suggesting that they might be regulated by ubiquitin-mediated degradation.

10.2.5 AP2 domain proteins

The AP2 DNA binding domain was discovered in *APETALA2* (*AP2*), a gene functioning within the ABC model of flower organogenesis in *Arabidopsis* (Jofuku *et al.*, 1994; Weigel, 1995). This DNA-binding domain, which is unique to plants, has been found subsequently in many other plant transcription factors (Okamuro *et al.*, 1997). In *Arabidopsis*, these proteins include TINY (Wilson *et al.*, 1996), AINTEGUMENTA (ANT) (Elliot *et al.*, 1996), the ERE binding protein from *Arabidopsis* (AtEBP) (Buttner and Singh, 1997), *C*-repeat/ dehydration-responsive element (DRE) binding factor 1 (CBF1) (Stockinger *et al.*, 1997), DRE-binding proteins (DREBs) (Liu *et al.*, 1998), abscisic acid (ABA)-insensitive 4 (ABI4) (Finkelstein *et al.*, 1998) and the ethylene-responsive factor 1 (ERF1) (Solano *et al.*, 1998). The ERE1–4 protein from tobacco, Glossy 15 from maize and Pti4, Pti5, and Pti6 from tomato also contain this DNA-binding domain (Ohta *et al.*, 2000; Moose and Sisco, 1996; Zhou *et al.*, 1997). These proteins can be divided into two subfamilies based on the number of DNA-binding domains. The AP2 subfamily includes AP2, ANT and Glossy and each contains two DNA-binding domains. Analysis of mutants has revealed that ANT plays a role in plant organogenesis and in regulating apical meristem activity in *Arabidopsis* (Elliot *et al.*, 1996), while Glossy regulates leaf epidermal cell identity in maize (Moose and Sisco, 1996). The second subfamily, the EREBP subfamily, includes the EREs from tobacco; Pti4, Pti5 and Pti6 from tomato; and TINY, CBF1, DREBs, ABI4, ERF1, AtEBP and other related EREBPs from *Arabidopsis*, each with one DNA-binding domain. Genomic sequencing of *Arabidopsis* has revealed that the double DNA-binding motif-containing AP2 subfamily has at least 12 members, while the EREBP subfamily with single DNA-binding domain is much larger, with more than 120 members (Riechman *et al.*, 2000).

The ERF1, AtEBP and EREBP proteins from *Arabidopsis* or tobacco, and Ptis from tomato bind to the G-C-C box in the ethylene-responsive element found in different promoters, while the CBF1 and DREB proteins bind to a dehydration-responsive element (C-repeat/DRE) that is involved in gene expression in response to drought and cold stress. Therefore, these factors play a role mostly in plant growth and development in response to environmental conditions. Increased expression of *TINY* suppresses cell proliferation and results in a diminutive phenotype (Wilson *et al.*, 1996). Ectopic expression of CBF1 or DREB1 in *Arabidopsis* improves freezing tolerance (Kasuga *et al.*, 1999). Pti4, Pti5 and Pti6 have been shown to interact directly with a tomato disease resistance *R* gene *Pto* product (Zhou *et al.*, 1997), supporting the idea that disease resistance *R* genes may regulate the expression of defense-related genes by regulating the function of DNA-binding proteins. The AtEBP protein has been shown to interact with an octopine synthase (*ocs*) element binding bZIP protein (Buttner and Singh, 1997), suggesting that cross-coupling between EREBP and bZIP transcription factors occurs and may be important in regulating gene expression during the plant defense response. Functional analysis of EREBPs reveals that some of these act as activators of G-C-C-box-dependent transcription in *Arabidopsis*, while others function as transcriptional repressors (Fujimoto *et al.*, 2000).

10.3 Combinatory interactions and the regulation of specific gene expression

Interactions between sequence-specific transcription factors to integrate cellular signals constitutes a major mechanism of regulation of spatial and temporal gene expression in eukaryotes. Many genes are controlled by multiple *cis*-acting promoter and enhancer elements that are specific binding sites of various transcription factors. Differential interactions between transcription factors bound to a given promoter, in response to environmental and cellular signals, lead to distinct expression profiles. As discussed above, many transcription activators can form homo- and/or heterodimers within members of the same family or between different transcription factor family members. Differential combinations of transcription factors bound to a promoter in a given time or a given cell type leads to specific regulation of transcription. Thus extensive duplication of plant transcription factors and their differential expression patterns increases the diversity of combinations of transcription factors.

10.3.1 Regulation of anthocyanin biosynthesis and trichrome development

The control of expression of genes coding for anthocyanin biosynthetic enzymes is a paradigm of synergistic interaction between plant transcription factors (Mol *et al.*, 1998). The biosynthesis of anthocyanins in maize is regulated by both developmental and environmental signals. High intensity light induces anthocyanin biosynthesis in the epidermis of leaves and petioles, whereas anthocyanin expression is developmentally induced in a number of organs such as kernels, cob, seedlings and leaves. Two families of transcription factors control anthocyanin biosynthesis in maize. The first is the C1 or Pl family, which comprise two closely related homologs, while the second is called R or B. The C1/Pl family are MYB domain-containing proteins, which are encoded by the *C1* and *Pl* loci (Paz-Ares *et al.*, 1987; Cone *et al.*, 1993). The R/B family of transcription factors, which are encoded by the *R* and *B* loci, contain the bHLH DNA-binding motif and are highly homologous to each other (Ludwig and Wessler, 1990). The members of each family exhibit different tissue-specific expression patterns. Genetic studies and transient expression assays reveal that individual family members alone are not sufficient to induce anthocyanin biosynthetic gene expression, but that it also requires the presence of a member from each family in the cell. Combinatory interactions between differentially expressed members of these two distinct classes of factors, therefore, define developmentally regulated anthocyanin production profiles. The combination of C1 and R induces pigmentation in the kernel, while B and Pl together are responsible for pigmentation in mature tissues. DNA sequences required by C1/Pl- and R/B-induced expression have been identified in the promoters of anthocyanin biosynthetic genes. Using the yeast two-hybrid system, it was found that a B factor could interact with the DNA-binding domain of C1. This interaction is mediated by the MYB domain of C1 and the *N*-terminal region of B (Goff *et al.*, 1992). Residues in the first helix of the R3 MYB repeat of C1 are sufficient for the specificity of interaction with R/B proteins (Grotewold *et al.*, 2000). However, this interaction seems not to be required to increase the DNA-binding specificity of C1, since C1 alone is able to bind to a specific site within the promoter of the *A1* gene that encodes an anthocyanin biosynthetic enzyme (Sainz *et al.*, 1997). One possibility is that the C1/R or B interaction is needed to constitute a transactivation activity either by recruiting a co-activator or by relieving inhibition of the transactivation activity within C1 protein which has been shown to have a strong activation domain (Sainz *et al.*, 1997).

The maize *P* gene (not to be confused with the *Pl* gene) controls the accumulation of 3-deoxy flavonoids and red phlobaphene pigments by activating a subset of the anthocyanin biosynthetic genes that are also controlled by *C1* and *R/B* genes. The P protein is homologous to C1 and Pl MYB proteins. P and C1 activate the expression of some common genes of anthocyanin biosynthesis such as *A1* by binding to same promoter elements of the *A1* gene (Grotewold *et al.*, 1994; Sainz *et al.*, 1997). In contrast, C1, but not P, binds to and activates (in the presence of the R/B protein) transcription of the *Bz1* gene that is specific to anthocyanin biosynthesis (Grotewold *et al.*, 1994; Sainz *et al.*, 1997). However, activation of the *A1* gene by P is independent of the presence of the R/B proteins. Indeed, P can not interact with R or B. Substitution of several predicted solvent-exposed residues in the first helix of the second MYB repeat of P with corresponding residues from C1 is sufficient to confer onto P the ability to physically interact with the bHLH protein R and, as a consequence, to activate the *Bz1* promoter that is normally regulated by C1 and R (but not by P) (Grotewold *et al.*, 2000). These data indicate that the MYB domains determine regulatory specificity and play an important role in combinatory interactions with other transcription factors.

Genetic analysis of trichome development in *Arabidopsis* has led to the identification of several loci involved in leaf trichome initiation and morphogenesis (Marks, 1997; Szymanski *et al.*, 2000). Among those loci are *GLABROUS1 (GL1), GLABROUS3 (GL3)*, and *TRANSPARENT TESTA GLABRA1 (TTG1)*. Loss-of-function mutations in *GL1* and *TTG1* genes result in a nearly complete loss of leaf trichome initiation, while mutations in *GL3* are less severe. These mutants still produce trichomes, but trichome branching is inhibited. These key genes of trichome development encode a MYB transcription factor (GL1) (Oppenheimer *et al.*, 1991), a bHLH protein (GL3) (Payne *et al.*, 2000) and a WD-40 repeat-containing protein (TTG1) (Walker *et al.*, 1999) that resembles the regulatory triad of genes that controls anthocyanin biosynthesis in petunia (de Vetten *et al.*, 1997; Mol *et al.*, 1998). Almost all of the *TTG1* coding sequence consists of WD-40 repeats; therefore, the function of TTG1 is likely to be the recruitment or assembly of other regulatory factors including GL1 and GL3. The GL1 MYB transcription factor is expressed transiently in developing trichomes. Overexpression of GL1 alone is not sufficient for substantial ectopic trichome formation, but can cause widespread trichome initiation if the maize bHLH-containing *R* gene is co-expressed. This is consistent with the finding that *GL1* and *R* physically interact *in vitro* (Szymanski *et al.*, 1998).

Recent experiments using the yeast two-hybrid assay have detected interactions between GL3 and GL1, and TTG1 with itself. GL1 and TTG1 do not interact, while GL3 has the potential to act as a bridging molecule between GL1 and TTG1 (Payne *et al.*, 2000). Different *N*-terminal domains of GL3 can bind to GL1 and TTG1 (Payne *et al.*, 2000). Therefore, the assembly of the GL1–GL3–TTG1 complex is needed for normal trichome initiation. It is suggested that interaction between this complex with additional factors leads to distinct activities that also negatively regulate trichome initiation (Szymanski *et al.*, 2000).

10.3.2 *Abscisic acid-regulated gene expression*

The phytohormone abscisic acid (ABA) regulates a number of processes in plants in response to a number of abiotic stresses, as well as regulating the expression of many genes during seed development (Bonetta and McCourt, 1998; McCarty, 1995; Busk and Pages, 1998). Promoter analysis has revealed *cis*-acting promoter sequences required for ABA-induced gene transcription. These *cis*-acting sequences are called ABREs (for ABA-responsive elements) (reviewed in Busk and Pages, 1998). The first ABRE to be identified was an A-C-G-T-containing sequence in the wheat *Em* gene (Guiltinan *et al.*, 1990).

The ABREs have now been identified in the promoters of many other ABA-responsive genes. There are also ABA-responsive genes that do not have A-C-G-T-containing ABREs and other promoter elements have been shown to be responsive to ABA in gene transcription.

The bZIP protein EmBP1 that specifically binds to the ABRE was identified initially (Guiltinan *et al.*, 1990). Subsequent gel retardation experiments have revealed that the heterodimer formed by EmBP1 and osZIP1, a rice bZIP protein, shows enhanced binding to the *Em* promoter, but the formation of heterodimers between EmBP1 and osZIP-2a or osZIP-2b prevents binding of EmBP1 to the promoter sequence (Nantel and Quatrano, 1996). In addition to osZIP-1, histone H1 also enhances the binding of EmBP1 to its binding sites *in vitro* (Schultz *et al.*, 1996). This suggests that the DNA-binding activity of EmBP1 is regulated differentially by interacting with other DNA-binding proteins. Furthermore, EMBP1 may interact with other proteins to form a large complex on the ABRE to regulate *Em* gene transcription. This complex includes VIVIPAROUS1 (VP1) and the GF14 proteins (Schultz *et al.*, 1998). GF14s are plant 14-3-3 proteins that were identified initially in a G-box binding complex (see chapter 3). GF14, which is unable to bind DNA, may provide a structural link in the ABA-responsive protein–DNA complex. It is suggested that transcription factors interact with a GF14 *N*-terminal dimerization domain, and it has been shown that GF14 proteins interact directly with general transcription factors such as TATA binding protein (TBP), hTAFII32 and TFIIB (Pan *et al.*, 1999). GF14 proteins may therefore participate in the regulation of transcription as a co-activator by bridging sequence-specific DNA binding proteins to the pre-initiation complex.

VP1 is a key regulator of seed maturation and germination by mediating certain ABA responses (McCarty, 1995). Severe *vp1* mutations cause premature germination of seed on the maize cob (McCarty *et al.*, 1991). All *vp1* mutants are insensitive to ABA. Analyses of truncated forms of VP1 have shown that VP1 contains a DNA-binding domain (the B3 domain that has been identified in other plant transcription factors such as ARFs, Suzuki *et al.*, 1997). However, the B3 domain is not required for gene activation mediated through the ABRE. Indeed, VP1 is unable to bind directly to the ABRE, but it can transactivate the *Em* promoter through the ABRE sequences, presumably via interaction with other proteins such as EmBP1 (Schultz *et al.*, 1998). A peptide consisting only of the B2 domain of VP1 enhances *in vitro* the DNA-binding activity of several bZIP proteins including EmBP1 (Hill *et al.*, 1996). Therefore, VP1 appears to act by facilitating the binding of EmBP1 to the ABREs. However, *in vivo* footprinting studies have shown strong protein binding to the ABREs of a promoter independent of VP1, suggesting that VP1 action does not involve changes in protein–DNA interactions (Busk and Pages, 1997). The deletion of the B2 domain from VP1 abolishes transactivation of the *Em* promoter (Hill *et al.*, 1996). Analysis of mutant alleles of the *Arabidopsis* ortholog of VP1, namely

ABI3, suggests that the B2 domain is involved in the activation of *Arabidopsis* seed protein genes (Bies-Etheve *et al.*, 1999). It has been suggested that the B2 domain is likely to be involved in tethering VP1/ABI3 to the complex at the ABRE, and may mediate protein–protein interactions. Using a yeast two-hybrid screen, a rice bZIP protein named transcription factor responsive for ABA regulation 1 (TRAB1) is shown to interact with VP1. TRAB1 interacts with a segment of VP1 containing the B1 and B2 domains (Hobo *et al.*, 1999).

In addition to the ABRE-containing *Em* promoters, VP1 can activate the expression of the maize *C1* gene apparently with a distinct mechanism. The Sph-promoter element of the *C1* gene is sufficient for VP1-mediated activation (Kao *et al.*, 1996). In fact, VP1 recognizes the Sph-element through the DNA-binding B3 domain and activates transcription via an activation domain (the A domain) located in the *N*-terminus (Suzuki *et al.*, 1997). This activation domain is required for the Sph-dependant activation of VP1, since its deletion abolishes transcriptional activation in transient expression assays (McCarty *et al.*, 1991). However, the full-length VP1 is unable to bind to the Sph element. This suggests that other regions of VP1 inhibit the DNA binding activity of the B3 domain and that this inhibition may be eliminated by interaction with other factors. Mutations of *vp1* disrupting the B3 DNA-binding domain block expression of the Sph-controlled *C1* gene, but undergo normal seed maturation and express the *Em* genes. Thus, VP1 activates the expression of *C1* through direct promoter-binding and transcriptional activation, while it activates *Em* gene expression by interacting with an ABRE-binding protein.

Recently, a screen with the two-hybrid system in yeast has led to the identification of several proteins that interact with an ABI3 derivative containing B2 and B3 domains (Kurup *et al.*, 2000). These ABI3-interacting proteins include ZnF proteins and the RNA polymerase II subunit RPB5. It has been shown that yeast RPB5 affects transcriptional activation at specific promoters. Human RPB5 has been shown to bind to transcriptional activators, TFIIB and TAF$_{II}$, and to interact with the hepatitis B virus X activator protein HBx (for references, see Kurup *et al.*, 2000). Therefore ABI3 would act as a cofactor bridging DNA-binding transcription to components of the RNA polymerase II containing initiation complex.

10.3.3 Regulation of storage protein gene transcription

Zeins are the major class of prolamin seed storage proteins of maize. These proteins are encoded by five distinct classes of genes distinguished on the basis of the molecular mass of their gene product. The expression of zein genes is limited to the endosperm during seed development. Mutations at the *opaque2* locus lead to a reduction of transcription of a specific set of zein genes, and a corresponding decrease of 22 kDa and 15 kDa zein proteins (Aukerman and Schmidt, 1994). *Opaque2* encodes a bZIP transcription factor that binds to a promoter element in the 22 kDa class of zein genes to activate their transcription (Schmidt *et al.*, 1990). The Opaque2 binding site also exists in the 15 kDa class of zein genes (Cord-Neto *et al.*, 1995). However, other zein genes do not contain the Opaque2 binding site, suggesting that additional regulatory elements exist that, in concert, control the activation of all classes of zein genes during storage protein accumulation. The prolamin box (P-box) would play such a role because it is present within the promoters of all zein genes (Vicente-Carbajosa *et al.*, 1997). The P-box in the 22 kDa zein gene promoter is located 20 nucleotides upstream of the Opaque2 binding site. Transient expression assays suggest that the P-box plays a positive role in the coordinated activation of zein gene expression during endosperm development (Quayle and Feix, 1992). A maize endosperm-specific transcription factor, named prolamin-box binding factor (PBF), binds specifically to the P-box (Vicente-Carbajosa *et al.*, 1997). This factor belongs to the Dof class of plant Cys_2-Cys_2 zinc-finger DNA binding proteins (Yanagisawa, 1996) and PBF interacts *in vitro* with Opaque2 (Vicente-Carbajosa *et al.*, 1997). However, it is not known whether this interaction leads to cooperative DNA-binding activity or an increase of transactivation potential of the target genes. Dof proteins have been shown to interact specifically with bZIP proteins and this interaction results in the stimulation of bZIP proteins binding to DNA target sequences in plant promoters (Chen *et al.*, 1996).

10.3.4 Regulation of flower development

Genetic and molecular analysis in *Arabidopsis* and *Antirrhinum majus* have identified several transcription factors that specify floral organ identity. These studies led to the proposal of the ABC model (Ma, 1994, Weigel and Meyerowitz, 1994). In this model, the A, B and C classes of genes function combinatorially in each of the four whorls of floral organs to specify the identity of sepals (whorl one, A functions alone), petals (whorl two, B functions), stamens (whorl three, B and C function) and carpels (whorl four, C functions alone). In *Arabidopsis*, the known A function genes are *APETALA1* (*AP1*), *APETALA2* (*AP2*) and *LEUNIG* (*LUG*), B function genes are *APETALA3* (*AP3*) and *PISTILLATA* (*PI*) and a C function gene has been identified as *AGAMOUS* (*AG*). On the basis of genetic and molecular data, the *Antirrhinum* floral identity genes *SQUAMOSA* (*SQUA*), *DEFICIENS* (*DEF*), *GLOBOSA* (*GLO*) and *PLENA* (*PLE*) are considered as the orthologs of *AP1*, *AP3*, *PI* and *AG*, respectively (Ma, 1994). *AP1* (*SQUA*), *AP3* (*DEF*), *PI* (*GLO*) and *AG* (*PLE*) are members of the MADS box gene family. The name MADS box was derived from the initially identified members: *M*CM1 (a transcription factor from yeast), *AG*, *DEF* and *S*RF (a serum-responsive transcription factor from human) (Schwarz-Sommer *et al.*, 1990). MADS box genes from *Arabidopsis* and *Antirrhinum* with the same floral identity function are more similar to each other than they are to other MADS box genes of different functions in the same species (Purugganan, 1995). MADS box genes have also been identified in monocotyledons and in gymnosperms (Ma and dePamphilis, 2000).

The MADS box is required for DNA binding (Schwarz-Sommer *et al.*, 1992). Plant MADS proteins have the MADS domain located at the *N*-terminus of the proteins. The MADS domain is followed by an intervening region, the I-box, and another moderately conserved region, the K-box, which is not present in the MADS proteins of other organisms (Ma *et al.*, 1991). This K-box is named for its similarity to the coiled-coil domain in keratin. The *C*-termini of the plant MADS box proteins are divergent among members and some of them may correspond to a transcriptional activation domain (Cho *et al.*, 1999). The specificity of heterodimerization of the different MADS proteins and the distinct phenotypes of loss-of-function alleles of several members of this protein family suggest that although DNA binding specificity *in vitro* seems to be conserved, the target genes recognized by different combinations of MADS box proteins *in vivo* are very different. This may be due to distinct interactions that lead to formation of homo- or heterodimers, and to additional interactions with other proteins.

The *Arabidopsis* genome encodes a large number of MADS box genes (Riechmann *et al.*, 2000, table 10.1), At least 28 of these have been studied, including the *AGL* (for AG-like) genes. Some of those MADS box genes are expressed in a temporally and/or spatially regulated manner, and they also appear to be involved in controlling floral development. Studies *in vivo* with *Arabidopsis* AG protein have indicated that the K-box is important for AG function (Mizukami *et al.*, 1996). Additional analysis showed that the K-box of AG alone is able to bind the K-box of other AGLs in yeast and *in vitro* (Fan *et al.*, 1997). It was suggested that the AG function requires interaction with at least one of those AGL proteins and such interactions contribute to the functional specificity of the AG protein. Selective heterodimerization observed between *Antirrhinum* MADS factors requires the K-box (Davies *et al.*, 1996). Exclusive interactions are detected between two factors, DEF and GLO, that together control petal and stamen development. In contrast, PLE, which is required for reproductive organ development, can interact with the products of MADS box genes expressed at early, intermediate and late stages, suggesting that direct interaction between MADS box proteins determines their specific function in controlling meristem and floral organ identity (Davies *et al.*, 1996). The K-box has also been shown to be required for interactions between rice MADS box proteins (Lim *et al.*, 2000). Therefore, MADS domain proteins cooperate with other MADS domain proteins by K-box-mediated interaction to control development in plants. The I-box region also plays a role in protein–protein interactions (Davies *et al.*, 1996).

Genetic analysis of double mutants between *SQUA* and *DEF* or *GLO* show that these MADS box proteins have a partially redundant function in determining the *Antirrhinum* floral organ identity (Egea-Cortines *et al.*, 1999). Protein interaction assays have shown that the three proteins interact to form a ternary complex containing a DEF–GLO heterodimer and a SQUA–SQUA homodimer. The ternary complex exhibits a significant increase in DNA-binding activity to the target binding sequences compared with DEF–GLO heterodimers or SQUA–SQUA homodimers. As indicated previously, the formation of homo- and heterodimers among plant MADS box proteins occurs through the I-region and the K-box, but ternary complex formation requires the C-terminal regions of the MADS box proteins. This domain is the most divergent among the different MADS proteins but displays conserved epitopes among members of the same subfamilies. These epitopes might confer the interaction specificity, and therefore determine the diverse biological functions of the MADS box proteins, possibly by creating a network of interactions with other members of the same family or with other factors. It has been shown recently that the C-terminal region of *Arabidopsis* AP1 also functions as a transcription activation domain (Cho *et al.*, 1999).

10.4 Interaction with signaling proteins

10.4.1 Light-regulated gene transcription

Light signals are perceived by a set of photoreceptors that regulate plant gene expression and growth. The phytochrome protein family monitors the red (R) and far red (FR) light wavelength through their capacity for switching between two light-induced and reversible forms: the R-absorbing, biologically inactive Pr form and the FR-absorbing, biologically active Pfr form (Smith, 2000). Light induced phytochromes regulate the expression of hundreds of genes including those encoding photosynthetic enzymes (Teraghi and Cashmore, 1995). These gene promoters contain various light responsive elements that integrate the control of gene transcription in response to light. Among these elements are the G-box, the G-A-T-A motif and the G-T elements, and biochemical and genetic approaches have isolated factors that bind to these elements. However, all light responsive genes do not contain these motifs, and they are also present in many other non-light regulated genes. For example, it has been shown that bZIP and bHLH proteins bind to the light-responsive G-box. Among the bZIP proteins are GBFs, CPRs and HY5 (Teraghi and Cashmore, 1995; Oyama *et al.*, 1997), while PIF3 and HFR1 contain a bHLH motif (Ni *et al.*, 1998; Fairchild *et al.*, 2000).

The activity of several transcription factors, implicated in light-regulated gene expression, has been shown to be regulated by casein kinase II (CKII) (Klimczak *et al.*, 1995). The DNA-binding activity of the *Arabidopsis* GBF1 to the G-box element, within many light-responsive genes, is stimulated by phosphorylation by CKII (Klimczak *et al.*, 1992). The transcription factor circadian clock-associated 1 (CCA1) is a MYB-related protein that binds specifically to the promoter of *Lhcb1* genes (Wang *et al.*, 1997). The transcription of these genes is regulated by light and also by circadian rhythms. Characterization of CCA1 has shown that it is involved in the light and circadian regulation of *Lhcb* genes. CCA1 mRNA and protein levels also exhibit circadian oscillations, and overexpression of CCA1 represses the expression of the endogenous *CCA1* gene (Wang and Tobin, 1998). A screen using a yeast two-hybrid system has led to the identification of the CKB3 protein that interacts specifically with CCA1 (Sugano *et al.*, 1999). CKB3 is a structural and functional homolog of the regulatory β-subunit of CKII. Other subunits of CKII also interact with CCA1 *in vitro*. The CKII β subunit stimulates binding of CCA1 to the CCA1 binding site on the *Lhcb1*3* promoter. CKII-like activity from *Arabidopsis* plant extracts can phosphorylate CCA1 *in vitro* and CKII phosphorylation is required for the formation of a DNA–protein complex containing CCA1. Overexpression of the *CKB3* gene in transgenic plants disturbs the light period-driven oscillation of the expression of *CCA1* and *LHY*, another MYB-related circadian clock protein gene. Protein kinase CK2 is also able to interact with and phosphorylate LHY *in vitro* and overexpression of the gene also reduced phytochrome induction of an *Lhcb* gene. Antisense inhibition of the catalytic subunit of CKII likewise disturbs light-regulated gene expression. Therefore, CKII is involved in light-regulated and circadian clock-regulated gene expression by directly interacting with DNA-binding transcription factors.

Phytochrome is a red-light activated serine–threonine kinase and several isoforms of the protein that have both overlapped and distinct physiological functions have been identified (Smith, 2000). Yeast two-hybrid screening for phytochrome-interacting proteins has identified a few candidate proteins including a nuclear-localized bHLH protein called PIF3 (Ni *et al.*, 1998). PIF3 interacts equally with the *C*-terminus of phytochrome A (phyA) and phytochrome B (phyB) of *Arabidopsis*, but preferentially with the intact phyB protein. Interestingly, PIF3 was simultaneously identified as a phytochrome-signaling component by a screen for T-DNA insertion mutants with a phenotype of an enhanced response to red light (Halliday *et al.*, 1999). The T-DNA insertion found in the promoter induces overexpression of the *PIF3* gene. Recent data have shown that PIF3 binds to sequences with a G-box element core (Martinez-Garcia *et al.*, 2000). However, PIF3 associates only with elements characteristically found in the promoters of certain genes whose expression is regulated by phytochrome. PIF3 antisense plants display reduced level of expression of some, but not all, phytochrome-induced genes (Martinez-Garcia *et al.*, 2000). Interestingly, both phyA and phyB bind reversibly to G-box-bound PIF3 upon red light-triggered conversion to the active Pfr forms (Martinez-Garcia *et al.*, 2000). However, the phyA binding affinity for PIF3 is tenfold lower when compared with phyB. This observation is consistent with *in vivo* data from PIF3-deficient plants, indicating that PIF3 plays a major role in phyB signaling pathway, but a minor role in phyA regulation (Halliday *et al.*, 1999). A segment of 37 amino acids present at the N-terminus of phyB, but absent in phyA, contributes to the higher affinity binding of phyB to PIF3. The interacting domain within PIF3 is localized to two segments at either side of the DNA-binding and dimerization bHLH domain. The two domains cooperate to bind to phyB. The segment on the N-terminal side of bHLH corresponds to the PAS domain of PIF3 (Ni *et al.*, 1998). PAS domains are proposed to function in protein–protein interactions (Dunlap, 1998). The segment on the C-terminal side of bHLH seems to be also involved in discriminating between the Pr and Pfr forms of phyB. PIF3 can also form, presumably via the bHLH domain, a heterodimer with HFR1, a protein that also belongs to the bHLH family (Fairchild *et al.*, 2000). HFR1 is involved in phyA signaling, as functional mutations lead to a reduction in seedling responsiveness specifically to continuous FR light (Fairchild *et al.*, 2000). In contrast to PIF3, HFR1 does not bind to either phyA or phyB, but the HFR1–PIF3 complex binds preferentially the Pfr form of both phytochromes. Since the expression of HFR1 itself is upregulated in FR light, this provides a basis for the specificity of HFR1 to phyA signaling. Thus, phytochromes can function as a component of a transcriptional regulator complex via interactions with a DNA-binding protein to activate a subclass of light inducible genes. The underling mechanism leading to transcriptional activation is unknown.

A bZIP protein, HY5, also has a role in the promotion of photomorphogenesis (Oyama *et al.*, 1997). Mutations in the *HY5* locus lead to light-insensitive responses with a high hypocotyl phenotype of seedlings grown in the light, suggesting that HY5 is a positive regulator of photomorphogenesis. Genetic analysis has suggested that HY5 acts downstream of multiple photoreceptor-mediated pathways and that it functionally interacts with the pleiotropic *COP1* gene, which is a negative regulator of photomorphogenesis (Ang and Deng, 1994; Ang *et al.*, 1998). Indeed, HY5 interacts physically with COP1 through its N-terminal domain. This interaction negatively regulates HY5 activity (Ang *et al.*, 1998). However, the activity of COP1 is inhibited by virtue of its light-induced sequestration in the cytoplasm (von Arnim and Deng, 1994). The abundance of HY5 protein level is light-dependent. In fact, the cellular HY5 level is regulated by a darkness-dependent degradation, probably mediated by the 26S proteosome (Osterlund *et al.*, 2000). Degradation of HY5 requires its COP1-interacting domain, supporting the hypothesis that COP1 targets HY5 for degradation. COP1 and HY5 interact through a COP1 WD-40 domain (Ang *et al.*, 1998). Expression *in vivo* of COP1 lacking this domain results in an increased HY5 level. Light negatively regulates the level of COP1 in the nucleus and so positively regulates the abundance of HY5 (von Arnim and Deng, 1994). Gel infiltration chromatography revealed that HY5 is present in a protein complex and exists in two isoforms, resulting from phosphorylation by a light-regulated kinase activity (Hardtke *et al.*, 2000). The phosphorylation site is within its COP1 interacting domain and unphosphrylated HY5 shows stronger interaction with COP1, indicating that the activity of HY5 is regulated at different levels.

10.4.2   Systemic acquired resistance signal induced transcription

Systemic acquired resistance (SAR) is a general plant-resistance response induced during a local infection by an avirulent pathogen (Ryals *et al.*, 1996). Salicylic acid (SA) is shown to be a necessary signal for SAR induction. An increase of endogenous SA level or exogenous application of SA not only results in an enhanced broad-spectrum resistance but also induces the expression of a set of genes known as pathogenesis-related (*PR*) genes (Ward *et al.*, 1991). *PR* genes may be directly involved in resistance because their expression coincides with the establishment of SAR and some of them encode antimicrobial activities. Research on regulators of *PR* gene expression has led to the identification of a locus *NPR1*. Plants with a mutation at this locus fail to express several *PR* genes and display enhanced susceptibility to infection after treatment with SA (Cao *et al.*, 1994). *NPR1* encodes a protein with several ankyrin repeats, which are found in proteins with diverse functions and are involved in protein–protein interactions (Cao *et al.*, 1997; Sedgwick and Smerdon, 1999). However, the NPR1 protein does not have obvious features of a DNA-binding transcription factor, although its overexpression in transgenic plants leads to enhanced induction of *PR* genes during an infection (Cao *et al.*, 1998). This suggests that NPR1 may function as a regulatory protein by interacting with other proteins. Using NPR1 protein as bait in a yeast two-hybrid screen, members of a subclass of the bZIP protein family (tomato NIF1, *Arabidopsis* AHBP-1b/TGA2, TGA5/OBP5, TGA3 and TGA7, but not TGA1 and TGA4) were shown to interact with NPR1 in yeast and *in vitro* (Zhang *et al.*, 1999; Després *et al.*, 2000; Zhou *et al.*, 2000). Mutant derivatives of NPR1 that abolish SAR in the plant failed to interact with two *Arabidopsis* tested TGA factors: TGA2 and TGA6. *Arabidopsis* TGA2 binds specifically to an SA-responsive promoter element of an *Arabidopsis* *PR* gene. NPR1 substantially increases the binding of *Arabidopsis* TGA2 to its cognate promoter element that corresponds to the *as-1* element (Després *et al.*, 2000). This element has been identified previously as a binding motif of tobacco bZIP transcription factors (TGA1a, TGA1b) and has been shown to be essential for SA-induced *PR-1* gene transcription *in planta* (Lebel *et al.*, 1998). Like *Arabidopsis* TGA1 and TGA4, tobacco TGA1a is unable to interact with NPR1, while two additional bZIP proteins, namely TGA2.1 and TGA2.2, do interact with NPR1 (Niggeweg *et al.*, 2000). This suggests that TGA factors respond to different signal transduction pathways.

The induction of *PR* gene expression by NPR1 is probably mediated through interaction with specific promoter-bound bZIP proteins. However, how SA induces the interacting activity of NPR1 to bZIP proteins and how the interaction leads to transcriptional activation of *PR-1* gene is, as yet, unknown. Unidentified repressor proteins under non-induced conditions, may inactivate those bZIP proteins and, upon induction, NPR1 is post-transcriptionally modified and relieves the bZIP proteins from such repression. Recent data have shown that the NPR1 protein is accumulated in the nucleus in response to activators of SAR and the nuclear localization of NPR1 is required for activation of *PR* gene expression (Kinkema *et al.*, 2000).

10.4.3 Disease resistance R gene mediated transcription

In tomato, the disease resistance *R* gene *Pto* confers resistance to *Pseudomonas syringae* pv *tomato* expressing the *avrPto* avirulence gene. *Pto* encodes an active serine–threonine protein kinase, while the bacterial arivulence *avrPto* gene encodes a small hydrophilic protein (Martin *et al.*, 1993; Ronald *et al.*, 1992). The Pto kinase interacts directly with the AvrPto protein, and mutations in *Pto* or *AvrPto* that disrupt this interaction also eliminate disease resistance (Tang *et al.*, 1996), indicating that the interaction determines the gene-for-gene specificity in this plant–pathogen interaction. This recognition event initiates multiple cellular reactions leading to the hypersensitive responses including induced expression of many *PR* genes. In addition to AvrPto, Pto protein interacts with many other cellular proteins. Among them, Pti4, Pti5 and Pti6 (for Pto-interacting) are transcription factors belonging to the EREBP subfamily (Zhou *et al.*, 1997). All these proteins bind to the GCC-box, a *cis*-element present in the promoter region of many *PR* genes (Zhou *et al.*, 1997). Pto kinase phosphorylates the Pti4 protein on at least four threonine residues. Phosphorylation of Pti4 increases its DNA binding activity *in vitro* to the GCC-box (Gu *et al.*, 2000). However, it is not known whether the phosphorylation affects other activities of the protein such as transactivation function. Pti4 transcripts accumulate earlier than the GCC-box containing genes whose induction requires *de novo* synthesis of new proteins, which may correspond to Pti4. This suggests that *Pti4* is an immediate–early response gene, the product of which is required to activate *PR* genes after interaction with and modification by the *R* gene product Pto.

10.5 Extensive protein interactions leading to the assembly of transcriptional initiation complexes

Promoters of eukaryotic structural genes contain two functional modules: a core promoter and a collection of specific transcription factor binding sequences. The core promoter comprises a T-A-T-A (TATA) box sequence that is located about 20–40 pb upstream from the transcription start site of many genes. Some genes have a TATA-less promoter with a loosely defined initiator sequence (Smale *et al.*, 1998). Transcription initiation by RNA polymerase II involves the assembly of general transcription factors on the core promoter to form a pre-initiation complex (Orphanides *et al.*, 1996). At least six general transcription factors are required for the assembly, which are named TFIIA, TFIIB, TFIID, TFIIE, TFIIF and TFIIH according to their chromatographic profiles and order of their discovery in mammalian cells. The first step of the pre-initiation complex assembly is the binding of the general transcription factor TFIID to the TATA box. TFIID is a multisubunit complex containing the TATA-box binding protein (TBP) and about 10–13 tightly linked TBP-associated factors ($TAF_{II}s$) (Verrijzer and Tjian, 1996). Plant genomes encode at least two TBP proteins that are likely to play a similar role in transcription (Li et al., 2001). Genes highly homologous to animal and yeast $TAF_{II}s$ have been found in the Arabidopsis genome. TBP can bind to the TATA box alone without association with $TAF_{II}s$, while its binding to TATA-less promoters requires an association with $TAF_{II}s$ or a form of TFIID (Goodrich et al., 1996). The TFIID (TBP)-TATA box complex then acts to nucleate the assembly, via protein–protein interactions, between the remaining general transcription factors and the RNA polymerase either through a stepwise assembly or a pre-assembly (or holoenzyme) model.

The ordered assembly of the transcription pre-initiation complex was origi-nally proposed on the basis of the formation of active transcription complexes in vitro (Buratowski, 1994). Essential steps leading to transcription by RNA polymerase II include: (i) formation of a stable complex containing TFIID–TFIIA–TFIIB bound to the TATA box; (ii) recruitment of RNA polymerase II escorted by TFIIF to the promoter through multiple protein–protein interactions with the TFIID–TFIIA–TFIIB complex; (iii) formation of an activated open promoter complex by the further addition of TFIIE and TFIIH, which stimulates a promoter-melting event; and (iv) synthesis of nascent RNA upon hyper-phosphorylation of the C-terminal domain of the largest subunit of the RNA polymerase II (Orphanides et al., 1996). Interaction between TBP and TFIIA first stimulates the formation and the stabilization of the TBP–TATA complex that is subsequently recognized by TFIIB. TFIIA has two (in yeast) or three (in animal cells) subunits that are encoded by two different genes (Orphanides et al., 1996). TFIIA genes are also duplicated in Arabidopsis (Li et al., 1999). The exact function of TFIIA in transcription is generally not clear. Biochemical evidence indicates that interaction between TFIIA and TFIID contributes directly to transcriptional activation (Kobayashi et al., 1995). Genetic analysis in yeast indicates that TFIIA may play a role both during and after the recruitment of TBP to the TATA box of a subset of genes, which is consistent with the evidence that TFIIA interacts directly with certain promoter-binding activators in animal and yeast systems (Liu et al., 1999). Direct interaction between a plant transcription activator and the TBP–TFIIA–TATA complex has been observed (Le Gourrierec et al., 1999). TFIIB is a single subunit protein that interacts directly with TBP and with DNA downstream to the TATA box (Buratowski and Zhou, 1993). Plant TFIIB has been characterized (Baldwin and Gurley, 1996) and its association with TBP supports both basal and activated transcription (Pan et al., 2000). Genes encoding general transcription factors are conserved in plants genomes (The Arabidopsis Genome Initiative, 2000), but their biochemical properties and function in plant gene transcription have not been studied.

The pre-assembly model was first proposed when certain preparations of RNA polymerase II were observed to co-purify with subsets of the general transcription factors (except TFIID and TFIIA), along with other regulators including chromatin remodeling factors (Parvin and Young, 1998). In this model, a minimum of two targeted steps are required to form an active initiation complex with the binding of TFIID or TBP to a promoter as a prerequisite for transcription.

Despite the complexity of this basal initiation transcriptional machinery with more than 40 polypeptides, its response to activators on specific target genes is still dependent on additional co-activators, which mediate and probably integrate the effects of transcriptional activators on the RNA polymerase II initiation complex (Roeder, 1998). Proteins with co-activator-like function have been discovered in plants, while co-activators with chromatin remodeling activities have not yet been functionally characterized, although homologous genes have been found in plants (The *Arabidopsis* Genome Initiative, 2000).

10.6 Conclusions

Despite significant advances in functional studies of plant transcription factors, the number of reported interactions among plant transcription factors is still small. High-throughput genomic approaches, such as the two-hybrid system, and proteomics coupled with the comprehensive phenotypic characterization of mutants for all transcription factor genes is required to identity functional protein–protein interactions between transcription factors genome-wide.

References

Abel, S. and Theologis, A. (1996) Early genes and auxin action. *Plant Physiol.*, **111**, 9-17.

Ang, L.H. and Deng, X.W. (1994) Regulatory hierarchy of photomorphogenic loci: allele-specific and light-dependent interaction between the HY5 and COP1 loci. *Plant Cell*, **6**, 613-628.

Ang, L.H., Chattopadhyay, S., Wei, N. *et al.* (1998) Molecular interaction between COP1 and HY5 defines a regulatory switch for light control of *Arabidopsis* development. *Mol. Cell*, **1**, 213-222.

Armstrong, G.A., Weisshaar, B. and Hahlbrock, K. (1992) Homodimeric and heterodimeric leucine zipper proteins and nuclear factors from parsley recognize diverse promoter elements with ACGT cores. *Plant Cell*, **4**, 525-543.

von Arnim, A.G. and Deng, X.W. (1994) Light inactivation of *Arabidopsis* photomorphogenic repressor COP1 involves a cell-specific regulation of its nucleocytoplasmic partitioning. *Cell*, **79**, 1035-1045.

Aukerman, M.J. and Schmidt, R.J. (1994) Regulation of alpha-zein gene expression during maize endosperm development, in *Results and Problems in Cell Differentiation 20: Plant promoters and Transcription Factors.* (ed. L. Nover) Springer-Verlag, Berlin, pp 209-233.

Baldwin, D.A. and Gurley, W.B. (1996) Isolation and characterization of cDNAs encoding transcription factor IIB from *Arabidopsis* and soybean. *Plant J.*, **10**, 561-568.

Bies-Etheve, N., da Silva Conceicao, A., Giraudat, J. *et al.* (1999) Importance of the B2 domain of the *Arabidopsis* ABI3 protein for Em and 2S albumin gene regulation. *Plant Mol. Biol.*, **40**, 1045-1054.

Bonetta, D. and McCourt, P. (1998) Genetic analysis of ABA signal transduction pathways. *Trends Plant Sci.*, **3**, 231-235.

Brown, C.E., Lechner, T., Howe, L. and Workman, J.L. (2000) The many hats of transcription coactivators. *Trends Biochem. Sci.*, **25**, 15-19.

Buratowski, S. (1994) The basics of basal transcription by RNA polymerase II. *Cell*, **77**, 1-3.

Buratowski, S. and Zhou, H. (1993) Functional domains of transcription factor TFIIB. *Proc. Natl Acad. Sci. USA*, **90**, 5633-5637.

Busk, P.K. and Pages, M. (1997) Protein binding to the abscisic acid-responsive element is independent of *VIVIPAROUS1 in vivo*. *Plant Cell*, **9**, 2261-2270.

Busk, P.K. and Pages, M. (1998) Regulation of abscisic acid-induced transcription. *Plant Mol. Bio.*, **37**, 425-435.

Buttner, M. and Singh, K.B. (1997) *Arabidopsis thaliana* ethylene-responsive element binding protein (AtEBP), an ethylene-inducible, GCC box DNA-binding protein interacts with an *ocs* element binding protein. *Proc. Natl Acad. Sci. USA*, **94**, 5961-5966.

Cao, H., Bowling, S.A., Gordon S. and Dong, X. (1994) Characterization of an *Arabidopsis* mutant that is nonresponsive to inducers of systemic acquired resistance. *Plant Cell*, **6**, 1583-1592.

Cao, H., Glazebrook, J., Clarke, J.D., Volko, S. and Dong, X. (1997) The *Arabidopsis NPR1* gene that controls systemic acquired resistance encodes a novel protein contaning ankyrin repeats. *Cell*, **88**, 57-63.

Cao, H., Li, X. and Dong, X. (1998) Generation of broad-spectrum disease resistance by overexpression of an essentail regulatory gene in systemic resistance. *Proc. Natl Acad. Sci. USA*, **95**, 6531-6536.

Chen, W., Chao, G. and Singh, K.B. (1996) The promoter of H_2O_2-inducible, *Arabidopsis* glutathione S-transferase gene contains closely linked OBF- and OBP1-binding sites. *Plant J.*, **10**, 955-966.

Cho, S., Jang, S., Chae, S. *et al.* (1999) Analysis of the *C*-terminal region of *Arabidopsis thalina* APETALA1 as a transcription activation domain. *Plant Mol. Biol.*, **40**, 419-429.

Cone, K.C., Cocciolone, S.M., Burr, F.A. and Burr, B. (1993) Maize anthocyanin regulatory gene *pl* is a duplicate of *c1* that functions in the plant. *Plant Cell*, **5**, 1795-1805.

Cord-Neto, G., Yunes, J.A., da Silva M.J., Vettore, A.L., Arruda, P. and Leite, A. (1995) The involvement of Opaque 2 on beta-prolamin gene regulation in maize and Coix suggests a more general role for this transcriptional activator. *Plant Mol. Biol.*, **27**, 1015-1029.

Davies, B., Egea-Cortines, M., de Andrade Silva, E., Saedler, H. and Sommer, H. (1996) Multiple interactions amongst floral homeotic MADS box proteins. *EMBO J.*, **15**, 4330-4343.

Després, C., Delong, C., Glaze, S., Liu, E. and Fobert, P.R. (2000) The *Arabidopsis* NPR1/NIM1 protein enhances the DNA binding activity of a subgroup of the TGA family of bZIP transcription factors. *Plant Cell*, **12**, 279-290.

de Vetten, N., Quattrocchio, F., Mol, J. and Koes, R. (1997) The *an11* locus controlling flower pigmentation in petunia encodes a novel WD-repeat protein conserved in yeast, plant and animals. *Genes Dev.*, **11**, 1422-1434.

Dunlap, J. (1998) Circadian rhythms. An end in the beginning. *Science*, **280**, 1548-1549.

Ellenberger, T.E., Brandl, C.J., Struhl, K. and Harrison, S.C. (1992) The GCN4 basic region leucine zipper binds DNA as a dimer of uninterrupted α-helices: crystal structure of the protein–DNA complex. *Cell*, **71**, 1223-1237.

Egea-Cortines, M., Saedler, H. and Sommer, H. (1999) Ternary complex formation between the MADS-box proteins SQUAMOSA, DEFICIENS and GLOBOSA is involved in the control of floral architecture in *Antirrhinum majus*. *EMBO J.*, **18**, 5370-5379.

Elliot, R.C., Betzner, A.S., Huttner, E. *et al.* (1996) *AINTEGUMENTA*, an *APETALA2*-like gene of *Arabidopsis* with pleiotropic roles in ovule development and floral organ growth. *Plant Cell*, **8**, 155-168.

Fairchild, C.D., Schumaker, M.A. and Quail, P.H. (2000) HFR1 encodes an atypical bHLH protein that acts in phytochrome A signal transduction. *Genes Dev.*, **14**, 2377-2391.

Fan, H.Y., Hu, Y., Tudor, M. and Ma, H. (1997) Specific interactions between the K domains of AG and AGLs, members of the MADS domain family of DNA binding proteins. *Plant J.*, **12**, 999-1010.

Ferl, R.J. (1996) 14-3-3 proteins and signal transduction. *Annu. Rev. Plant Physiol. Plant Mol. Biol.*, **47**, 49-73.

Finkelstein, R.R., Wang, M.L., Lynch, T.J. and Beach, D. (1998) The *Arabidopsis* abscisic acid response locus *ABI4* encodes an APETALA2 domain protein. *Plant Cell*, **10**, 1043-1054.

Fujimoto, S.Y., Ohta, M., Usui, A., Shinshi, H. and Ohme-Takagi, M. (2000) *Arabidopsis* ethylene-responsive element binding factors act as transcriptional activators or repressors of GCC box-mediated gene expression. *Plant Cell*, **12**, 393-404.

Giraudat, J., Hauge, B.M., Valon, C., Smalle, J., Parcy, F. and Goodman, H.M. (1992) Isolation of the *Arabidopsis* ABI3 gene by positional cloning. *Plant Cell*, **4**, 1251-1261.

Goff, S.A., Cone, K.C. and Chandler, V.L. (1992) Functional analysis of the transcriptional activator encoded by the maize *B* gene: evidence for a direct functional interaction between two classes of regulatory proteins. *Genes Dev.*, **6**, 864-875.

Goodrich, J.A., Hoey, T. and Tjian, R. (1996) Contacts in context: promoter specificity and macromolecular interactions in transcription. *Cell*, **84**, 825-830.

Gray, W.M. and Estelle, M. (2000) Function of the ubiquitin–proteasome pathway in auxin response. *Trends Biochem. Sci.*, **25**, 133-138.

Grotewold, E., Drummond, B.J., Bowen, B. and Peterson T. (1994) The *myb-Homologous P* gene controls phlobaphene pigmentation in maize floral organs by directly activating a flavonoid biosynthetic gene subset. *Cell*, **76**, 543-553.

Grotewold, E., Sainz, M.B., Tagliani, L., Hernandez, M., Bowen, B. and Chandler, V.L. (2000) Identification of the residues in the Myb domain of maize C1 that specify the interaction with the bHLH cofactor R. *Proc. Natl Acad. Sci. USA*, **97**, 13,579-13,584.

Gu, Y.-Q., Yang, C., Thara, V.K., Zhou, J. and Martin, G.B. (2000) Pti4 is induced by ethylene and salicylic acid, and its product is phosphorylated by the Pto kinase. *Plant Cell*, **12**, 771-785.

Guiltinan, M.J., Marcotte, W.R. and Quatrano, R.S. (1990) A plant leucine zipper protein that recognizes an abscisic acid response element. *Science*, **250**, 267-271.

Halliday, K.J., Hudson, M., Ni, M., Qin, M. and Quail, P.H. (1999) *poc1*: An *Arabidopsis* mutant perturbed in phytochrome signaling because of a T-DNA insertion in the promoter of *PIF3*, a gene encoding a phytochrome-interacting bHLH protein. *Proc. Natl Acad. Sci. USA*, **96**, 5832-5837.

Hardtke, C.S., Gohda, K., Osterlund, M.T., Oyama, T., Okada, K. and Deng, X.W. (2000) HY5 stability and activity in *Arabidopsis* is regulated by phosphorylation in its COP1 binding domain. *EMBO J.*, **19**, 4997-5006.

Hill, A., Nantel, A., Rock, C.D. and Quatrano, R.S. (1996) A conserved domain of the viviparous1 gene product enhances the DNA binding activity of the bZip protein EmBP1 and other transcription factors. *J. Biol. Chem.*, **271**, 3366-3374.

Hobo, T., Kowyama, Y. and Hattori, T. (1999) A bZIP factor, TRAB1, interacts with VP1 and mediates abscisic acid-induced transcription. *Proc. Natl Acad. Sci. USA*, **96**, 15348-15353.

Hurst, H.C. (1995) Transcription factors. 1. bZIP proteins. *Protein Profile*, **2**, 105-168.

Jin, H. and Martin, C. (1999) Multifunctionality and diversity within the plant *MYB*-gene family. *Plant Mol. Biol.*, **41**, 577-585.

Jofuku, K.D., Den Boer, B.G.W., Van Montague, M. and Okamuro, J.K. (1994) Control of *Arabidopsis* flower and seed development by the homeotic gene *APETALA2*. *Plant Cell*, **6**, 1211-1225.

Kao, C.-Y., Cocciolone, S.M., Vasil, I.K. and McCarty D.R. (1996) Localization and interaction of the *cis*-acting elements for abscisic acid, *VIVIPAROUS1* and light activation of the *C1* gene of maize. *Plant Cell*, **8**, 1171-1179.

Kasuga, M., Liu, Q., Miura, S., Yamaguchi-Shinozaki, K. and Shinozaki, K. (1999) Improving plant drought, salt and freezing tolerance by gene transfer of a single stress-inducible transcription factor. *Nature Biotechnol.*, **17**, 287-291.

Katagiri, F., Lam, E., and Chua N.H. (1989) Two tobacco DNA-binding proteins with homology to the nuclear factor CREB. *Nature*, **340**, 727-730.

Kim, J., Harter, K. and Theologis, A. (1997) Protein–protein interaction among the Aux/IAA proteins. *Proc. Natl Acad. Sci. USA*, **94**, 11786-11791.

Kinkema, M., Fan, W. and Dong, X. (2000) Nuclear localization of NPR1 is required for activation of PR gene expression. *Plant Cell*, **12**, 2339-2350.

Klimczak, L.J., Schindler, U. and Cashmore, A.R. (1992) DNA binding activity of the *Arabidopsis* G-box binding factor GBF1 is stimulated by phosphorylation by casein kinase from broccoli. *Plant Cell*, **4**, 87-98.

Klimczak, L.J., Collinge, M.A., Farini, D., Giuliano, G., Walker, J.C. and Cashmore, A.R. (1995) Reconstitution of *Arabidopsis* casein kinase II from recombinant subunits and phosphorylation of transcription factor GBF1. *Plant Cell*, **7**, 105-115.

Kobayashi, N., Boyer, T.G. and Berk, A.J. (1995) A class of activation domains interacts directly with TFIIA and stimulates TFIIA-TFIID-promoter complex assembly. *Mol. Cell Biol.*, **15**, 6465-6473.

Kranz, H.D., Denekamp, M., Greco, R. *et al.* (1998) Towards functional characterization of members of the *R2R3-MYB* gene family from *Arabidopsis thaliana*. *Plant J.*, **16**, 263-276.

Kurup, S., Jones, H.D. and Holdsworth, M.J. (2000) Intercation of the developmental regulator ABI3 with proteins identified from developing *Arabidopsis* seeds. *Plant J.*, **21**, 143-155.

Lebel, E., Heifetz, P., Thorne, L., Uknes, S., Ryals, J. and Ward, E. (1998) Functional analysis of regulatory sequences controlling *PR-1* gene expression in *Arabidopsis*. *Plant J.*, **16**, 223-233.

Le Gourrierec, J., Li, Y.F. and Zhou, D.X. (1999) Transcriptional activation by *Arabidopsis* GT-1 may be through interaction with TFIIA-TBP-TATA complex. *Plant J.*, **18**, 663-638.

Li, Y.F., Le Gourrierec, J., Torki, M., Kim, Y.-J., Guerineau, F. and Zhou, D.X. (1999) Characterization and functional analysis of *Arabidopsis* TFIIA reveal that the evolutionarily unconserved region of the large subunit has a transcription activation domain. *Plant Mol. Biol.*, **39**, 515-525.

Li, Y.-F., Dubois, F. and Zhou, D.X. (2001) Ectopic expression of TATA box-binding protein induces shoot proliferation in *Arabidopsis*. *FEBS Lett.*, **489**, 187-191.

Lim, J., Moon, Y.-H., An, G. and Jang, S.K. (2000) Two rice domain proteins interact with OsMADS1. *Plant Mol. Biol.*, **44**, 513-527.

Liu, Q., Kasuga, M., Sakuma, Y.*et al.* (1998) Two transcription factors, DREB1 and DREB2, with an EREBP/AP2 DNA binding domain separate two cellular signal transduction pathways in drought- and low-temperature-responsive gene expression, respectively, in *Arabidopsis*. *Plant Cell*, **10**, 1391-1406.

Liu, Q., Gabriel, S.E., Roinick, K.L., Ward, R.D. and Arndt, K.M. (1999) Analysis of TFIIA function *In vivo*: evidence for a role in TATA-binding protein recruitment and gene-specific activation. *Mol. Cell. Biol.*, **19**, 8673-8685.

Ludwig, S.R. and Wessler, S.R. (1990) Maize *R* gene family: tissue specific helix-loop-helix proteins. *Cell*, **66**, 895-905.

Luscombe, N.M., Austin, S.E., Berman, H.M., and Thornton J.M.(2000) http://genomebiology.com/2000/1/1/reviews/001/.

Ma, H. (1994) The unfolding drama of flower development: recent results from genetic and molecular analysis. *Genes Dev.*, **8**, 745-756.

Ma, H. and dePamphilis, C. (2000) The ABCs of the floral evolution. *Cell*, **101**, 5-8.

Ma, H., Yanofsky, M.F. and Meyerowitz, E.M. (1991) *AGL1-AGL6*, an *Arabidopsis* gene family with similarity to floral homeotic and transcription factor genes. *Genes Dev.*, **5**, 484-495.

McCarty, D.R. (1995) Genetic control and intergration of maturation and germination pathways in seed development. *Annu. Rev. Plant Physiol. Plant Mol. Biol.*, **46**, 71-93.

McCarty, D.R. and Chory, J. (2000) Conservation and innovation in plant signaling pathways. *Cell*, **130**, 201-209.

McCarty, D.R., Hattori, T., Carson, C.B., Vimla Vasil, V., Lazar, M. and Vasil, I.K. (1991) The *Viviparous-1* developmental gene of maize encodes a novel transcriptional activator. *Cell*, **66**, 895-905.

Marks, M.D. (1997) Molecular genetic analysis of trichome development in *Arabidopsis*. *Annu. Rev. Plant Physiol. Plant Mol. Biol.*, **48**, 137-163.

Martin, G.B., Brommonschenkel, S., Chunwongse, J. *et al.*(1993) Map-based cloning of a protein kinase gene conferring disease resistance in tomato. *Science*, **262**, 1432-1436.

Martinez-Garcia, J.F., Huq, E. and Quail, P.H. (2000) Direct targeting of light signals to a promoter element-bound transcription factor. *Science*, **288**, 859-863.

Menkens, A. E. and Cashmore A.R. (1994) Isolation and characterization of a fourth *Arabidopsis* thaliana G-box-binding factor, which has similarities to Fos oncoprotein. *Proc. Natl Acad. Sci. USA*, **91**, 2522-2526.

Menkens, A. E., Schindler, U. and Cashmore A.R. (1995) The G-box: a ubiquitous regulatory DNA element in plant bound by GBF family of bZIP proteins. *Trends Biochem. Sci.*, **20**, 506-510.

Mizukami, Y., Huang, H., Tudor, M., Hu, Y. and Ma H. (1996) Functional domains of the floral regulator AGAMOUS: characterization of the DNA binding domain and analysis of dominant negative mutations. *Plant Cell*, **8**, 831-845.

Mol, J., Grotewold, E. and Koes, R. (1998) How genes paint flowers and seeds. *Trends Plant Sci.*, **3**, 212-217.

Moose, S.P. and Sisco, P.H. (1996) *Glossy 15*, an *APETALA2*-like gene from maize that regulates leaf epidermal cell identity. *Genes Dev.*, **10**, 3018-3027.

Murre, C. and Baltimore, D. (1989) A new DNA binding and dimerization motif in immunoglobulin enhancer binding, daughterless, MyoD, and myc proteins. *Cell*, **56**, 777-783.

Nantel, A. and Quatrano, R.S. (1996) Characterization of three rice bZIP factors, including two inhibitors of EmBP-1 DNA-binding activity. *J. Biol. Chem.*, **271**, 31,296-31,305.

Ng, H.H. and Bird, A. (2000) Histone deacetylase: silencers for hire. *Trends Biochem. Sci.*, **25**, 121-126.

Ni, M., Tepperman, M.J. and Quail, P.H. (1998) PIF3, a phytochrome-interacting factor necessary for normal photoinduced signal transduction, is a nocel basic helix-loop-helix protein. *Cell*, **96**, 657-667.

Niggeweg, R., Thurow, C., Weigel, R., Pfitzner, U. and Gatz, C. (2000) Tobacco TGA factors differ with respect to interaction with NPR1, activation potential and DNA-binding properties. *Plant Mol. Biol.*, **42**, 775-788.

Niu, X., Renshaw-Gegg, L., Miller, L. and Guiltinan, M.J. (1999) Bipartite determinants of DNA-binding specificity of plant basic leucine zipper proteins. *Plant Mol Biol.*, **41**, 1-13

Ogata, K., Morikawa, S., Akamura, H. *et al.* (1995) Comparison of the free and DNA complexed forms of the DNA-binding domain of c-MYB. *Nature Struct. Biol.*, **2**, 309-320.

Ohta, M., Ohme-Takagi, M. and Sinshi, H. (2000) Three ethylene-responsive transcription factors in tobacco with distinct transactivation function. *Plant J.*, **22**, 29-38.

Okamuro, J.K., Caster, B., Villarroel, R., Van Montague, M. and Jofuku, K.D. (1997) The AP2 domain of APETALA2 defines a large new family of DNA binding proteins in *Arabidopsis*. *Proc. Natl Acad. Sci. USA*, **94**, 7076-7081.

Oppenheimer, D.G., Herman, P.L., Sivakumaran, S., Esch, J. and Marks, M.D. (1991) A *myb*-related gene required for leaf trichome differentiation in *Arabidopsis* is expressed in stipules. *Cell*, **67**, 483-493.

Orphanides, G., Lagrange, T. and Reinberg, D. (1996) The general transcription factors of RNA polymerase II. *Genes Dev.*, **10**, 2657-2683.

Osterlund, M.T., Hardtke, C.S., Wei, N. and Deng, X.W. (2000) Targeted destabilization of HY5 in light development of *Arabidopsis*. *Nature*, **405**, 462-466.

Oyama, T., Shimura, Y. and Okada, K. (1997) The *Arabidopsis HY5* gene encodes a bZIP protein that regulates stimulus-induced development of root and hypocotyl. *Genes Dev.*, **11**, 2983-2995.

Pan, S., Sehnke, P.C., Ferl, R.J. and Gurley, W.B. (1999) Specific interactions with TBP and TFIIB *in vitro* suggest that 14-3-3 proteins may participate in the regulation of transcription when part of a DNA binding complex. *Plant Cell*, **11**, 1591-1602.

Pan, S., Czarnerka-Verner, E. and Gurley, W.B. (2000) Role of the TATA binding protein-transcription factor IIB interaction in supporting basal and activated transcription in plant cells. *Plant Cell*, **12**, 125-135.

Parvin, D.J. and Young, R.A. (1998) Regulatory targets in the RNA polymerase II holoenzyme. *Curr. Opin. Genet. Dev.*, **8**, 565-570.

Payne, C.T., Zhang, F. and Lloyd, A. (2000) GL3 encodes a bHLH protein that regulates trichome development in *Arabidopsis* through interaction with GL1 and TTG1. *Genetics*, **156**, 1349-1362.

Paz-Ares, J., Ghosal, D., Wienand, U., Peterson, P.A. and Saedler, H. (1987) The regulatory c1 locus of *Zea mays* encodes a protein with homology to myb proto-oncogene products with structural similarities to transcriptional activators. *EMBO J.*, **6**, 3553-3558.

Purugganan, M.D., Rousley, S.D., Schmidt, R.J. and Yanofsky, M.F. (1995) Molecular evolution of flower development: diversification of the plant MADS box regulatory family. *Genetics*, **140**, 345-356.

Qin, X.-F., Holuigue, L., Horvath, D.M. and Chua, N.-H. (1994) Immediate early transcription activation by salicylic acid via the cauliflower mosaic virus *as-1* element. *Plant Cell*, **6**, 863-874.

Quayle, T. and Feix, G. (1992) Functional analysis of the −300 region of maize zein genes. *Mol. Gen. Genet.*, **231**, 369-374.

Riechmann, J.L., Hears, J., Martin, G. *et al.* (2000) *Arabidopsis* transcription factors: genome-wide comparative analysis among eukaryotes. *Science*, **290**, 2105-2110.

Roeder, R.G. (1998) Role of general and gene-specific cofactors in the regulation of eukaryotic transcription. *Cold Spring Harb. Symp. Quant. Biol.*, **63**, 201-208.

Romero, I., Fuertes, A., Benito, M.J., Malpica, J.M., Leyva, A. and Paz-Ares, J. (1998) More than 80 *R2R3–MYB* regulatory genes in the genome of *Arabidopsis thaliana*. *Plant J.*, **14**, 273-284.

Ronald, P.C., Salmeron, J.M., Oldroyd, G.E.D. and Staskawicz, B.J. (1992) The cloned avirulence gene *avrPto* induces disease resistance in tomato cultivars containing the *Pto* resistance gene. *J. Bacteriol.*, **174**, 1604-1611.

Rouse, D., Mackay, P., Stirnberg, P., Estelle, M. and Leyser, O. (1998) Changes in auxin response from mutations in an *Aux/IAA* gene. *Science*, **279**, 1371-1373.

Rosinski, J.A. and Atchley, W.R. (1998) Molecular evolution of the Myb family of transcription factors: evidence for polyphyletic origin. *J. Mol. Evol.*, **46**, 74-83.

Ryals, J.A., Neuenschwander, U.H., Willits, M.G., Molina, A., Steiner, H.Y. and Hunt, M.D. (1996) Systemic acquired resistance. *Plant Cell*, **8**, 1809-1819.

Ryseck, R.P. and Bravo, R. (1991) c-JUN, JUN B, and JUN D differ in their binding affinities to AP-1 and CRE consensus sequences: effect of FOS proteins. *Oncogene*, **6**, 533-542.

Sainz, M.B., Grotewold, E. and Chandler, V.L. (1997) Evidence for direct activation of an anthocyanin promoter by the maize C1 protein and comparison of DNA binding by related Myb domain protein. *Plant Cell*, **9**, 611-625.

Schindler, U., Menkens, A.E., Beckmann, H., Ecker, J.R. and Cashmore, A.R. (1992) Heterodimerization between light-regulated and ubiquitously epressed *Arabidopsis* GBF bZIP proteins. *EMBO J.*, **11**, 1261-1273.

Schmidt, R.J., Burr, F., Aukeman, M.J. and Burr, B. (1990) Maize regulatory gene opaque-2 encodes a protein with a leucine-zipper motif that binds to zein DNA. *Proc. Natl Acad. Sci. USA*, **87**, 46-50.

Schultz, T.F., Spiker, S. and Quatrano, R.S. (1996) Histone H1 enhances the DNA binding activity of the transcription factor EmBP1. *J. Mol. Biol.*, **271**, 25,742-25,745.

Schultz, T.F., Medina, J., Hill, A. and Quatrano, R.S. (1998) 14-3-3 proteins are part of an abscisic acid-viviparous1 (VP1) response complex in Em promoter and interact with VP1 and EmBP1. *Plant Cell*, **10**, 837-847.

Schwarz-Sommer, Z., Huijser, P., Nacken, W., Saedler, H. and Sommer, H. (1990) Genetic control of flower development:homeotic genes in *Antirrhinum majus*. *Science*, **250**, 931-936.

Schwarz-Sommer, Z., Hue, I., Huijser, P. *et al.* (1992) Characterisation of the *Antirrhinum* floral homeotic MADS box gene *deficiens*: evidence for DNA binding and autoregulation of its persistent expression throughout flower development. *EMBO J.*, **11**, 251-263.

Sedgwick, S.G. and Smerdon, S.J. (1999) The ankyrin repeat: A diversity of interactions on a common structural framework. *Trends Biochem. Sci.*, **24**, 311-316.

Smale, S.T., Jain, A., Kaufmann, J., Emami, K.H., Lo, K. and Garraway, I.P. (1998) The initiator element: a paradigm for core promoter heterogeneity within metazoan protein-coding genes. *Cold Spring Harb. Symp. Quant. Biol.*, **63**, 21-31.

Smith, H. (2000) Phytochromes and light signal perception by plants—an emerging synthesis. *Nature*, **407**, 585-591.

Solano, R., Stepanova, A., Chao, Q. and Ecker, J. (1998) Nuclear events in ethylene signalling: a transcriptional cascade mediated by ETHYLENE6INSENSITIVE3 and ETHYLENE-RESPONSIVE-FACTOR1. *Genes Dev.*, **12**, 3703-3714.

Stockinger, E.J., Gilmour, S.J. and Thomashow, M.F. (1997) *Arabidopsis thaliana* CBF1 encodes an AP2 domain-containing transcription activator that binds to the *C*-repeat/DRE, a *cis*-acting DNA regulatory element that stimulates transcription in response to low temperature and water deficit. *Proc. Natl Acad. Sci. USA*, **94**, 1035-1040.

Sugano, S., Andronis, C., Green, R.M., Wang, Z.Y. and Tobin, E.M. (1998) Protein kinase CK2 interacts with and phosphorylates the *Arabidopsis* circadian clock-associated 1 protein. *Proc. Natl Acad. Sci. USA*, **95**, 11,020-11,025.

Sugano, S., Andronis, C., Ong, M.S., Green, R.M. and Tobin, E.M. (1999) The protein kinase CK2 is involved in regulation of circadian rhythms in *Arabidopsis*. *Proc. Natl Acad. Sci. USA*, **96**, 12,362-12,366.

Suzuki, M., Kao, C.Y. and McCarty, D.R. (1997) The conserved B3 domain of *VIVIPAROUS1* has a cooperative DNA binding activity. *Plant Cell*, **9**, 799-807.

Szymanski, D.B., Jilk, R.A., Pollock, S.M. and Marks, M.D. (1998) Control of *GL2* expression in *Arabidopsis* leaves and trichome. *Development*, **125**, 1161-1171.

Szymanski, D.B., Lloyd, A.M. and Marks M.D. (2000) Progress in the molecular genetic analysis of trichome initiation and morphogenesis in *Arabidopsis*. *Trends Plant Sci.*, **5**, 214-219.

Tang, X., Frederick, R.D., Zhou, J., Halterman, D.A., Jia, Y. and Martin, G.B. (1996) Initiation of plant disease resistance by physical interaction of AvrPto and thePto kinase. *Science*, **274**, 2060-2063.

Teraghi, W.B. and Cashmore, A.R. (1995) Light-regulated transcription. *Annu. Rev. Plant Physiol. Plant Mol. Biol.*, **46**, 445-474.

The *Arabidopsis* Genome Initiative (2000) Analysis of the genome sequence of the flowering *Arabidopsis thaliana*. *Nature*, **408**, 796-815.

Ulmasov, T., Hagen, G. and Guifoyle, T.J. (1997a) AFR1, a transcription factor that binds auxin response elements. *Science*, **276**, 1865-1868.

Ulmasov, T., Murfett, J., Hagen, G. and Guifoyle, T.J. (1997b) Aux/IAA proteins repress expression of reporter genes containing natural and highly active synthetic auxin response elements. *Plant Cell*, **9**, 1963-1971.

Ulmasov, T., Hagen, G. and Guilfoyle, T.J. (1999a) Activation and repression of transcription by auxin-response factors. *Proc. Natl Acad. Sci. USA*, **96**, 5844-5849.

Ulmasov, T., Hagen, G. and Guifoyle, T.J. (1999b) Dimerization and DNA binding of auxin response factors. *Plant J.*, **19**, 309-319.

Verrijzer, C.P. and Tjian, R. (1996) TAFs mediate transcriptional activation and promoter selectivity. *Trends Biochem. Sci.*, **21**, 338-342.

Vicente-Carbajosa, J., Moose, S.P., Parsons, R.L. and Schmidt, R.J. (1997) A maize zinc-finger protein binds the prolamin box in zein gene promoters and interacts with the basic leucine zipper transcriptional activator Opaque2. *Proc. Natl Acad. Sci. USA*, **94**, 7685-7690.

Wada, T., Tachinana, T., Shimura, Y. and Okada, K. (1997) Epidermal cell differentiation in *Arabidopsis* determined by a Myb homolog, CPC. *Science*, **277**, 1113-1116.

Walker, A.R., Davison, P.A., Bolognesi-Winfield, A.C. *et al.* (1999) The *TRANSPARENT TESTA GLABRA1* locus, which regulates trichome differentiation and anthocyanin biosynthesis in *Arabidopsis*, encodes a WD40 repeat protein. *Plant Cell*, **6**, 1065-1076.

Wang, Z.Y. and Tobin, E.M. (1998) Constitutive expression of the *CIRCADIAN CLOCK ASSOCIATED 1 (CCA1)* gene disrupts circadian rhythms and suppresses its own expression. *Cell*, **93**, 1207-1217.

Wang, Z.Y., Kenigsbuch, D., Sun, L., Harel, E., Ong, M.S. and Tobin, E.M. (1997) A Myb-related transcription factor is involved in the phytochrome regulation of an *Arabidopsis Lhcb* gene. *Plant Cell*, **9**, 491-507.

Ward, E., Uknes, S., Williams, S.C. *et al.* (1991) Coordinate gene activity in response to agents that induce systemic acquired resistance. *Plant Cell*, **3**, 1085-1094.

Weigel, D. (1995) The APETALA2 domain is regulated to a novel type of DNA binding domain. *Plant Cell*, **7**, 388-389.

Weigel, D. and Meyerowitz, E.M. (1994) The ABCs of floral homeotic genes. *Cell*, **78**, 203-209.

Weston, K. (1999) MYB proteins in life, death and differentiation. *Curr. Opin. Genet. Dev.*, **8**, 76-81.

Wilson, K., Long, D., Swinburne, J. and Coupland, G. (1996) A dissociation insertion causes a semidominant mutation that increases expression of *TINY*, an *Arabidopsis* gene related to *APETALA2*. *Plant Cell*, **8**, 659-671.

Yanagisawa, S. (1996) Dof DNA-binding proteins contain a novel zinc finger motif. *Trends Plant Sci.*, **1**, 213.

Zawel, L. and Reinberg, D. (1995) Common themes in assembly and function of eukaryotic transcription factors. *Annu. Rev. Biochem.*, **64**, 533-561.

Zhang, B. and Singh, K.B. (1994) ocs element promoter sequences are activated by auxin and salicylic acid in *Arabidopsis*. *Proc. Natl Acad. Sci. USA*, **91**, 2507-2511.

Zhang, Y., Fan, W., Kinkema, M., Li, X. and Dong, X. (1999) Interaction of NPR1 with basic leucine zipper protein transcription factors that bind sequences required for salicylic acid induction of the *PR-1* gene. *Proc. Natl Acad. Sci. USA*, **96**, 6523-6528.

Zhou, J., Tang, X. and Martin, G.B. (1997) The Pto kinases conferring resistance to tomato bacterial speck disease interacts with proteins that bind a *cis*-element of pathogenesis-related genes. *EMBO J.*, **16**, 3207-3218.

Zhou, J.M., Trifa, Y., Silva, H. *et al.* (2000) NPR1 differentially interacts with members of the TGA/OBF family of transcription factors that bind an element of the *PR-1* gene required for induction by salicylic acid. *Mol. Plant Microbe Interact.*, **13**, 191-202.

Zhu, Y., Tepperman, J.M., Fairchild, C.D. and Quail, P.H. (2000) Phytochrome B binds with greater apparent affinity than phytochrome A to the basic helix-loop-helix factor PIF3 in a reaction requiring the PAS domain of PIF3. *Proc. Natl Acad. Sci. USA*, **97**, 13,419-13,424.

11   Calmodulin

Teerapong Buaboocha and Raymond E. Zielinski

11.1   Introduction

Plant cells respond to a variety of stimuli via changes in their intracellular Ca^{2+} concentration ($[Ca^{2+}]_i$), which are perceived primarily by calmodulin (CaM), CaM-like domain protein kinases and a family of structurally related Ca^{2+}-binding proteins, the EF-hand family of Ca-modulated proteins (Harmon *et al.*, 2000; Zielinski, 1998). A considerable body of evidence has demonstrated that, in plants, increases in $[Ca^{2+}]_i$ are involved in transducing environmental signals produced by osmotic and temperature shocks, mechanical perturbation, light and symbiotic and pathogenic microorganisms (Gilroy and Trewavas, 2001; Trewavas and Malho, 1998). CaM transduces the signal of increased $[Ca^{2+}]_i$ by binding to and altering the activities of a variety of proteins (CaM-binding proteins or CaM-interacting proteins) under either the positive or negative influence of Ca^{2+}. CaM and CaM-interacting proteins may, in turn, cross-talk with other cellular signaling pathways. The activities of these interlaced, Ca^{2+}-dependent and Ca^{2+}-independent signaling pathways effect physiological responses to the vast array of specific stimuli received by plants.

The study of Ca^{2+} signaling in biological systems is both fascinating and frustrating. The use of Ca^{2+}-specific dyes and genetically encoded Ca^{2+} sensors have revealed that Ca^{2+} signals are rich in both spatial and temporal information (Berridge *et al.*, 2000; Gilroy and Trewavas, 2001; Trewavas and Malho, 1998). Frustratingly, however, the mechanisms by which this information is transduced are poorly understood. This is a consequence of several factors: Ca^{2+} signals are generated in response to virtually every stimulus tested; the primary non-enzymatic Ca^{2+} receptor, CaM, is a pleotropic regulatory molecule that interacts with a wide array of proteins and is encoded by a surprisingly large gene family. This has made genetic and reverse genetic analyses in multicellular eukaryotes ambiguous. Finally, CaM-interacting proteins cannot be identified by simple bioinformatic analyses, as their CaM-binding domains do not share a common sequence motif.

The capacity for transducing the information in Ca^{2+} signals is at least twofold. Firstly, although CaM is ubiquitously expressed, its interacting proteins often accumulate in tissue or organ-specific patterns. Thus Ca^{2+}-binding by CaM takes place in conjunction with these target proteins. Secondly, different CaM-target protein complexes have distinct affinities and kinetics for Ca^{2+}

binding and release. Protein–protein interaction involving CaM and CaM-interacting proteins, therefore, represent crucial components of the signaling pathways initiated by increases in $[Ca^{2+}]_i$.

This chapter will focus on the mechanisms by which CaM interacts with target proteins. Several aspects of CaM will be discussed, including structures, conformational changes upon Ca^{2+} binding, interactions with target proteins (including mechanisms and kinetics) and the characteristics of CaM-binding structures in target proteins. The possible physiological roles of plant CaM-binding proteins have been reviewed extensively in the recent past (Snedden and Fromm, 1998; Zielinski, 1998) and therefore will not be emphasized here. There are a number of examples of Ca^{2+}-independent CaM-binding proteins in animals (recently reviewed by Jurado et al., 1999), but little information is available on apo-CaM (the Ca^{2+}-free form of CaM)-binding proteins in plants. Therefore, this review will focus on Ca^{2+}-dependent interactions involving CaM. Although much of the mechanistic information on CaM and CaM-target interactions discussed here is derived from animal systems, all indications are that similar mechanisms operate in plants and relevant examples of plant proteins are included as much as possible.

11.2 Calmodulin sequences and structures

Calmodulin (CaM) is a small (148 residues), multifunctional protein that acts as a Ca^{2+} sensor in eukaryotic cells. Primary structures of CaM are generally conserved throughout evolution, especially in the functional motifs among various organisms including yeast, animals and plants. One striking difference is that numerous isoforms of CaM may occur within a single plant species. Figure 11.1a comprises a comparison of the primary structures of *Arabidopsis* CaM isoforms, as a representative of plants, with vertebrate and yeast CaM proteins (Gawienowski et al., 1993; Ling et al., 1991; Ling and Zielinski, 1993; Perera and Zielinski, 1992; Zielinski, 2001). It should be noted that gene families encoding CaM isoforms have also been reported and characterized in potato (Takezawa et al., 1995), soybean (Lee et al., 1995) and petunia (Rodriguez-Concepcion et al., 1999). Plant CaM proteins share about 90% and 60% identity with those from vertebrates and yeast, respectively. The residues directly responsible for Ca^{2+}-binding are highly conserved, as well as non-Ca^{2+}-coordinating residues that are important for CaM function. Four regions of CaM are delineated in figure 11.1a, each encompassing an EF-hand Ca^{2+}-binding motif. These regions will be referred to here as domains I to IV, with I closest to the *N*-terminus of the protein. The amino acid sequence similarity of these regions has been interpreted as indicating that the four-domain structure of CaM arose from two gene duplication events (Kawasaki and Kretsinger, 1994). These domains are paired (I with II and III with IV) to form two globular

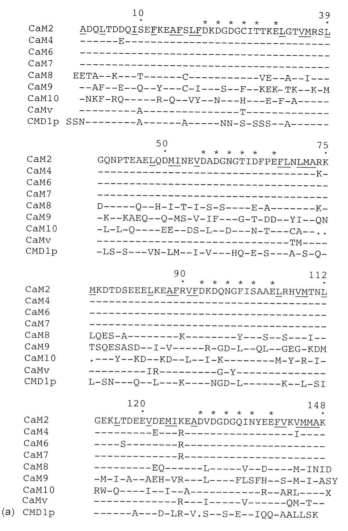

Figure 11.1(a) Amino acid sequence alignment of *Arabidopsis* CaM isoforms and CaM proteins from vertebrates (CaMv) and *Saccharomyces cerevisiae* (CMD1p). The sequences are compared with *Arabidopsis* CaM2 as a standard; identical residues in other sequences are indicated by a dash (−), and gaps introduced for alignment purposes are indicated by dots (·). The sequences are arranged to show the relationships among the four Ca^{2+}-binding domains of the molecules. Residues serving as Ca^{2+}-binding ligands are marked with asterisks (∗). Strongly hydrophobic amino acids are underlined. Note that strongly hydrophobic patches flank each of the four Ca^{2+}-binding loops. The residue labeled X in the CaM10 sequence represents a 45-residue extension unrelated to other CaM sequences that was omitted from the alignment. GenBank accession numbers for the sequences used in the alignment are: CaM2 (M38380); CaM4 (Z12022); CaM6 (Z12024); CaM7 (AF178073); CaM8 (AF178074); CaM9 (AF178075); CaM10 (Z12136); CaMv (M65156); and CMD1p (M14760).

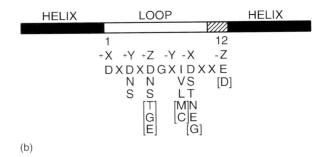

(b)

Figure 11.1(b) Schematic representation of EF-hand structure. Amino acids most commonly found in CaM EF-hands are indicated in one-letter nomenclature, where X represents any amino acid. The Ca^{2+}-binding residues are indicated in Arabic numerals and given in Cartesian coordinates. Alternative residues in each position commonly found in CaM are indicated; bracketed residues are ones commonly found in EF-hand proteins other than CaM.

regions, which, for convenience, will be termed the *N*- and *C*-terminal globular domains. The sequences most commonly found in the EF-hands of CaM are indicated in figure 11.1b. Specific roles of the EF-hand residues in Ca^{2+} binding are discussed in section 11.2.1.

The sequence alignment shown in figure 11.1a also suggests the importance of hydrophobic interaction in the mechanism of CaM function. The positions of the phenylalanine (Phe or F), and methionine (Met or M) residues are conserved in virtually all CaMs, with the exception of the Met–Met dipeptide at positions 145–146 in plant CaMs, which is displaced one residue when compared with the vertebrate proteins. With the exception of Phe_{99}, the location of all Phe residues is also conserved between plant, vertebrate, and yeast CaMs. Furthermore, regions in plant CaM comprising the remaining hydrophobic residues [alanine (A), leucine (L), isoleucine (I) and valine (V); CaM generally does not contain tryptophan (W)] are highly conserved functionally in their relative locations in all CaMs, including the highly diverged *Arabidopsis* proteins CaM8, CaM9 (Zielinski, 2001) and CaM10 (Ling and Zielinski, 1993). These residues are highlighted in figure 11.1a. This pattern of conservation is consistent with the finding that about 80% of the contacts between CaM and its target proteins are hydrophobic interactions rather than charge-to-charge interactions (reviewed in Crivici and Ikura, 1995).

11.2.1 EF-hands, Ca^{2+}-binding and intermolecular tuning by target proteins

Metal binding by CaM is an important consideration because Ca^{2+} enhances the many protein–protein interactions involving CaM and in some cases it negatively regulates CaM–target protein interaction. A portion of the ability of CaM to bind Ca^{2+} specifically is derived from its primary structural motif, the EF-hand.

Amino acid residues within and immediately surrounding EF-hands modulate this intramolecular affinity for Ca^{2+} (Falke *et al.*, 1994; Linse and Forsen, 1995). A crucial, but often unappreciated, aspect of Ca^{2+}-binding by CaM is that it is sensitized and tuned to a specific macroscopic affinity by the interaction of CaM with its target proteins. This cooperative mode of action has been termed 'intermolecular tuning' (Peersen *et al.*, 1997), and will be discussed below and in conjunction with the structural constraints governing CaM–target protein interaction.

Structures of apo-CaM and Ca^{2+}-loaded CaM have been determined by NMR and X-ray crystallography (Figure 11.2a) (Babu *et al.*, 1988; Chattopadhyaya *et al.*, 1992). When crystallized under these conditions, the protein resembles a dumb-bell, with the ends consisting of the *N*- and *C*-terminal globular domains and the handle consisting of an extended central α-helix formed by residues 65 to 92. Each globular domain contains two Ca^{2+}-binding sites, which allow the protein to bind a total of four ions. The Ca^{2+}-binding sites are composed of a characteristic helix-loop-helix motif called an EF-hand (Kretsinger and Nockholds, 1973), illustrated schematically in figure 11.1b. Each loop, including

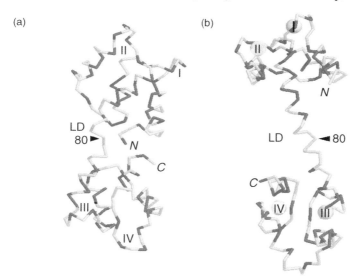

Figure 11.2 Comparison of the structures of (a) apo-CaM determined by NMR spectroscopy and (b) Ca^{2+}-CaM determined by X-ray crystallography. The polypeptide backbones of the molecules are shown in which hydrophobic residues are shaded to emphasize their self-association in (a) and solvent exposure in (b). Liganded Ca^{2+} ions are represented by shaded spheres and the EF-hands are numbered (I–IV) beginning with the *N*-terminus. The amino (*N*) and carboxy (*C*) termini of the protein and the flexible linker domain (LD) are indicated, as is residue Asp_{80}, which is approximately the center of the flexible region. Structure codes 1 CFD and 3CLN were used in (a) and (b), respectively (available at http://www.pdb.bnl.gov/) and were modeled using the program Protein Explorer (available at http://www.umass.edu/microbiol/chime/explorer/).

the end of the second flanking helix, provides seven ligands for binding Ca^{2+} with a pentagonal bipyramid geometry. The residues of the loop are either numbered 1 to 12 or described as a series of points in a Cartesian coordinate system with Ca^{2+} at the center to designate the Ca^{2+}-binding residues. Three ligands for Ca^{2+} coordination are provided by carboxylate oxygens from residues 1 (+X), 3 (+Y) and 5 (+Z), one from a carbonyl oxygen in residue 7 (−Y), and two from carboxylate oxygens in residue 12 (−Z), which is a highly conserved Glu. The seventh ligand is provided either by a carboxylate side chain from residue 9 (−X), or from a water molecule recruited via the side chain or carbonyl oxygen of residue 9. CaM binds Ca^{2+} preferentially over Mg^{2+} (even though Mg^{2+} is present at 10^3- to 10^4-fold excess in the cytosol) for two reasons: coordination number and ion-binding cavity size (Falke $et\ al.$, 1994; Cates $et\ al.$, 1999; Peersen $et\ al.$, 1997). Mg^{2+} prefers sixfold coordination while Ca^{2+} favors sevenfold coordination. Thus the role of the E residue invariably found in position 12 (−Z) of each EF-hand of CaM is crucial. This residue may rotate to give bidentate or monodentate metal ion chelation, favoring Ca^{2+} or Mg^{2+}, respectively. Mutation of Glu_{12} to Asp reduced the affinity of EF-hands for Ca^{2+} in parvalbumin by 100-fold and raised the affinity for Mg^{2+} by tenfold (Cates $et\ al.$, 1999). Consistent with this finding is the observation that non-regulatory EF-hands, which are constantly occupied, usually by Mg^{2+}, possess Asp in the 12th (−Z) position (Johnson and Potter, 1981). Thus it is likely that the EF-hand in domain II of $Arabidopsis$ CaM9 binds Mg^{2+} rather than Ca^{2+}. The overall sizes of the EF-hands of CaM also favor the larger ionic radius of Ca^{2+} (ca. 1.06 Å) over that of Mg^{2+} (ca. 0.6 Å). Here again, E in the 12th position plays an important role by allowing a larger ion-binding cavity size favorable for Ca^{2+} binding when compared with the shorter D. Although Mg^{2+}-binding by CaM has been characterized, its physiological relevance is questionable. Ohki $et\ al.$ (1997) used non-physiologically high levels of Mg^{2+} in studies of ion binding and target protein recognition, but more recent studies of ion effects on target protein binding studied by surface plasmon resonance (Ozawa $et\ al.$, 1999) indicate that physiological levels of Mg^{2+} have little or no effect on target binding by CaM.

In spite of the fact that the geometry of CaM's EF-hands are optimized for binding Ca^{2+} specifically over Mg^{2+}, its affinity for Ca^{2+} is only modest. Saturation of CaM's four EF-hands with Ca^{2+} $in\ vitro$ requires nonphysiologically high (> 10 µM) levels of Ca^{2+} (Falke $et\ al.$, 1994; Linse and Forsen, 1995). Resolution of this paradox was initially provided by Storm and colleagues (Olwin $et\ al.$, 1984; Olwin and Storm, 1985), who observed that in the presence of a CaM-binding protein, Ca^{2+} binding by CaM saturates at physiologically relevant concentrations of the ion (ca. 1 µM). Similar effects have been reported on the Ca^{2+}-dependence of glutamate decarboxylase (Snedden $et\ al.$, 1996) and NAD kinase (Liao $et\ al.$, 1996) activation by plant CaM, which indirectly reflect the affinity of the target protein–CaM complex for Ca^{2+}. Similarly,

CaM-binding peptides derived from CaM-regulated enzymes enhance CaM's ability to bind Ca^{2+} (Haiech *et al.*, 1991; Peerson *et al.*, 1997). This cooperative binding effect is sufficiently strong to suppress mutations to Ca^{2+} coordinating residues in EF hands and to restore the macroscopic K_d for Ca^{2+} by CaM to near wild-type affinity (Findlay *et al.*, 1995; Haiech *et al.*, 1991). Since $K_d = k_{off}/k_{on}$, the effect of a target protein on Ca^{2+} binding by CaM could be to increase k_{on} or decrease k_{off}. Several studies indicate that CaM-binding peptides and proteins can alter either or both kinetic parameters (Brown *et al.*, 1997; Johnson *et al.*, 1996; Kasturi *et al.*, 1993; Olwin and Storm, 1985; Peersen *et al.*, 1997; Persechini *et al.*, 1996). Thus the general picture that has emerged is that the second messenger Ca^{2+} is perceived by CaM acting in concert with its target proteins. These interactions serve as an intermolecular tuning mechanism that transduces informationally complex signals into diverse responses by selective activation and inactivation of different CaM–target protein complexes.

11.2.2 Hydrophobic interaction: Met puddles and conserved Phe residues

The extensive conservation of many hydrophobic residues observed in the comparison between amino acid sequences of CaM from different organisms (underlined in figure 11.1a) implies the importance of hydrophobic interaction in the mechanism of CaM–target protein complex formation. This inference is supported by the X-ray structure of Ca^{2+}-loaded CaM (Babu *et al.*, 1988; Chattopadhyaya *et al.*, 1992), which revealed large, surface exposed hydrophobic patches in both globular domains of the molecule. Each hydrophobic patch is formed by the side chains of 14 residues, including many highly conserved Met and Phe residues that are dispersed throughout the primary structure of CaM. These patches can be seen in figure 11.2, which compares the structures of apo-CaM and Ca^{2+}-loaded CaM. In apo-CaM, the hydrophobic residues (indicated by shading in figure 11.2a) are generally sequestered in globular structures by self-association. In contrast, Ca^{2+}-binding by CaM results in important tertiary structure changes in the globular domains of the molecule (Zhang *et al.*, 1995; Kuboniwa *et al.*, 1995; Finn *et al.*, 1995). A major difference is in the arrangement of EF-hands. In Ca^{2+}-free CaM, the four helices in each globular domain form a compact bundle, termed a 'closed form', whereas the analogous structure in Ca^{2+}-loaded CaM appears not as tightly packed, and is termed an 'open form'. In concert with this tertiary structural change, another obvious difference is the opening of a central deep hydrophobic cavity in CaM induced by Ca^{2+}-binding. This can be seen most clearly in figure 11.2b for the *C*-terminal globular domain where the shaded hydrophobic residues face a central, solvent-exposed cavity. Upon Ca^{2+}-binding, the two helices in each EF-hand become more perpendicular, which can be systematically determined by the degree of change in their interhelical angles. This movement leads to the exposure of interacting hydrophobic residues, which are buried in the tightly bundled helices in the closed Ca^{2+}-free state (Finn *et al.*, 1995). The hydrophobic residues are derived mainly from the Ca^{2+}-binding loop and the second helix of the EF-hand (Ikura, 1996). Together with several acidic residues from the first helix of the EF-hand, these residues construct a hydrophobic patch with a large hydrophobic cavity in the middle flanked by some acidic amino acids in each globular domain. Met is particularly important because of the observation that the solvent accessibility of several Met residues is significantly reduced by the removal of Ca^{2+} (Zhang *et al.*, 1995). All these changes are initiated by rearrangements in the EF-hands when the negatively charged side chains and backbone carbonyl groups coordinate Ca^{2+}. The electrostatic interactions with Ca^{2+} ions neutralize the negative charges of the Ca^{2+}-binding loops and provide the driving force for helix movement and hydrophobic patch exposure (Linse and Forsen, 1995; Zhang *et al.*, 1995).

Met residues are unusually numerous in CaM and are estimated to contribute nearly half of the accessible surface area of the hydrophobic patches in regions referred to as Met puddles or patches (O'Neil and DeGrado, 1990). Met puddles are likely to be the major reason CaM can interact with target proteins in a sequence-independent manner (O'Neil and DeGrado, 1990; Gellman, 1991). This situation appears to be the result of two properties of this amino acid. First, Met has considerable conformational flexibility compared with the aliphatic amino acids, Leu, Ile and Val. Flexibility arises because the C_γ–S bond of the Met side-chain is slightly longer and has a lower energy barrier for rotation than the C_γ–C_ϵ bond of the aliphatic amino acids (Gellman, 1991). Consistent with this idea is the variety of configurations of the C_γ–S bond observed in the crystal structures of many proteins (Janin *et al.*, 1978). Secondly, sulfur atoms are more easily polarized when compared with carbon atoms. Weakly polarized Met residues allow for the exposure of Met-rich hydrophobic patches on CaM's surface by facilitating interaction with highly polarized solvent water more readily than would hydrophobic amino acids with aliphatic side-chains. This prediction is supported by comparisons of the structures of apo- and Ca^{2+}-loaded CaM (figure 11.2 and Zhang *et al.*, 1995) and by studies of spin labeling of apo-CaM, Ca^{2+}-loaded CaM and CaM–target peptide complexes (Yuan *et al.*, 1999b). The role of Met in mediating protein–protein interactions in CaM was tested by systematically changing Met to Leu (Zhang *et al.*, 1994). While Met-to-Leu mutants retained their overall structure as evidenced by NMR spectroscopy, many of the mutants perturbed the ability to CaM to activate cyclic nucleotide phosphodiesterase. Consistent with the importance of Met residues in target recognition by CaM are the observations that NAD kinase activation was reduced substantially by substitution of charged residues for the Met–Met dipeptide in the *C*-terminal domain of *Arabidopsis* CaM (Liao *et al.*, 1996). Further, Met-to-Glu substitutions in the *C*-terminal globular domain of CaM impair the activation of protein kinases

(Chin and Means, 1996). The flexibility of Met residues in mediating sequence-independent protein–protein interaction is also observed in the system by which signal peptides are recognized and targeted to the endoplasmic reticulum for co-translational translocation by signal recognition particles (Bernstein *et al.*, 1989). Signal peptides, like most CaM-binding domains, are amphiphilic α-helices with no common amino acid sequence motif. The 54 kDa component of the signal recognition particle contains surface exposed Met residues that mediate sequence-independent docking of the signal recognition particle with apolar domains of signal peptides to facilitate their association with the endoplasmic reticulum.

Means and co-workers extensively studied the role of hydrophobic interaction in CaM by mutating each of the nine Met residues to Gln, a more polar, but similarly sized, amino acid. The ability of each mutant protein to bind and activate smooth muscle myosin light chain kinase (smMLCK), CaM-dependent protein kinase IIα (CaM-KII), and CaM-dependent protein kinase IV (CaM-KIV) was determined (Chin and Means, 1996). Most of the Gln mutants of the N-terminal domain produced maximal enzyme activity and small changes in the affinity for all enzymes tested. The exception was the loss in maximal activity of smMLCK by the substitutions at Met_{71} and Met_{72} without affecting the affinity, which may be attributed to the more extensive contacts in the CaM-smMLCK complex than in the CaM–CaM–KII complex (Meador *et al.*, 1992, 1993). These results indicated the necessity of the N-terminal domain of CaM in smMLCK activation but lesser involvement in determining binding affinity. In contrast, the Gln mutants of the C-terminal domain showed decreased affinity for smMLCK, especially in residues Met_{109} and Met_{124}. This indicates that Met_{109} and Met_{124} are involved in generating a high affinity interaction between the C-terminal domain of CaM and different target proteins. In the structure of CaM-target peptide complexes, Met_{124} interacts with a conserved, large hydrophobic side chain in the CaM-binding domains of three CaM-dependent kinases: Leu_{229} in CaM-KII (Meador *et al.*, 1993), Trp_{800} in smMLCK (Meador *et al.*, 1992) and Trp_{580} in skMLCK (Ikura *et al.*, 1992). Interaction of these residues with the hydrophobic patch on CaM, including Met_{124}, was proposed to be a significant factor in defining the specificity between CaM and CaM-binding proteins (Chin and Means, 1996). From these studies, a model of CaM-dependent activation of a target enzyme was proposed in which residues in the C-terminal domain of CaM including Met_{124} initially interact with a large hydrophobic side chain of the CaM-binding domain of the target protein at resting Ca^{2+} concentrations. As Ca^{2+} level increases, both domains become occupied by Ca^{2+} and move closer, which is facilitated by the flexible central helix described in section 11.2.3 below. Finally, the N-terminal domain interacts with the target protein and leads to enzyme activation. This aspect of the model is supported by mutational analyses of both the smMLCK CaM-binding domain and in the N-terminal globular domain of CaM (Chin *et al.*, 1997) and is elaborated in section 11.3.3 below.

Phenylalanine residues in CaM are particularly striking because they are absolutely conserved in the primary structure of CaM across eukaryotic phylogeny (with the exception that Phe_{99} in the Ca^{2+}-binding loop of EF-hand III in plant and vertebrate CaM is replaced by Leu in yeast). This arrangement suggested a functional role to Ohya and Botstein (1994a,b) who systematically constructed a series of 33 combinatorial mutants of the conserved Phe residues in yeast CaM (CMD1p). They used these site directed mutants to complement a yeast CaM null allele (*cmd1*) and tested the transformants for temperature sensitive (*ts*) phenotypes. In the *N*-terminal globular domain, all Phe_{16} and Phe_{19} mutants (*cmd1B*) failed to localize CaM to the bud and all Phe_{65} and Phe_{68} mutants (*cmd1C*) had nuclear division defects. In the *C*-terminal globular domain, Phe_{92} (*cmd1A*) mutants had disrupted actin organization and Phe_{89} and Phe_{140} (*cmd1D*) displayed bud emergence (i.e. growth polarization) defects. The Phe_{12} mutants appeared to enhance the effects of other Phe to Ala mutations in the molecule. Together, these results provide a simple and elegant *in vivo* demonstration that hydrophobic residues play a key role in CaM function and support the idea that different regions of CaM are critically required for interaction with different groups of target proteins. These studies are currently being extended by biochemical and cell biology analyses to develop a better understanding of which specific targets of CaM regulation are responsible for the phenotypes associated with the *cmd1A*, *B*, *C* and *D* complementation groups (Okano *et al.*, 1998; Sekiya-Kawasaki *et al.*, 1998).

11.2.3   *The central helix: a flexible linker domain*

The structure of Ca^{2+}-loaded CaM, revealed by X-ray crystallography (figure 11.2b), suggests the presence of a most unusual structure in CaM; a long, solvent exposed central α-helix (Babu *et al.*, 1988). In the crystal structure, this helix, which includes the extreme ends of an EF-hand motif from each globular domain, was reasonably well defined. However, the authors commented that the ϕ and ψ dihedral angles of the peptide bonds between residues 79/80 and 80/81 deviate from ideal α-helical geometry and noted that H-bonds between the carbonyl oxygens and main chain nitrogen atoms in this region are longer than usual. The significance of these observations was revealed when the secondary structure of *Drosophila* CaM was determined by NMR spectroscopy (Ikura *et al.*, 1991). In solution, residues Asp_{78} to Ser_{81} of the central helix adopt a non-helical conformation with considerable flexibility. The flexibility of this region is key to the mechanism of action of CaM and had been suggested by biochemical analyses of CaM function.

In a critical study, Persechini and Kretsinger (1988) demonstrated by chemical cross-linking that the *N*- and *C*-terminal globular domains of CaM lie in close proximity to one another when CaM interacts with skeletal muscle myosin light chain kinase (skMLCK). This result prompted the suggestion that the central helix of CaM acts as a flexible linker, thereby allowing CaM to interact with peptides of different lengths. Consistent with this hypothesis, CaM with one (Glu_{84}), two (Glu_{83} and Glu_{84}) or four (Ser_{81} to Glu_{84}) deleted residues in the central helix showed little or no difference in activating or binding skMLCK, NAD kinase and calcineurin compared with *wt* protein (Persechini *et al.*, 1989). Likewise, substitution of Lys for Glu in positions 82 to 84 did not alter the ability of CaM to activate MLCK (Craig *et al.*, 1987). The most dramatic demonstration of the function of this region, however, was provided by experiments in which the central helix of CaM was replaced by the analogous region of troponin C (TnC) and yielded a chimeric protein capable of activating CaM-KII (George *et al.*, 1993). TnC is a component of the troponin complex of striated and cardiac muscle that sensitizes muscle tension to changes in cytosolic Ca^{2+}. It is an EF-hand protein that shares ca. 55% amino acid sequence identity with CaM and has a similar overall conformation. CaM can substitute for TnC *in vitro*, but TnC cannot activate all CaM-regulated proteins (Kawasaki and Kretsinger, 1994). These observations are consistent with the idea that the central, highly charged region of CaM does not interact with target proteins directly. Rather, it acts as a flexible linker domain that bends to position the *N*- and *C*-terminal globular domains of CaM properly for docking with CaM binding domains of variable sizes in different CaM-target proteins. Confirmation of this role for the central region of CaM came from studies of the structures of CaM–peptide complexes (Ikura *et al.*, 1992; Meador *et al.*, 1992, 1993), including those that were formed using CaM deletion or substitution mutants in the flexible linker domain (Meador *et al.*, 1995). CaM mutants with deletions in the central linker domain also functionally complement CaM function in yeast (Persechini *et al.*, 1991).

11.2.4 Additional structure–function relationships of CaM

There are additional residues that confer CaM's ability to activate target proteins but which do not interact directly with the CaM-binding domain. To identify those critical residues, functional domains and subdomains between CaM and TnC were exchanged to produce chimeric mutant proteins and their abilities to activate target proteins were examined. Because of the inability of TnC to activate most of CaM-binding proteins, determining the binding and activation ability of these chimeras was reasoned to define where in the CaM molecule critical residues for each target protein reside. One target protein analyzed by such an approach was smMLCK. CaM substituted with the first (George *et al.*, 1990) or the third (George *et al.*, 1993) Ca^{2+}-binding domain from TnC exhibited severely impaired activation of smMLCK. Individual substitutions directed by the sequence of TnC were further used to identify specific activating residues on each domain of CaM, including Thr_{34} and Ser_{38} from helix 2, and Thr_{110}, Leu_{112} and Lys_{115} from helix 6 (Su *et al.*, 1994; Van Berkum and Means, 1991). Comparison of the structures of CaM and the CaM–smMLCK complex showed that even though these residues are far apart in CaM, peptide binding brings them into close proximity. Su *et al.* (1994) hypothesized that these residues together might function as an activating domain for smMLCK when the protein binds to CaM, unwinding the central helix and bringing the *N*- and *C*-terminal domains of CaM in a proper orientation on the smMLCK CaM-binding domain. However, it is likely that CaM does not activate all target enzymes in precisely the same way, probably because the variations in sequence and size of CaM-binding domains (described in section 11.3.1) causes CaM to be positioned differently when it is bound to different target enzymes. This idea is supported by similar, but more limited analyses of other CaM-regulated enzymes (George *et al.*, 1996).

11.2.5   CaM isoforms in plants: what are the limits of the CaM gene family?

One of the defining characteristics of CaM in plants is the expression of multiple CaM isoforms. In *Arabidopsis* these proteins vary between 49% and 99% in amino acid sequence identity (corresponding to 75 to 1 sequence substitutions, respectively) with the most highly expressed CaM isoform represented by the CaM2 sequence shown in figure 11.1a, which is encoded by a family of three genes (Zielinski, 1998). Animals and yeast express EF-hand CaM-like proteins that serve specialized functions *in vivo* (Kawasaki and Kretsinger, 1994). Thus a logical question is what defines a 'true' CaM and distinguishes it from a CaM-like protein that serves a distinct role *in vivo*? Two general approaches have been taken to address this question.

11.2.5.1   Biochemical analyses

The broad significance of multiple CaM isoforms in plants is not understood. However, a frequently proposed hypothesis is that diverged CaM isoforms may activate selected subsets of target proteins involved in Ca^{2+}-mediated signal transduction (Heo *et al.*, 1999; Liao *et al.*, 1996; Snedden and Fromm, 1998). A number of biochemical studies demonstrating differential regulation of target proteins by plant CaM isoforms support this idea. CaM from *Arabidopsis* was shown to stimulate differentially the activity of NAD kinase (Liao *et al.*, 1996), phosphodiesterase (Reddy *et al.*, 1999) and the binding of the transcription factor TGA3 to a region of CaM3 gene (*Cam3*) promoter (Szymanski *et al.*, 1996). CaM isoforms from soybean also differentially activate NAD kinase. The enzyme was activated by highly conserved SCaM1, but not by the divergent soybean CaM isoform, SCaM4, which is the apparent ortholog of CaM8 from *Arabidopsis* (Lee *et al.*, 1995; Zielinski, 2001). Domains responsible for this differential activation were determined by testing chimeric SCaMs generated by exchanging functional domains between SCaM1 and SCaM4. Domain I was found to play a key role in the differential activation of NAD kinase.

Single residue substitution mutants in domain I, evaluated for their ability to activate NAD kinase, showed that Lys_{30} and Glu_{40} were critical for the activation (Lee *et al.*, 1997). In addition, SCaM1 and SCaM4 were shown to activate mammalian nitric oxide synthase (NOS) differentially *in vitro*, as well as the protein phosphatase calcineurin, and this regulation was reciprocal. SCaM4 half-maximally activated NOS at 180 nM while SCaM1 served as a competitive antagonist (K_i ca. 120 nM) of this activation. Calcineurin was half-maximally activated by SCaM1 at ca. 12 nM and SCaM4 competitively antagonized (K_i ca. 70 nM) its activation. This could allow selectivity of activating target proteins by competitive binding of SCaM isoforms to target proteins *in vivo* (Cho *et al.*, 1998). This approach was extended to a wider array of CaM-regulated enzymes, where activation by SCaM4 generally required higher levels of the EF-hand protein or higher Ca^{2+} concentrations than did activation by the more conventional SCaM1 (Lee *et al.*, 2000). Limited experiments *in vivo* also support the idea of different functional roles for CaM isoforms. SCaM1 and the diverged SCaM4 and SCaM5 mRNAs are differentially induced in response to pathogens, elicitors and treatments that elevate cytosolic Ca^{2+}. Further, ectopic expression of SCaM4 in transgenic plants triggered the formation of spontaneous lesions reminiscent of hypersensitive cell death on leaves and induced the expression of several resistance-associated genes (Heo *et al.*, 1999). Finally, Köhler and Neuhaus (2000) tested the abilities of CaM8 and CaM9 to interact with the CaM-binding domain of the cyclic nucleotide-gated channel CNGC2. In contrast to more conventional CaM isoforms, however, CaM8 and CaM9 showed no ability to bind the channel.

In addition, differential expression of CaM isoforms may play an essential role in selective target activation in various tissues. In *Arabidopsis*, Northern blot and reverse transcriptase PCR assays indicated that *Cam1*, *2* and *3* mRNAs were expressed in similar but distinct patterns in organs including root, floral stalk, leaf, flower and silique (Perera and Zielinski, 1992). In contrast, *Cam4*, *5* and *6* mRNAs accumulated in *Arabidopsis* leaf RNA fractions, but only *Cam4* and *5* mRNAs were detected in silique total RNA (Gawienowski *et al.*, 1993). Overlapping but distinct and quantitatively different patterns of expression were also found for *Cam7*, *8* and *9* mRNAs (Zielinski, 2001). However, in soybean, immunolocalization of SCaM isoforms showed that the highly conserved SCaM isoforms (SCaM1, SCaM2 and SCaM3) and the divergent SCaM isoforms (SCaM4 and ScaM5) had no difference in specific tissue distribution (Cho *et al.*, 1998).

11.2.5.2 *Genetic analyses of CaM function: is yeast a valid paradigm?*

The expression of multiple *Cam* sequences in even the most genetically tractable plant species has greatly hampered mutational analysis of CaM function *in planta*. As an alternative approach, the ability of *Arabidopsis* CaM isoforms

(Zielinski, 2001) and a diverged CaM isoform from petunia (Rodriguez-Concepcion *et al.*, 1999) to complement a yeast *Cam* null (*cmd1*) has been used. Surprisingly, even the highly diverged CaM isoforms encoded by *Cam8* and *Cam9* functionally substituted for yeast CaM (CMD1p). However, there were subtle differences in the efficiencies with which the diverged sequences complemented the loss of CMD1p and the rates of growth they supported. In contrast, the protein encoded by *Arabidopsis Cam10* (figure 11.1a, Ling and Zielinski, 1993) was unable to complement the yeast mutant (R.E. Zielinski, unpublished data). Petunia *Cam53*, a 184-amino acid protein with a 150-residue *N*-terminal CaM domain and a 34-residue prenylated *C*-terminal domain, also complemented *cmd1* (Rodriguez-Concepcion *et al.*, 1999). Unfortunately, no information on the growth rates of the strains complemented by *Cam53* was presented, nor was the complementation efficiency of *Cam53* reported. Thus it is difficult to compare these experiments with the *Arabidopsis* CaM isoform complementation results. These studies suggest that genetic analyses in yeast, coupled with reverse genetics in *Arabidopsis*, may provide a powerful method for dissecting CaM function that have previously been elusive in higher organisms.

In addition to the plant complementation studies cited above, studies of CaM function by Ohya and Botstein (1994a,b), described in section 11.2.2, have provided critical insight into the role of hydrophobic residues in CaM function *in vivo*. In spite of these successes, however, questions are raised about the appropriateness of *Saccharomyces cerevisiae* for studies on Ca^{2+} signaling based on the surprising results reported by Davis and co-workers (Geiser *et al.*, 1991). In this study, mutants of yeast CMD1p having Glu to Val substitutions at the critical 12th position and Asp to Ala in the first position of the Ca^{2+}-binding loops, were used to complement *cmd1*. Each of the CMD1p mutants was shown to be deficient in binding Ca^{2+} *in vitro*. Quite unexpectedly, however, they all complemented the null allele and supported wild-type (*wt*) growth rates. The results were interpreted to indicate that CMD1p is likely to function without binding Ca^{2+} *in vivo*. The assumption underlying this conclusion, however, is that the interaction between CaM, Ca^{2+} and Ca^{2+}/CaM and target proteins is strictly ordered. Although this sequence of events is often used to describe Ca^{2+}/CaM-mediated signaling, a wide variety of evidence indicates that CaM's ability to bind Ca^{2+} depends on association with its interaction partners, as described in section 11.2.1. On the other hand, CMD1p is the most divergent CaM identified thus far and, owing to a deleted Asp residue, it binds only three Ca^{2+} instead of the usual four (figure 11.1a); CMD1p may, indeed, function differently from other CaM proteins. The clearest answer to this inconsistency, which will either validate or rule out the further use of yeast complementation by plant CaM proteins, will be a direct analysis of the mutants used by Geiser *et al.* (1991) to bind Ca^{2+} in the presence of CaM-binding proteins or peptides.

11.3   Structures of Ca^{2+}/CaM–target peptide complexes

The main mechanism by which CaM interacts with its target proteins is Ca^{2+}-dependent, and therefore this form of interaction is the focus of this chapter. Most of what is understood at the atomic level of CaM–target protein interaction has been derived from studies of CaM complexed with peptides comprising the CaM-binding domain of a larger protein. Indications are that these model complexes faithfully recapitulate many of the physiologically relevant interactions between CaM and the proteins whose activities it regulates, although with some qualifications (Olwin and Storm, 1985; Peersen *et al.*, 1997). First, the affinities of some of these peptides for CaM or the affinities of the CaM–peptide complex for Ca^{2+} may be higher than they are in the intact target protein. Second, the structures of the CaM-binding peptides may vary from their conformation in the intact protein prior to their interaction with CaM.

The structures of Ca^{2+}/CaM bound to peptides corresponding to the CaM-binding domains of skMLCK (Ikura *et al.*, 1992), smMLCK (Meador *et al.*, 1992), CaM-KII (Meador *et al.*, 1993), the plasma membrane Ca^{2+} pump (Elshorst *et al.*, 1999) and CaM-KK (Osawa *et al.*, 1999) have been solved. In addition, high-resolution structures of CaM-dependent protein kinase I (CaM-KI, Goldberg *et al.*, 1996) and CaM-regulated adenylyl cyclase (Zhang *et al.*, 1997) have been solved in the absence of CaM. These structures support a general mechanism of action of CaM, with some variations. This mechanism is likely to describe in principle the interaction between CaM and many of its protein targets including those in plants. As a representative of this canonical interaction, figure 11.3 shows two views of the structure of the skMLCK peptide (also known as the M13 peptide)–CaM complex that was constructed from the coordinates deposited in the Brookhaven protein database (Ikura *et al.*, 1992). The overall structure of the complex is compact and has an ellipsoid shape, with the *N*- and *C*-terminal globular domains of CaM much closer in proximity than they are in the crystal structure of Ca^{2+}/CaM shown in figure 11.1b. This dramatic rearrangement is facilitated by an adjustable bend created by the central linker domain. Flexible distortion of the linker domain allows the globular domains of CaM to engulf the CaM-binding domain peptide, forming a tunnel diagonal to CaM's long axis. This can be seen most clearly in the end-on view of the complex shown in figure 11.3b. About 80% of the contacts between CaM and the skMLCK peptide in this complex are van der Waals interactions, which are formed between the surface-exposed hydrophobic patches on CaM and hydrophobic residues in the target peptide (Crivici and Ikura, 1995). Salt bridges are also formed in the complex involving Glu residues in CaM and basic residues, Lys and Arg, in the skMLCK CaM-binding domain. These structures dramatically confirmed the biochemical analyses and predicted mechanism of CaM–target protein interaction proposed by Persechini and Kretsinger (1988).

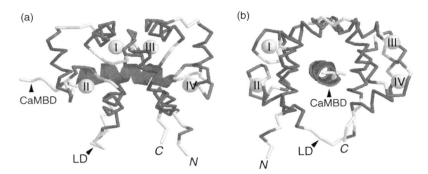

Figure 11.3 Structure of the CaM–skMLCK complex determined by NMR spectroscopy. The polypeptide backbone of CaM is shown with the hydrophobic residues shaded to emphasize their location relative to the skMLCK CaM-binding peptide. The skMLCK CaM-binding peptide is shown in ribbon format to distinguish it from the backbone of CaM. Panels (a) and (b) illustrate two views of the same complex rotated 90° about a vertical axis. Bound Ca^{2+} ions are represented by shaded spheres and the EF-hands are numbered (I–IV) beginning with the N-terminus. The amino- (N) and carboxy- (C) termini of the protein and the flexible linker domain (LD) are indicated. Structure code 2BBN (available at http://www.pdb.bnl.gov/) was modeled as described in the legend to figure 11.2.

The α-helical conformation and antiparallel orientation of the target peptide in the CaM–M13 complex (that is, CaM's C-terminal domain associates with the N-terminal portion of the target peptide) is illustrated in figure 11.3. This arrangement is typical of CaM-binding domains complexed with CaM, but not necessarily representative of their conformations in the absence of CaM. Many CaM-binding domains possess a pair of aromatic or long-chain aliphatic residues (or both) separated by a stretch of 12 residues, such as M13 peptide from skMLCK shown in figure 11.3. From the structure of its complex with CaM, M13 uses Trp_4 and Leu_{17} (equivalent to Trp_{580} and Leu_{593} in holo-skMLCK, and Trp_{800} and Leu_{813} in holo-smMLCK) to anchor the peptide to the C- and N-terminal hydrophobic patches of CaM, respectively. There are also two minor hydrophobic anchors at positions 8 and 11 on the target peptide. However, CaM-KII, another well-characterized CaM-binding protein, has a CaM-binding domain with the major hydrophobic anchors separated by only 8 residues (Leu_{299} and Leu_{308}). How can CaM recognize the difference in the CaM-binding domains? To answer this question, the structure of the CaM–CaM–KII complex was solved (Meador et al., 1993) and compared with the CaM–smMLCK complex (Meador et al., 1992). The overall structures are similar, with the two domains of CaM forming a hydrophobic tunnel that wraps around the target peptides. In spite of the difference in the separating distance between the two anchor residues, the binding of the N- and C-terminal lobes with respect to their positions in the structure of CaM–smMLCK is maintained, and the central helix is expanded into a bend and serves as a flexible joint, as predicted by the Persechini and Kretsinger (1988) model. The central helix of CaM in the CaM–CaM–KII complex, however, is further unraveled in its C-terminal end and loses more of its helical structure than it does in the complex with smMLCK. By lengthening and stretching the central helix, the N-terminal globular domain of CaM can be shifted and positioned adjustably on the second anchor residue, thereby accommodating variances in the sizes of CaM-binding domains in different target proteins.

There are at least two variations on the mechanism of CaM–target protein interaction described above. These were dramatically revealed in the structures of the plasma membrane Ca^{2+} pump (Elshorst *et al.*, 1999), where only the C-terminal half of CaM physically interacts with a peptide derived from the CaM-binding domain of the enzyme. This observation, although somewhat unexpected, is consistent with biochemical studies that demonstrated only the C-terminal but not the N-terminal half of CaM can activate the Ca^{2+}-pump. This complex resembles an intermediate in the folding pathway typical of most CaM-target protein complexes (described below). The CaM–KK complex with CaM (Osawa *et al.*, 1999) illustrates a second important variant of the canonical CaM–target protein interaction. In this structure, CaM folds and engulfs the CaM-binding domain, typical of the canonical interaction. However, the CaM-binding domain is oriented in the opposite polarity with respect to other known CaM–target peptide complexes. That is, the C-terminal half of CaM interacts with the C-terminal half of the CaM-binding domain of CaM-KK. A second surprising feature of this complex is that the CaM–KK CaM-binding domain forms an α-helix and a hairpin loop whose C-terminus folds back on itself and is accommodated by the hydrophobic tunnel formed by CaM. Relative to the numbers of known CaM-binding target proteins, relatively few structures of CaM–target peptide complexes have been solved. This is particularly true for CaM-binding proteins having a lower binding affinity for CaM. Therefore, it would not be surprising if more variations on the theme outlined above will be forthcoming as more structures are elucidated.

11.3.1 *Structures of CaM-binding domains*

Sequences representing CaM-binding domains have been identified by deletion mutagenesis and chemical methods. Ca^{2+}-dependent CaM-binding regions can frequently be reduced to less than 20 residues while retaining their high affinity binding capability (K_ds are typically in the range of 10^{-7} to 10^{-9} M). CaM-binding domains, however, share remarkably little sequence identity. The first conceptual breakthrough in identifying CaM-binding domains came from the work of DeGrado and co-workers who identified basic amphiphilic α-helices as high affinity targets for CaM binding with K_ds approaching 10^{-10} M (O'Neil and DeGrado, 1990). In this conformation the aggregated hydrophobic residues make contact with the hydrophobic patches of Met and Phe exposed in the *N*- and *C*-terminal domains of CaM. The rest of the contact is via electrostatic interaction from basic, positively charged residues (Arg, Lys, His) in the target peptide with negatively charged Glu and Asp residues in CaM. A further refinement on this theme came from the observation that many CaM-binding peptides possess a near-zero hydrophobicity and a relatively high hydrophobic moment (Erickson-Viitanan and DeGrado, 1987). Combined with a net positive charge above a defined minimum value, these parameters provided a rapid way to search for a CaM-binding region in target proteins. This prediction appeared to be dramatically confirmed when the first structures of CaM–target peptide complexes were deduced (described in section 11.3). However, as the sequences of a broader cross-section of CaM-binding proteins have become available, it is clear that CaM-binding 'motifs' vary even more than was originally proposed.

Rhoads and Friedberg (1997) classified Ca^{2+}-dependent CaM-binding sequences into two related groups, which they termed 1-8-14 and 1-5-10 motifs, emphasizing the positions of the conserved, strongly hydrophobic residues. Alignment of some selected plant CaM-binding domains with these consensus motifs is shown in table 11.1. This has also been expanded to include the

Table 11.1 Amino acid sequence motifs of CaM-binding domains from selected plant CaM-regulated proteins[a]

Protein	GenBank accession	CaM-binding sequence		
1-(4)5-10 motif		1	5	10
CaM-kinase I*	L24907	SK WKQAFNATAV VRHMR		
V-type ATPase		SI VKNRARRFRM ISNL		
SAUR1	AF148498	AF LRSFLGAKQI IRRES		
CaM-kinase II-like	AF289237	KR LALKALSKAL SED		
Ethylene-regulated	AF253511	II WSVGILEKVI LRWRR		
P-type Ca^{2+}-ATPase	AF025842	EV LEHWRNLCGV VKNPK		
CIP111	AF217546	SL WTPLKSVAMF LRRHI		
CaM-activated kinase	U24188	SR LRSFNARRKL RAAAI		
1-8-14 motif		1	8	14
skMLCK*	A25830	RR WKKNFIAVSAANRF KK		
Glutamate decarboxylase 1	U10034	SD IDKQRDIITGWKKF VA		
1-12 motif		1	12	
Cation channel CNGC2	Y16328	RE LKRTARYYSSNW KTW		
Glutamate decarboxylase 2	U49937	KE ILMEVIVGWRKF VKE		

[a]Motifs are aligned with known, well-characterized CaM-binding domains from animal sources (designated by *) and are grouped according to the CaM-binding motifs characterized by the placement of strongly hydrophobic residues identified by Rhoads and Friedberg (1997), with the addition of the 1-4-10 and 1-12 motifs identified by Ikura and co-workers (http://calcium.oci.utoronto.ca/).

1-4-10 and 1-12 motifs identified by Ikura and co-workers (http://calcium.oci.utoronto.ca/). It should be noted that no attempt is made here to compile an exhaustive list of plant CaM-binding proteins. However, plant CaM-binding proteins are not simply a collection of orthologs of animal CaM-regulated proteins. Some are shared with animal species while others appear to be unique to plants. Most target proteins have two flanking hydrophobic residues at position 1, and either 14 or 10. The 1-8-14 motif, which is the most highly represented motif in animal CaM-binding proteins, has one internal hydrophobic residue at position eight with an additional internal hydrophobic residue at position five in some proteins. Although an exhaustive search, such as the one conducted by Rhoads and Friedberg (1997), has not been conducted for plant CaM-binding protein motifs, there appear to be relatively few plant proteins that possess the 1-8-14 motif. The 1-(4)5-10 motif, which is much less common in animals but well-represented in plants, has one internal hydrophobic residue at position five and is present in the animal CaM-dependent protein kinase family as well as the related kinases from plants. This motif is commonly found in the various classes of CaM-regulated Ca^{2+}-transporting ATPases (Sze et al., 2000).

Most CaM-binding domains in target proteins have been assumed to comprise regions that form basic, amphiphilic α-helices, or that can be induced to assume such a conformation. However, a growing list of CaM-binding peptides have been shown to include acidic as well as basic residues. This arrangement is particularly noteworthy in the CaM-binding domain of plant glutamate decarboxylase isoenzymes (Baum et al., 1993; Zik et al., 1998), which possesses several acidic residues in short peptides that were shown to bind CaM (table 11.1). These proteins are also noteworthy because their CaM-binding domains are representative of different CaM-binding motif families. The second modification now recognized in CaM-binding domains is that they are not always in an α-helical conformation prior to binding CaM. This seems particularly counterintuitive based on the structures of CaM–target peptide complexes described in section 11.3, and the observation that systematic Ala substitutions in the CaM-binding domain of skMLCK, which tend to increase their propensity for helix formation, increase the affinity of CaM binding (Montigiana et al., 1996). However, a particularly striking example of this apparent paradox was revealed by the high-resolution structure of CaM-KI (Goldberg et al., 1996). The C-terminal portion of the CaM-KI CaM-binding domain is in a helical conformation in the intact protein, but Trp_{303} in the N-terminal portion of the domain is in a random coil region of the protein. As shown in table 11.1, Trp_{303} resides in position 1 of a 1-5-10 CaM-binding motif, which is generally the initial, high affinity anchoring point for CaM binding. This residue is oriented away from the protein into the medium ready to be bound by the exposed hydrophobic residues in the C-terminal globular domain of CaM. This suggests that at least a portion of many CaM-binding domains are induced to form helices by Ca^{2+}/CaM and the degree of coil to helix transition, that must take place during CaM binding, may play an important role in determining the affinity for binding.

11.3.2　Energetics of target peptide binding by CaM

From the discussion in sections 11.2.2, 11.3, and 11.3.1 above, it is clear that hydrophobic interactions play a major role in governing CaM–target protein interaction. Hydrophobic interactions involve the burial of non-polar residues in a manner that reduces their interaction with water. It is often assumed that interactions between non-polar residues at a protein–protein interface are accurately modeled by the properties of non-polar compounds moving from an aqueous to a liquid organic phase (Privalov and Gill, 1988). By this reasoning, burial of exposed hydrophobic residues, such as those exposed by Ca^{2+}-binding in CaM, are generally considered to proceed with a decrease in free energy (ΔG) and to be predominantly entropy driven. The question is: does entropy change associated with this hydrophobic interaction provide sufficient free energy to drive the binding of CaM to its protein targets? The free energy change associated with the formation of most Ca^{2+}-dependent CaM-target protein complexes is highly favorable, ranging between about -30 and -50 kJ mol^{-1} at 20°C. This is calculated from

$$\Delta G = -RT \ln (K_a) \tag{11.1}$$

assuming that the K_as for CaM-target protein complexes lie in the range of 10^6 to 10^9 M^{-1} (O'Neil and DeGrado, 1990). Initial isothermal microcalorimetric measurements of the association of smMLCK and CaM, surprisingly, indicated that binding was enthalpy driven (Wintrode and Privalov, 1997). However, more recent calorimetric studies with a broader range of CaM-binding peptides have demonstrated that the variations in enthalpies of binding are tremendous and indicate that CaM binding may be entropically or enthalpically driven, depending on the particular target protein substrate (Brokx et al., 2001). The conclusion of this study was that the most important determining factor in the energetics of binding and relative affinity for CaM is the degree of helix folding that takes place in the target sequence; the less the target must rearrange to form a helix, the more the binding is driven by entropic effects, which is consistent with hydrophobic residue burial. Brokx et al. (2001) also pointed out that, by the energetic criteria they defined, inhibitors designed to perturb CaM activity should be most effective if they are not only amphiphilic, but also have a strong propensity for helix formation. This suggests that the synthetic helical peptides originally designed by DeGrado and co-workers to model the structure of CaM-binding domains (O'Neil and DeGrado, 1990) may be particularly potent inhibitors of CaM function. Targeted expression of these peptides, therefore, may be useful for dissecting CaM functions in future experiments in vivo.

11.3.3 Mechanisms of CaM-mediated activation

Biochemical studies of CaM structure–function relationships (section 11.2.4), in conjuction with studies of concerted Ca^{2+}-binding by CaM-target protein complexes (section 11.2.1), provide a global view of how CaM-binding may regulate enzyme activity. Furthermore, this information provides clues on how the functions of these proteins may be examined *in vivo*. As illustrated in figure 11.4, autoinhibition is a common mechanism of regulation in proteins whose activities are modulated by CaM (reviewed in Crivici and Ikura, 1995). The CaM-binding domains of these proteins either overlap or reside in close proximity to a region, known as the autoinhibitory domain, that either blocks or whose interaction with the CaM-binding domain blocks the active site. This could either occur directly as shown in figure 11.4 or indirectly by influencing the

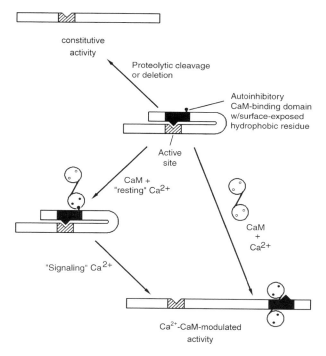

Figure 11.4 Model for alternative modes of Ca^{2+}-CaM-mediated activation of a target enzyme. CaM typically interacts with a 15 to 30-residue basic amphiphilic peptide domain, altering its interaction with the remainder of the target protein. Interaction may proceed at resting levels of Ca^{2+} with apo- or partially liganded CaM, followed by complex formation or dissociation when Ca^{2+} levels are elevated in response to a stimulus (lower left pathway). Alternatively, interaction may occur by an ordered mechanism in response to increases in Ca^{2+} (lower right pathway). Proteolytic removal, mutation or deletion of the autoinhibitory CaM-binding domain or interaction of the domain with a monoclonal antibody may yield a constitutively activated enzyme (upper pathway).

conformation of the active site. Limited proteolysis, mutation or deletion of the autoinhibitory domain, or perturbation of the domain's function by monoclonal antibody binding, frequently render the enzyme active in the absence of Ca^{2+} and CaM and insensitive to exogenous CaM. These observations experimentally support the autoinhibitory model. This mode of regulation has been demonstrated in plants for a plasma membrane localized Ca^{2+}-ATPase (Rasi-Caldogno et al., 1993) and glutamate decarboxylase (GAD) (Snedden et al., 1996). It appears, in vivo, that there are several mechanisms by which autoinhibition may be relieved by CaM, which is illustrated in the lower portion of figure 11.4. Some proteins, such as GAD (Arazi et al., 1995), appear to interact with CaM even at resting levels of CaM, consistent with the observation that most or all CaM does not freely diffuse in unstimulated cells (reviewed in Zielinski, 1998). The semi-open conformation of CaM that forms at resting Ca^{2+} levels (Swindells and Ikura, 1996) seems to be the likely form of CaM that would bind to an exposed hydrophobic residue, corresponding to position 1 in the CaM-binding motifs shown in table 11.1, although other modes of association not occurring through the hydrophobic anchor may also occur (Yuan et al., 1999b). When cellular Ca^{2+} levels rise, the partially formed CaM–target protein complex would then saturate with Ca^{2+} and the activity of the target enzyme would be modulated either up or down as shown in the lower left side of figure 11.4. In contrast, other proteins may interact poorly or not at all with CaM at resting levels of Ca^{2+}. These proteins would bind CaM only when signaling levels of Ca^{2+} are sufficiently high to promote their interaction with CaM as shown in the lower right portion of figure 11.4. Different combinations of CaM-binding proteins, each with a slightly different affinity for apo-CaM and each sensitizing CaM to bind Ca^{2+} with slightly different affinity would provide a cell with a diverse array of responses to a second messenger signal.

The relief of autoinhibition by CaM regulation mechanism also provides a potential means for perturbing enzyme activities in vivo. By expressing a protein whose CaM-binding domain has either been deleted or mutated to change its affinity for CaM, control over the protein's function may be altered sufficiently to produce a measurable phenotype that can be used to assess its physiological role. However, this method is not without pitfalls. When Baum et al. (1996) expressed GAD whose CaM-binding domain had been deleted in transgenic plants, a variety of phenotypic effects were observed that made straightforward interpretation of GAD's role difficult. Nevertheless, as more plant CaM-binding proteins are identified, this approach is likely to be a valuable complement to gene knockouts as a means to assess physiological function.

11.4   Prospects for monitoring calmodulin function *in vivo*

Green fluorescent protein (GFP) from *Aequorea victoria* is a spontaneously fluorescent molecule that has gained wide popularity for molecular and cell biological analyses as a reporter of gene expression, subcellular localization and cell function (Tsien, 1998). Three important developments involving GFP and GFP mutants with altered fluorescence properties have directly or potentially influenced the field of Ca^{2+} signaling mediated through CaM in recent years. GFP–CaM fusion proteins (Moser *et al.*, 1997; Li *et al.*, 1999), cameleon Ca^{2+} sensors (Miyawaki *et al.*, 1997, 1999), and fluorescent sensors of Ca^{2+}-dependent CaM binding (Romoser *et al.*, 1997; Persechini and Cronk, 1999) have provided sensitive, genetically encoded markers that offer the possibility for monitoring Ca^{2+} and CaM function *in vivo* and better understanding of how Ca^{2+} signals are transduced. Development of these sensors for studies *in vivo* has been facilitated by the detailed dissection of CaM-target protein interaction described previously.

CaM is expressed in all eukaryotic cell types and it accumulates to relatively high levels in rapidly growing plant tissues where its concentration may reach as high as 1 to 2 μM (Zielinski, 1998). Studies of CaM localization and possible redistribution in response to Ca^{2+} signals previously relied on either immunofluorescence microscopy or on imaging cells microinjected with fluorescently labeled CaM. These methods have revealed changes in subcellular CaM localization (reviewed in Zielinski, 1998) in response to various stimuli, but their usefulness has been limited by the low affinity of anti-CaM antibodies and the perturbations associated with delivering labeled molecules into cells by microinjection. Recently, however, it has been possible to image GFP–CaM fusion proteins after transfection or transformation in live cells. Genetically encoded fusion protein reporters eliminate many potential artifacts associated with sample preparation but suffer from the potential shortcoming that the protein of interest may not retain all its biological properties when fused with a reporter protein. In the case of CaM, full or nearly full activity is retained by GFP–CaM fusion proteins as evidenced by their ability to complement CaM null mutations in budding and fission yeast (Moser *et al.*, 1997). In yeast, GFP–CaM localizes to the sites of cell growth, where it overlaps with patches of actin in buds. In addition, the fusion protein simultaneously associates with spindle pole bodies during cell division. Similar association of CaM with the spindle was observed in HeLa cells, together with a submembrane fraction that aggregated at the equatorial region of the cell marking the future location of the cleavage furrow (Li *et al.*, 1999). No such imaging has been reported in a plant system at this time, but the observations that plant, vertebrate and yeast CaM proteins are functionally interchangeable (Persechini *et al.*, 1991; Zielinski, 2001) suggests that GFP–plant CaM fusions should prove equally useful for cell biological studies in plants.

In addition to marking the locations of CaM itself, imaging CaM binding to target proteins has recently been possible in live cells using fusion proteins consisting of blue or cyan mutant GFPs joined to green or yellow mutant GFPs by a short peptide comprising a CaM-binding sequence derived from smMLCK (Persechini and Cronk, 1999; Romoser *et al.*, 1997). These indicators normally transfer energy between the lower wavelength GFP variant and the higher wavelength variant by fluorescence resonance energy transfer (FRET) because the short CaM-binding domain linking the reporter proteins tethers them within the Förster radius (about 30 to 50 Å for BFP/GFP and CFP/YFP donor/acceptor pairs; Tsien, 1998). Free CaM interacting with the CaM-binding linker, however, physically separates the tethered fluorescent proteins, reducing their efficiency for FRET. As a result, relative levels of free CaM can be estimated by monitoring the ratio of the blue and green fluorescence emitted by cells expressing the indicator protein. By varying the sequence of smMLCK linker protein, it is possible to monitor the binding of CaM to targets having different affinities for CaM. One of the surprising findings of these studies is that, although CaM is a relatively abundant protein, CaM-binding proteins outnumber CaM on a molar basis by a factor of about two, and there are only minute amounts of *free* Ca^{2+}-CaM produced even at the highest levels of Ca^{2+}. These measurements highlight the coupling effect between Ca^{2+} and target binding in CaM–target protein complexes described in section 11.2.1; the effective K_d of CaM for Ca^{2+} *in vivo* is estimated to be about 20-fold lower than the K_d of purified CaM for Ca^{2+} *in vitro* (Persechini and Cronk, 1999). Perhaps the most interesting result to emerge from these studies from the perspective of cellular signaling, however, is the finding that only the largest Ca^{2+} signals appear to be sufficient to drive the formation of CaM–target protein complexes with targets of moderate affinity for CaM, while high affinity target–CaM complexes are formed in response to lower Ca^{2+} concentrations. This result suggests that a key area in which knowledge is lacking in plant systems is not simply the identities of CaM-regulated proteins but their affinities for CaM and subcellular locations, which will dictate specific patterns of Ca^{2+}-CaM-mediated signaling.

Cameleons are multiple fusion proteins consisting of a blue or cyan mutant of GFP, CaM, the CaM-binding domain of skMLCK, and a green or yellow mutant of GFP. Elevated Ca^{2+} causes CaM to bind the skMLCK portion of the fusion protein. As a result of the Ca^{2+}-dependent binding, the fusion protein adopts a compact conformation, which places the blue or cyan chromophore in close proximity to the green or yellow chromophore. This conformational change increases the efficiency of FRET between the shorter and the longer wavelength mutant GFP, thus providing a ratiometric fluorescence signal that is proportional to the level of Ca^{2+}. Cameleons act as a closed system in that they appear neither to bind endogenous CaM nor CaM target proteins and only report local changes in Ca^{2+} concentration. One cameleon variant, the split cameleon, however, was shown to report CaM interaction with the skMLCK CaM-binding domain *in vivo* (Miyawaki *et al.*, 1997). This observation raises the possibility that split cameleons may be used to evaluate the physiological relevance of CaM interaction with potential target proteins or peptides derived from them in an *in vivo* environment.

References

Arazi, T., Baum, G., Snedden, W.A., Shelp, B.J. and Fromm, H. (1995) Molecular and biochemical analysis of calmodulin interactions with the calmodulin-binding domain of plant glutamate decarboxylase. *Plant Physiol.*, **108**, 551-561.

Babu, Y.S., Bugg, C.E. and Cook, W.J. (1988) Structure of calmodulin refined at 2.2 Å resolution. *J. Mol. Biol.*, **204**, 191-204.

Baum, G., Chen, Y., Arazi, T., Takatsuji, H. and Fromm, H. (1993) A plant glutamate decarboxylase containing a calmodulin-binding domain: cloning, sequence and functional analysis. *J. Biol. Chem.*, **268**, 19,610-19,617.

Baum, G., Lev-Yadun, S., Fridmann, Y. *et al.* (1996) Calmodulin binding to glutamate decarboxylase is required for regulation of glutamate and GABA metabolism and normal development in plants. *EMBO J.*, **15**, 2988-2996.

Bernstein, H.D., Poritz, M.A., Strub, K., Hoben, P.J., Brenner, S. and Walter, P. (1989) Model for signal sequence recognition from amino-acid sequence of 54K subunit of signal recognition particle. *Nature*, **340**, 482-486.

Berridge, M.J., Lipp, P. and Bootman, M.D. (2000) The versatility and universality of calcium signaling. *Nature Rev.*, **1**, 11-21.

Brokx, R.D., Lopez, M.M., Vogel, H.J. and Makhatadze, G.L. (2001) Energetics of target peptide binding by calmodulin reveals different modes of binding. *J. Biol. Chem.*, **276**, 14,083-14,091.

Brown, S.E., Martin, S.R. and Bayley, P.M. (1997) Kinetic control of the dissociation pathway of calmodulin-peptide complexes. *J. Biol. Chem.*, **272**, 3389-3397.

Cates, M.S., Berry, M.B., Ho, E.L., Li, Q., Potter, J.D. and Phillips Jr, G.N. (1999) Metal-ion affinity and specificity in EF-hand proteins: coordination geometry and domain plasticity in parvalbumin. *Structure*, **7**, 1269-1278.

Chattopadhyaya, R., Meador, W.E., Means, A.R. and Quiocho, F.A. (1992) Calmodulin structure refined at 1.7 Å resolution. *J. Mol. Biol.*, **228**, 1177-1192.

Chin, D. and Means, A.R. (1996) Methionine to glutamine substitutions in the *C*-terminal domain of calmodulin impair the activation of three protein kinases. *J. Biol. Chem.*, **271**, 30,465-30,471.

Chin, D., Sloan, D.J., Quiocho, F.A. and Means, A.R. (1997) Functional consequences of truncating amino acid side chains located at a calmodulin-peptide interface. *J. Biol. Chem.*, **272**, 5510-5513.

Cho, M.J., Vaghy, P.L., Kondo, R. *et al.* (1998) Reciprocal regulation of mammalian nitric oxide synthase and calcineurin by plant calmodulin isoforms. *Biochemistry*, **37**, 15,593-15,597.

Craig, T.A., Watterson, D.M., Prendergast, F.G., Haiech, J. and Roberts, D.M. (1987) Site-specific mutagenesis of the α-helices of calmodulin. Effects of altering a charge cluster in the helix that links the two halves of calmodulin. *J. Biol. Chem.*, **262**, 3278-3284.

Crivici, A. and Ikura, M. (1995) Molecular and structural basis of target recognition by calmodulin. *Ann. Rev. Biophys. Biomol. Struct.*, **24**, 85-116.

Elshorst, B., Hennig, M., Forsterling, H. *et al.* (1999) NMR solution structure of a complex of calmodulin with a binding peptide of the Ca^{2+} pump. *Biochemistry*, **38**, 12,320-12,332.

Erickson-Viitanen, S. and DeGrado, W.F. (1987) Recognition and characterization of calmodulin-binding sequences in peptides and proteins. *Methods Enzymol.*, **139**, 455-478.

Falke, J.J., Drake, S.K., Hazard, A.L. and Peersen, O.B. (1994) Molecular tuning of ion binding to calcium signaling proteins. *Q. Rev. Biophys.*, **27**, 219-290.

Findlay, W.A., Martin, S.R., Beckingham, K. and Bayley, P.M. (1995) Recovery of native structure by calcium binding site mutants of calmodulin upon binding of sk-MLCK target peptides. *Biochemistry*, **34**, 2087-2094.

Finn, B.E., Evenas, J., Drakenberg, T., Waltho, J.P., Thulin, E. and Forsen, S. (1995) Calcium-induced structural changes and domain autonomy in calmodulin. *Nature Struct. Biol.*, **2**, 777-783.

Gawienowski, M.C., Szymanski, D., Perera, I.Y. and Zielinski, R.E. (1993) Calmodulin isoforms in *Arabidopsis* encoded by multiple divergent mRNAs. *Plant Mol. Biol.*, **22**, 215-225.

Geiser, J.R., Tuinen, D., Brockerhoff, S.E., Neff, M.M. and Davis, T.N. (1991) Can calmodulin function without binding calcium? *Cell*, **65**, 949-959.

Gellman, S.H. (1991) On the role of methionine residues in the sequence-independent recognition of nonpolar protein surfaces. *Biochemistry*, **30**, 6633-6636.

George, S.E., VanBerkum, M.F.A., Ono, T. *et al.* (1990) Chimeric calmodulin-cardiac troponin C proteins diffentially activate calmodulin target enzymes. *J. Biol. Chem.*, **265**, 9228-9235.

George, S.E., Su, Z., Fan, D. and Means, A.R. (1993) Calmodulin-cardiac troponin C chimeras. *J. Biol. Chem.*, **268**, 25,213-25,220.

George, S.E., Su, Z., Fan, D., Wang, S. and Johnson, J.D. (1996) The fourth EF-hand of calmodulin and its helix-loop-helix components: impact on calcium binding and enzyme activation. *Biochemistry*, **35**, 8307-8313.

Gilroy, S. and Trewavas, A. (2001) Signal processing and transduction in plant cells: the end of the beginning? *Nature Rev. Mol. Cell Biol.*, **2**, 307-314.

Goldberg, J., Nairn, A.C. and Kuriyan, J. (1996) Structural basis for the autoinhibition of calcium/calmodulin-dependent protein kinase I. *Cell*, **84**, 875-887.

Haiech, J., Kilhoffer, M.C., Lukas, T.J., Craig, T.A., Roberts, D.M. and Watterson, D.M. (1991) Restoration of the calcium binding activity of mutant calmodulins toward normal by the presence of a calmodulin binding structure. *J. Biol. Chem.*, **266**, 3427-3431.

Harmon, A.C., Gribskov, M. and Harper, J.F. (2000) CDPKs—a kinase for every Ca^{2+} signal? *Trends Plant Sci.*, **5**, 154-159.

Heo, W.D., Lee, S.H., Kim, M.C. *et al.* (1999) Involvement of specific calmodulin isoforms in salicyclic acid-independent activation of plant disease resistance responses. *Proc. Natl. Acad. Sci. USA.*, **96**, 766-771.

Ikura, M. (1996) Calcium binding and conformational response in EF-hand proteins. *Trends Biochem. Sci.*, **21**, 14-17.

Ikura, M., Spera, S., Barbato, G., Kay, L.E., Krinks, M. and Bax, A. (1991) Secondary structure and side-chain ^1H and ^{13}C resonance assignments of calmodulin in solution by heteronuclear multidimentional NMR spectroscopy. *Biochemistry*, **30**, 9216-9228.

Ikura, M., Clore, G.M., Gronenborn, A.M., Zhu, G., Klee, C.B. and Bax, A. (1992) Solution structure of a calmodulin-target peptide complex by multidimensional NMR. *Science*, **256**, 632-638.

Janin, J., Wodak, S., Levitt, M. and Maigret, B. (1978) Conformation of amino acid side-chains in proteins. *J. Mol. Biol.*, **125**, 357-386.

Johnson, J.D. and Potter, J.D. (1981) Detection of two classes of Ca^{2+}-binding sites in troponin C with circular dichroism and tyrosine fluorescence. *J. Biol. Chem.*, **253**, 3675-3677.

Johnson, J.D., Snyder, C., Walsh, M. and Flynn, M. (1996) Effects of myosin light chain kinase and peptides on Ca^{2+} binding sites of calmodulin. *J. Biol. Chem.*, **271**, 761-767.

Jurado, L.A., Chockalingam, P.S. and Jarrett, H.W. (1999) Apocalmodulin. *Physiol. Rev.*, **79**, 661-682.

Kasturi, R., Vasulka, C. and Johnson, J.D. (1993) Ca^{2+}, caldesmon, and myosin light chain kinase exchange with calmodulin. *J. Biol. Chem.*, **268**, 7958-7964.

Kawasaki, H. and Kretsinger, R.H. (1994) Calcium-binding proteins 1: EF-hands. *Protein Profile*, **1**, 343-517.

Köhler, C. and Neuhaus, G. (2000) Characterisation of calmodulin binding to cyclic nucleotide-gated ion channels from *Arabidopsis thaliana*. *FEBS Lett.*, **471**, 133-136.

Kretsinger, R.H. and Nockholds, C.E. (1973) Carp muscle calcium-binding protein. II. Structure determination and general description. *J. Biol. Chem.*, **248**, 3313-3326.

Kuboniwa, H., Tjandra, N., Grzesiek, S., Ren, H., Klee, C.B. and Bax, A. (1995) Solution structure of calcium-free calmodulin. *Nat. Struct. Biol.*, **2**, 768-776.

Lee, S.H., Johnson, J.D., Walsh, M.P. *et al.* (2000) Differential regulation of Ca^{2+}/calmodulin-dependent enzymes by plant calmodulin isoforms and free Ca^{2+} concentration. *Biochem. J.*, **350**, 299-306.

Lee, S.H., Kim, K.C., Lee, M.S. *et al.* (1995) Identification of a novel divergent calmodulin isoform from soybean which has differential ability to activate calmodulin-dependent enzymes. *J. Biol. Chem.*, **270**, 21,806-21,812.

Lee, S.H., Seo, H.Y., Kim, J.C. *et al.* (1997) Differential activation of NAD kinase by plant calmodulin isoforms. *J. Biol. Chem.*, **272**, 9252-9259.

Li, C.-J., Heim, R., Lu, Y., Tsien, R.Y. and Chang, D.C. (1999) Dynamic redistribution of calmodulin in HeLa cells during cell division as revealed by a GFP-calmodulin fusion protein technique. *J. Cell Sci.*, **112**, 1567-1577.

Liao, B., Gawienowski, M.C. and Zielinski, R.E. (1996) Differential stimulation of NAD kinase and binding of peptide substrates by wild-type and mutant plant calmodulin isoforms. *Arch. Biochem. Biophys.*, **327**, 53-60.

Ling, V. and Zielinski, R.E. (1993) Isolation of an *Arabidopsis* cDNA sequence encoding a 22 kDa calcium-binding protein (CaBP-22) related to calmodulin. *Plant Mol. Biol.*, **22**, 207-214.

Ling, V., Perera, I.Y. and Zielinski, R.E. (1991) Primary structures of *Arabidopsis* calmodulin isoforms deduced from the sequences of cDNA clones. *Plant Physiol.*, **90**, 1196-1202.

Linse, S. and Forsen, S. (1995) Determinants that govern high-affinity calcium binding. *Adv. Second Messenger Phosphoprot. Res.*, **30**, 88-151.

Meador, W.E., Means, A.R. and Quiocho, F.A. (1992) Target enzyme recognition by calmodulin: 2.4 Å structure of a calmodulin-peptide complex. *Science*, **257**, 1251-1255.

Meador, W.E., Means, A.R. and Quiocho, F.A. (1993) Modulation of calmodulin plasticity in molecular recognition on the basis of x-ray structures. *Science*, **262**, 1718-1721.

Meador, W.E., George, S.E., Means, A.R. and Quiocho, F.A. (1995) X-ray analysis reveals conformational adaptation of the linker in functional calmodulin mutants. *Nat. Struct. Biol.*, **2**, 943-945.

Miyawaki, A., Griesbeck, O., Heim, R. and Tsien, R.Y. (1999) Dynamic and quantitative Ca^{2+} measurements using improved cameleons. *Proc. Natl Acad. Sci. USA*, **96**, 2135-2140.

Miyawaki, A., Liopis, J., Heim, R., McCaffery, J.M., Adams, J.A., Ikura, M. and Tsien, R.Y. (1997) Fluorescent indicators for Ca^{2+} based on green fluorescent proteins and calmodulin. *Nature*, **388**, 882-887.

Montigiani, S., Neri, G., Neri, P. and Neri, D. (1996) Alanine substitutions in calmodulin-binding peptides result in unexpected affinity enhancement. *J. Mol. Biol.*, **258**, 6-13.

Moser, M.J., Flory, M.R. and Davis, T.N. (1997) Calmodulin localizes to the spindle pole body of *Schizosaccharomyces pombe* and performs an essential function in chromosome segregation. *J. Cell Sci.*, **110**, 1805-1812.

Ohki, S., Ikura, M. and Zhang, M. (1997) Identification of Mg^{2+}-binding sites and the role of Mg^{2+} on target recognition by calmodulin. *Biochemistry*, **36**, 4309-4316.

Ohya, Y. and Botstein, D. (1994a) Diverse essential functions revealed by complementing yeast calmodulin mutants. *Science*, **263**, 963-966.

Ohya, Y. and Botstein, D. (1994b) Structure-based systematic isolation of conditional-lethal mutations in the single yeast calmodulin gene. *Genetics*, **138**, 1041-1054.

Okano, H., Cyert, M.S. and Ohya, Y. (1998) Importance of phenylalanine residues of yeast calmodulin for target binding and activation. *J. Biol. Chem.*, **273**, 26,375-26,382.

Olwin, B.B. and Storm, D.R. (1985) Calcium binding to complexes of calmodulin and calmodulin binding proteins. *Biochemistry*, **24**, 8081-8086.

Olwin, B.B., Edelman, A.M., Krebs, E.G. and Storm, D.R. (1984) Quantitation of energy coupling between Ca^{2+}, calmodulin, skeletal muscle myosin light chain kinase, and kinase substrates. *J. Biol. Chem.*, **259**, 10,949-10,955.

O'Neil, K.T. and DeGrado, W.F. (1990) How calmodulin binds its targets: sequence independent recognition of amphiphilic α-helices. *Trends Biochem. Sci.*, **15**, 59-64.

Osawa, M., Tokumitsu, H., Swindells, M.B. *et al.* (1999) A novel target recognition revealed by calmodulin in complex with Ca^{2+}-calmodulin-dependent kinase kinase. *Nat. Struct. Biol.*, **6**, 819-824.

Ozawa, T., Sasaki, K. and Umezawa, Y. (1999) Metal ion selectivity for formation of the calmodulin-metal-target peptide ternary complex studied by surface plasmon resonance spectroscopy. *Biochim. Biophys. Acta*, **1434**, 211-230.

Peersen, O.B., Madsen, T.S. and Falke, J.J. (1997) Intermolecular tuning of calmodulin by target peptides and proteins: differential effects on Ca^{2+} binding and implications for kinase activation. *Prot. Sci.*, **6**, 794-807.

Perera, I.Y. and Zielinski, R.E. (1992) Structure and expression of the *Arabidopsis* CaM-3 calmodulin gene. *Plant Mol. Biol.*, **19**, 649-664.

Persechini, A. and Cronk, B. (1999) The relationship between the free concentrations of Ca^{2+} and Ca^{2+}-calmodulin in intact cells. *J. Biol. Chem.*, **274**, 6827-6830.

Persechini, A. and Kretsinger, R.H. (1988) The central helix of calmodulin functions as a flexible tether. *J. Biol. Chem.*, **263**, 12,175-12,178.

Persechini, A., Blumenthal, D.K., Jarrett, H.W., Klee, C.B., Hardy, D.O. and Kretsinger, R.H. (1989) The effects of deletions in the central helix of calmodulin on enzyme activation and peptide binding. *J. Biol. Chem.*, **264**, 8052-8058.

Persechini, A., Kretsinger, R.H. and Davis, T.N. (1991) Calmodulins with deletions in the central helix functionally replace the native protein in yeast cells. *Proc. Natl Acad. Sci. USA*, **88**, 449-452.

Persechini, A., White, H.D. and Gansz, K.J. (1996) Different mechanisms for Ca^{2+} dissociation from complexes of calmodulin with nitric oxide synthase or myosin light chain kinase. *J. Biol. Chem.*, **271**, 62-67.

Privalov, P.L. and Gill, S.J. (1988) Stability of protein structure and hydrophobic interactions. *Adv. Protein Chem.*, **39**, 191-234.

Rasi-Caldogno, F., Carnelli, A. and De Michelis, M.I. (1993) Controlled proteolysis activates the plasma membrane Ca^{2+} pump of higher plants. *Plant Physiol.*, **103**, 385-390.

Reddy, V.S., Safadi, F., Zielinski, R.E. and Reddy, A.S.N. (1999) Interaction of a kinesin-like protein with calmodulin isoforms from *Arabidopsis*. *J. Biol. Chem.*, **274**, 31,727-31,733.

Rhoads, A.R. and Friedberg, F. (1997) Sequence motifs for calmodulin recognition. *FASEB J.*, **11**, 331-340.

Rodriguez-Concepcion, M., Yalovsky, S., Zik, M., Fromm, H. and Gruissem, W. (1999) The prenylation status of a novel plant calmodulin directs plasma membrane or nuclear localization of the protein. *EMBO J.*, **18**, 1996-2007.

Romoser, V.A., Hinkle, P.M. and Persechini, A. (1997) Detection in living cells of Ca^{2+}-dependent changes in the fluorescence emission of an indicator composed of two green fluorescent protein variants linked by a calmodulin-binding sequence: A new class of fluorescent indicators. *J. Biol. Chem.*, **272**, 13270-13274.

Sekiya-Kawasaki, M., Botstein, D. and Ohya, Y. (1998) Identification of functional connections between calmodulin and the yeast actin cytoskeleton. *Genetics*, **150**, 43-58.

Snedden, W.A. and Fromm, H. (1998) Calmodulin, calmodulin-related proteins and plant responses to the environment. *Trends Plant Sci.*, **3**, 299-304.

Snedden, W.A., Koutsia, N., Baum, G. and Fromm, H. (1996) Activation of a recombinant petunia glutamate decarboxylase by calcium/calmodulin or by a monoclonal antibody which recognizes that calmodulin binding domain. *J. Biol. Chem.*, **271**, 4148-4153.

Su, Z., Fan, D. and George, S.E. (1994) Role of domain 3 of calmodulin in activation of calmodulin-stimulated phosphodiesterase and smooth muscle myosin light chain kinase. *J. Biol. Chem.*, **269**, 16,761-16,765.

Swindells, M.B. and Ikura, M. (1996) Pre-formation of the semi-open conformation by the apo-calmodulin C-terminal domain and implications binding IQ-motifs. *Nature Struct. Biol.*, **3**, 501-504.

Sze, H., Liang, F., Hwang, I., Curran, A.C. and Harper, J.F. (2000) Diversity and regulation of plant Ca^{2+} pumps: insights from expression in yeast. *Annu. Rev. Plant Physiol. Plant Mol. Biol.*, **51**, 433-462.

Szymanski, D.B., Liao, B. and Zielinski, R.E. (1996) Calmodulin isoforms differentially enhance the binding of cauliflower nuclear proteins and recombinant TGA3 to a region derived from the *Arabidopsis CaM3* promoter. *Plant Cell*, **8**, 1069-1077.

Takezawa, D., Liu, Z.H., An, G. and Poovaiah, B.W. (1995) Calmodulin gene family in potato: developmental and touch-induced expression of the mRNA encoding a novel isoform. *Plant Mol. Biol.*, **27**, 693-703.

Trewavas, A.J. and Malho, R. (1998) Ca^{2+} signaling in plant cells: the big network! *Curr. Opin. Plant Biol.*, **1**, 428-433.

Tsien, R.Y. (1998) The green fluorescent protein. *Annu. Rev. Biochem.*, **67**, 509-544.

Van Berkum, M.F.A. and Means, A.R. (1991) Three amino acid substitutions in domain I of calmodulin prevent the activation of chicken smooth muscle myosin light chain kinase. *J. Biol. Chem.*, **266**, 21,488-21,495.

Wintrode, P.L. and Privalov, P.L. (1997) Energetics of target peptide recognition by calmodulin: a calorimetric study. *J. Mol. Biol.*, **266**, 1050-1062.

Yuan, T. and Vogel, H. (1998) Calcium-calmodulin-induced dimerization of the carboxyl-terminal domain from petunia glutamate decarboxylase. *J. Biol. Chem.*, **273**, 30,328-30,335.

Yuan, T., Ouyang, H. and Vogel, H. (1999a) Surface exposure of the methionine side chains of calmodulin in solution. *J. Biol. Chem.*, **274**, 8411-8420.

Yuan, T., Walsh, M.P., Sutherland, C., Fabian, H. and Vogel, H.J. (1999b) Calcium-dependent and -independent interactions of the calmodulin-binding domain of cyclic nucleotide phosphodiesterase with calmodulin. *Biochemistry*, **38**, 1446-1455.

Zhang, M. and Vogel, H.J. (1994) Two-dimensional NMR studies of selenomethionyl calmodulin. *J. Mol. Biol.*, **239**, 545-554.

Zhang, M., Li, M., Wang, J.H. and Vogel, H.J. (1994) The effect of Met → Leu mutations on calmodulin's ability to activate cyclic nucleotide phosphodiesterase. *J. Biol. Chem.*, **269**, 15,546-15,552.

Zhang, M., Tanaka, T. and Ikura, M. (1995) Calcium-induced conformational transition revealed by the solution structure of apo calmodulin. *Nat. Struct. Biol.*, **2**, 758-767.

Zhang, G., Liu, Y., Ruoho, A.E. and Hurley, J.H. (1997) Structure of the adenylyl cyclase catalytic core. *Nature*, **386**, 247-253.

Zielinski, R.E. (1998) Calmodulin and calmodulin-binding proteins in plants. *Annu. Rev. Plant Physiol. Plant Mol. Biol.*, **49**, 697-725.

Zielinski, R.E. (2001) Characterization of three new members of the *Arabidopsis thaliana* calmodulin gene family: conserved and diverged members of the gene family functionally complement a yeast calmodulin null. *Planta* (in press). Available electronically (DOI 10.1007/S004250/00636).

Zik, M., Arazi, T., Snedden, W.A. and Fromm, H. (1998) Two isoforms of glutamate decarboxylase in *Arabidopsis* are regulated by calcium/calmodulin and differ in organ distribution. *Plant Mol. Biol.*, **37**, 967-975.

Index